AF595006

Modeling and Simulation of Lipid Membranes

Modeling and Simulation of Lipid Membranes

Editors

Jordi Martí
Carles Calero

MDPI • Basel • Beijing • Wuhan • Barcelona • Belgrade • Manchester • Tokyo • Cluj • Tianjin

Editors
Jordi Martí
Technical University of Catalonia-Barcelona Tech
Spain

Carles Calero
University of Barcelona
Spain

Editorial Office
MDPI
St. Alban-Anlage 66
4052 Basel, Switzerland

This is a reprint of articles from the Special Issue published online in the open access journal *Membranes* (ISSN 2077-0375) (available at: https://www.mdpi.com/journal/membranes/special_issues/Model_Simulation_Lipid_Membranes).

For citation purposes, cite each article independently as indicated on the article page online and as indicated below:

LastName, A.A.; LastName, B.B.; LastName, C.C. Article Title. *Journal Name* **Year**, *Volume Number*, Page Range.

ISBN 978-3-0365-4937-8 (Hbk)
ISBN 978-3-0365-4938-5 (PDF)

Contents

About the Editors

Jordi Martí

Jordi Martí is an Associate Professor of Physics at the Department of Physics of the Polytechnic University of Catalonia-Barcelona Tech (UPC). He got his Ph.D. degree in Physics at the University of Barcelona and has been a visiting professor at the University of California-Berkeley, the Lawrence Berkeley National Laboratory, the Atomic Energy Commission of Argentina and the Atomic Energy Research Institute of the Hungarian Academy of Sciences, among others. He is a scientific coordinator of the research group on "Computer Simulation in Condensed Matter" at UPC.

He has made significant contributions to the study of microscopic dynamics of water, especially on the interpretation of molecular vibrations and has authored pioneering works on single and collective dynamics of hydrogen-bonded liquids. His present research interests include structure and dynamics of aqueous and ionic systems, intermolecular proton transfer in confined liquids, nucleation of helium in liquid metals and the molecular modeling and simulation of biomembranes, with especial attention paid to the interaction of oncogenic proteins with cell membranes and to the design of suitable new drugs for tumor treatment. He has published 90 scientific articles in international peer-reviewed journals with about 4000 citations and has presented around 60 communications in international conferences and workshops (h-index 38). He is a member of the editorial board of journals: MDPI *Materials* (Soft Matter section), *Frontiers in Nanotechnology and Graphene*.

Carles Calero

Dr. Calero earned his Ph.D. in Physics at the City University of New York (CUNY, USA) under the supervision of Dist. Prof. E. M. Chudnovsky. He held post-doctoral appointments at the Instituto de Ciencia de Materiales de Barcelona (was awarded a JAE-doc fellowship), at the Universitat Politècnica de Catalunya and at Boston University in the group of H. E. Stanley. Since 2016 he works in the Condensed Matter Physics Department of the University of Barcelona.

His main research line addresses the theoretical description of colloidal systems, focusing on the dynamics of aggregation of colloids and the interfacial phenomena which determine the stability conditions of colloidal suspensions. For his research, Dr. Calero employs tools based on statistical mechanics and a range of computer simulation techniques. His research has an interdisciplinary character, with publications in physics, materials science, physical chemistry and biophysics journals.

Preface to "Modeling and Simulation of Lipid Membranes"

This Special Issue of *Membranes* on "Modeling and Simulation of Lipid Membranes" discusses recent progress in the study of membrane systems by means of computational or experimental tools, paying special attention to the interplay between the two main techniques. It contains eight research articles and one editorial. This book is addressed to all scientists who are interested in new insights into the knowledge of physiological function of lipid cell membranes and their relationship with other components of cells such as proteins, vesicles and organelles. We express our grateful acknowledgment to all authors, reviewers and MDPI editorial staff for their excellent contributions and continuous support.

This publication has been supported by funds from the I+D+i project Reference PID2021-124297NB-C32, founded by MCIN/ AEI/10.13039/501100011033/ and "FEDER Una manera de hacer Europa".

Jordi Martí and Carles Calero
Editors

MDPI

Editorial

Modeling and Simulation of Lipid Membranes

Jordi Martí [1,†] and Carles Calero [2,*,†]

1 Department of Physics, Polytechnic University of Catalonia-Barcelona Tech, B5-209 Northern Campus UPC, 08034 Barcelona, Spain; jordi.marti@upc.edu
2 Department of Condensed Matter Physics, University of Barcelona, Carrer de Martí i Franquès, 1, 08028 Barcelona, Spain
* Correspondence: carles.calero@ub.edu; Tel.: +34-93403-9212
† These authors contributed equally to this work.

Citation: Martí, J.; Calero, C. Modeling and Simulation of Lipid Membranes. *Membranes* **2022**, *12*, 549. https://doi.org/10.3390/membranes12060549

Received: 18 May 2022
Accepted: 23 May 2022
Published: 25 May 2022

Cell membranes separate the interior of cells and the exterior environment, providing protection, controlling the passage of substances, and governing the interaction with other biomolecules and signalling processes. They are complex structures that, mainly driven by the hydrophobic effect [1], are based upon phospholipid bilayer assemblies containing sterols, glycolipids, and a wide variety of proteins located both at the exterior surface and spanning the membrane [2,3]. There exist a large number of different types of phospholipids, each with a given function, although we understand only a small fraction of them [4]. Recently, studies of the physical and biochemical characteristics of lipid molecules as been referred to as lipidomics [5] in recognition of their fundamental importance for the understanding of cell biology.

Over the years, a great variety of experimental techniques have been developed to investigate the structure, dynamics and function of phospholipid membranes. These include nuclear magnetic resonance [6], X-ray scattering [7], small angle and quasi-elastic neutron scattering spectroscopy [8], scanning tunneling microscopy [9], and more recently new techniques to probe previously unaccessible length- and time-scales, such as stimulated emission depletion microscopy-fluorescence correlation spectroscopy [10], terahertz time-domain spectroscopy [11], or microfluidic techniques [12], to mention just a few. In parallel, in recent decades the increase of computer power and the development of new modeling and simulation techniques have allowed a significant improvement in the theoretical description of lipid membranes. As a consequence, plenty of papers have been devoted to the modeling and simulation of cell membranes, from pioneering works at the atomic level of description [13–15] to a multiplicity of coarse-grained approaches [16], the latter allowing to run for long simulations over larger and larger time and distance scales and to study processes such as lipid rafts [17] or full membrane dynamics [18]. Indeed, computer simulations provide relevant information on the structure and dynamics of lipid membranes, and can be used to complement and interpret the experimental data, which is limited by the length and time resolution of the experiment.

This Special Issue of *Membranes* discusses recent progress in the study of membrane systems mainly using computational (usually molecular dynamics) or mixed methodologies. It contains eight research articles. The complete description of each study and the main results are presented in more detail in the full manuscript, which the reader is invited to read. A brief summary of the articles is presented as follows.

Sessa et al. [19] investigate with a combination of permeability experiments and molecular dynamics simulations the crucial issue of the interaction between proteins and phospholipid membranes. The authors compare the effects on a model lipid bilayer of a natural peptide and an analog synthetic peptide which contains a highly hydrophobic azobenzene group. Their computer simulations suggest that the affinity of the peptide is significantly enhanced by the inclusion of such residue. In addition, simulations and experiments on the entrapment capacity of large vesicles show that the modified peptide induces a larger perturbance on the structure of the lipid bilayer, increasing its permeability.

Understanding this effect may be important for the design of new peptides with specific functionalities with potential therapeutic applications.

The article by Lu and Marti [20] highlights the influence of cholesterol in the orientations and structural conformations of the oncogene KRas-4B. This protein is well known for its extended presence in a wide variety of cancers and because of its undruggability. The authors have performed microsecond molecular dynamics simulations using the CHARMM36 force field to observe that high cholesterol contents in the cell membrane favor a given orientation with the protein exposing its effector-binding loop for signal transduction and helping KRas-4B mutant species to remain in its active state. This suggests that high cholesterol intake will increase mortality of some cancer patients.

The next contribution was due to Aragon-Muriel et al. [21] and it reports a study of a newly designed Schiff base derivative from 2-(m-aminophenyl)benzimidazole and 2,4-dihydroxybenzaldehyde interacting with two synthetic membrane models prepared with pure 1,2-dimyristoyl-sn-glycero-3-phosphocholine and a 3:1 mixture of this lipid with 1,2-dimyristoyl-sn-glycero-3-phosphoglycerol, in order to mimic eukaryotic and prokaryotic membranes. The study was performed by means of a combined in vivo-in silico study using differential scanning calorimetry, spectroscopic and spectrometric techniques and molecular dynamics simulations. The main results indicate that the Schiff derivative induces higher fluidity at the mixed membrane. As a second part of their study, the authors modeled an erythrocyte membrane model formed by 1-palmitoyl-2-oleoyl-sn-glycero-3-phosphoethanolamine, N-(15Z-tetracosenoyl)-sphing-4-enine-1-phosphocholine and 1-palmitoyl-2-oleoyl-sn-glycero-3-phosphocholine and observed that the Schiff derivative showed high affinity to the different membranes due to hydrophobic interactions or hydrogen bonds.

The interplay between scattering experiments and molecular dynamics simulations to obtain information on the structure of model phospholipid membranes is discussed in the article [22]. Zec and co-workers provide a detailed comparison between the results of scattering experiments (neutron and X-ray reflectometry and small angle scattering measurements) and calculated values obtained from standard all-atom MD simulations of bilayers composed of popular phospholipids (1,2-dimyristoyl-sn-glycero-3-phosphocholine (DMPC) and 1,2-dilinoleoyl-sn-glycero-3-phosphocholine. The authors show that MD simulations can be used to interpret from a nanoscopic perspective the results from scattering experiments, which prove larger length and time scales. Their analysis also identifies the uncertainties and sources of error from scattering experiments and simulations, which need to be considered in order to draw significant conclusions from their comparison.

In the paper by Radhakrishnan et al. [23] the authors used molecular dynamics techniques in order to study the permeation of membranes by several relevant solutes, such as Withaferin A, Withanone, Caffeic Acid Phenethyl Ester and Artepillin C when they are at the interface of a cell membrane model formed by phosphatidylserine lipids. Their results indicated that exposure of phosphatidylserine can favor the permeation of Withaferin A, Withanone and of Caffeic Acid Phenethyl Ester through a cancer cell membrane when compared to a normal membrane. The authors showed the ability of phosphatidylserine exposure-based models for analyzing how cancer cells are able to perform drug selectivity.

In Reference [24], Trejo and co-workers review the main properties of red blood cells' (RBC) membranes and their effect on blood rheology. The authors describe the mechanical properties of RBC membranes and the mesoscopic theory to model their relevant elastic features, as well as the resulting membrane dynamics. They also discuss the interaction of RBCs with the constituents of blood plasma through the membrane, of great importance to understand RBCs mutual interactions and the formation of RBCs aggregates. The consequences of RBCs properties on fluid dynamics of blood in the circulatory system (hemodynamics) are also reviewed, giving an account of recent advancements in numerical and experimental techniques which have provided new information on the subject. In particular, Trejo et al. review in detail the use of recent microfluidic techniques to obtain information on the properties of single RBCs as well as on collective effects which determine the rheological properties of blood (hemorheology). Finally, a review of the disorders which alter the hemodynamics and rheological properties of blood is provided, and an account is given of the microfluidic techniques developed for their diagnostic.

In the work of Hu and Marti [25], the authors reported a molecular dynamics study on the atomic interactions of a lipid bilayer membrane formed by dioleoylphosphatidylcholine, 1,2-dioleoyl-sn-glycero-3-phosphoserine and cholesterol with a series of derivatives of the drug benzothiadiazine, designed in silico, all within a potassium chloride aqueous solution. The benzothiadiazine derivatives were obtained by single-hydrogen site substitution and it has been revealed that all them have strong affinity to remain at the cell membrane interface, with variable residence times in the range 10-70 ns. The authors observed that benzothiadiazine derivatives can bind lipids and cholesterol chains with single and double hydrogen-bonds of rather short characteristic lengths.

The influence of the membrane on the properties of transmembrane proteins is investigated by Asare and co-workers using numerical simulations [26]. The authors perform MD simulations of KCNE3, a transmembrane protein associated with several potassium channels, inserted in different phospholipid bilayers: DMPC, 1-palmitoyl-2-oleoyl-sn-glycero-3-phosphocholine (POPC), and a mixture of POPC and POPG (1-Palmitoyl-2-oleoyl-sn-glycero-3-phosphatidylglycerol) in a 3:1 proportion, to study how such environments determine its structural and dynamical properties. Their simulations indicate that the central part of the protein immersed in the membrane, the transmembrane domain, is more rigid and stable than the two ends of the protein which are surrounded by the electrolyte. The results reported by Asare and co-workers can help complement the information extracted from experiment on KCNE3's function in its native membrane environment.

Despite studies of model lipid membranes have been carried out for long time, there are still many aspects and theoretical findings that have not been yet verified experimentally and for which the existing results are incomplete or inconsistent. Conversely, there are also experimental results which still lack of appropriate microscopical interpretation. Therefore, the main objective of this Special Issue was to collect a sample of recent scientific works on the modeling and simulation lipid membranes, with special aim in the interactions of the two principal techniques (theory-simulation vs. experiments) and their mutual benefit. The techniques presented here, from purely computational to the mixture of simulation and experimental methods in some cases, have helped us to understand essential physical properties as the structure and dynamics of specific lipid membranes and solutes. These studies will provide new insights into the fundamental principles underlying physiological functions of cell membranes and their relationship with other components of cells and tissues. We believe that this objective has been successfully achieved, for which we express our heartfelt appreciation to all authors and reviewers for their excellent contributions.

Author Contributions: Writing—original draft preparation, review and editing, J.M. and C.C. All authors have read and agreed to the published version of the manuscript.

Funding: This contribution received no external funding.

Institutional Review Board Statement: Not applicable.

Informed Consent Statement: Not applicable.

Data Availability Statement: Not applicable.

Acknowledgments: We thank all authors and reviewers who contributed to this Special Issue for their excellent works and accurate revisions.

Conflicts of Interest: The authors declare no conflict of interest.

References

1. Nagle, J.F.; Tristram-Nagle, S. Structure of lipid bilayers. *Biochim. Biophys. Acta (BBA)-Rev. Biomembr.* **2000**, *1469*, 159–195. [CrossRef]
2. Tien, H.T.; Ottova-Leitmannova, A. *Membrane Biophysics: As Viewed from Experimental Bilayer Lipid Membranes*; Elsevier: Amsterdam, The Netherlands, 2000.
3. Mouritsen, O.G. *Life-As a Matter of Fat*; Springer: Berlin/Heidelberg, Germany, 2005.
4. Van Meer, G.; Voelker, D.R.; Feigenson, G.W. Membrane lipids: Where they are and how they behave. *Nat. Rev. Mol. Cell Biol.* **2008**, *9*, 112–124. [CrossRef] [PubMed]
5. Shevchenko, A.; Simons, K. Lipidomics: Coming to grips with lipid diversity. *Nat. Rev. Mol. Cell Biol.* **2010**, *11*, 593–598. [CrossRef]
6. Stockton, G.W.; Smith, I.C. A deuterium nuclear magnetic resonance study of the condensing effect of cholesterol on egg phosphatidylcholine bilayer membranes. I. Perdeuterated fatty acid probes. *Chem. Phys. Lipids* **1976**, *17*, 251–263. [CrossRef]
7. Pan, J.; Tristram-Nagle, S.; Nagle, J.F. Effect of cholesterol on structural and mechanical properties of membranes depends on lipid chain saturation. *Phys. Rev. E* **2009**, *80*, 021931. [CrossRef] [PubMed]
8. Pabst, G.; Kučerka, N.; Nieh, M.P.; Rheinstädter, M.; Katsaras, J. Applications of neutron and X-ray scattering to the study of biologically relevant model membranes. *Chem. Phys. Lipids* **2010**, *163*, 460–479. [CrossRef] [PubMed]
9. Woodward IV, J.; Zasadzinski, J. High-resolution scanning tunneling microscopy of fully hydrated ripple-phase bilayers. *Biophys. J.* **1997**, *72*, 964–976. [CrossRef]
10. Hedde, P.N.; Dörlich, R.M.; Blomley, R.; Gradl, D.; Oppong, E.; Cato, A.C.; Nienhaus, G.U. Stimulated emission depletion-based raster image correlation spectroscopy reveals biomolecular dynamics in live cells. *Nat. Commun.* **2013**, *4*, 1–8. [CrossRef] [PubMed]
11. Tielrooij, K.; Paparo, D.; Piatkowski, L.; Bakker, H.; Bonn, M. Dielectric relaxation dynamics of water in model membranes probed by terahertz spectroscopy. *Biophys. J.* **2009**, *97*, 2484–2492. [CrossRef]
12. Trejo-Soto, C.; Costa-Miracle, E.; Rodríguez-Villarreal, I.; Cid, J.; Alarcón, T.; Hernández-Machado, A. Capillary filling at the microscale: Control of fluid front using geometry. *PLoS ONE* **2016**, *11*, e0153559.
13. Bassolino-Klimas, D.; Alper, H.E.; Stouch, T.R. Mechanism of solute diffusion through lipid bilayer membranes by molecular dynamics simulation. *J. Am. Chem. Soc.* **1995**, *117*, 4118–4129. [CrossRef]
14. Feller, S.E. Molecular dynamics simulations of lipid bilayers. *Curr. Opin. Colloid Interface Sci.* **2000**, *5*, 217–223. [CrossRef]
15. Berkowitz, M.L.; Bostick, D.L.; Pandit, S. Aqueous solutions next to phospholipid membrane surfaces: Insights from simulations. *Chem. Rev.* **2006**, *106*, 1527–1539. [CrossRef] [PubMed]
16. Orsi, M.; Haubertin, D.Y.; Sanderson, W.E.; Essex, J.W. A quantitative coarse-grain model for lipid bilayers. *J. Phys. Chem. B* **2008**, *112*, 802–815. [CrossRef] [PubMed]
17. Simons, K.; Toomre, D. Lipid rafts and signal transduction. *Nat. Rev. Mol. Cell Biol.* **2000**, *1*, 31–39. [CrossRef] [PubMed]
18. Giacomello, M.; Pyakurel, A.; Glytsou, C.; Scorrano, L. The cell biology of mitochondrial membrane dynamics. *Nat. Rev. Mol. Cell Biol.* **2020**, *21*, 204–224. [CrossRef]
19. Sessa, L.; Concilio, S.; Walde, P.; Robinson, T.; Dittrich, P.S.; Porta, A.; Panunzi, B.; Caruso, U.; Piotto, S. Study of the interaction of a novel semi-synthetic peptide with model lipid membranes. *Membranes* **2020**, *10*, 294. [CrossRef]
20. Lu, H.; Martí, J. Influence of cholesterol on the orientation of the farnesylated GTP-bound KRas-4B binding with anionic model membranes. *Membranes* **2020**, *10*, 364. [CrossRef]
21. Aragón-Muriel, A.; Liscano, Y.; Morales-Morales, D.; Polo-Cerón, D.; Oñate-Garzón, J. A study of the interaction of a new benzimidazole schiff base with synthetic and simulated membrane models of bacterial and mammalian membranes. *Membranes* **2021**, *11*, 449. [CrossRef]
22. Zec, N.; Mangiapia, G.; Hendry, A.C.; Barker, R.; Koutsioubas, A.; Frielinghaus, H.; Campana, M.; Ortega-Roldan, J.L.; Busch, S.; Moulin, J.F. Mutually beneficial combination of molecular dynamics computer simulations and scattering experiments. *Membranes* **2021**, *11*, 507. [CrossRef]
23. Radhakrishnan, N.; Kaul, S.C.; Wadhwa, R.; Sundar, D. Phosphatidylserine Exposed Lipid Bilayer Models for Understanding Cancer Cell Selectivity of Natural Compounds: A Molecular Dynamics Simulation Study. *Membranes* **2022**, *12*, 64. [CrossRef] [PubMed]
24. Trejo-Soto, C.; Lázaro, G.R.; Pagonabarraga, I.; Hernández-Machado, A. Microfluidics approach to the mechanical properties of red blood cell membrane and their effect on blood rheology. *Membranes* **2022**, *12*, 217. [CrossRef] [PubMed]
25. Hu, Z.; Marti, J. In silico drug design of benzothiadiazine derivatives interacting with phospholipid cell membranes. *Membranes* **2022**, *12*, 331. [CrossRef] [PubMed]
26. Asare, I.K.; Galende, A.P.; Garcia, A.B.; Cruz, M.F.; Moura, A.C.M.; Campbell, C.C.; Scheyer, M.; Alao, J.P.; Alston, S.; Kravats, A.N.; et al. Investigating Structural Dynamics of KCNE3 in Different Membrane Environments Using Molecular Dynamics Simulations. *Membranes* **2022**, *12*, 469. [CrossRef]

Article

Mutually Beneficial Combination of Molecular Dynamics Computer Simulations and Scattering Experiments

Nebojša Zec [1], Gaetano Mangiapia [1], Alex C. Hendry [2], Robert Barker [3], Alexandros Koutsioubas [4], Henrich Frielinghaus [4], Mario Campana [5], José Luis Ortega-Roldan [2], Sebastian Busch [1,*] and Jean-François Moulin [1,*]

1 German Engineering Materials Science Centre (GEMS) at Heinz Maier-Leibnitz Zentrum (MLZ), Helmholtz-Zentrum Hereon, Lichtenbergstr. 1, 85748 Garching bei München, Germany; nebojsa.zec@hereon.de (N.Z.); gaetano.mangiapia@hereon.de (G.M.)
2 School of Biosciences, University of Kent, Canterbury CT2 7NJ, UK; ach49@kent.ac.uk (A.C.H.); J.L.Ortega-Roldan@kent.ac.uk (J.L.O.-R.)
3 School of Physical Sciences, University of Kent, Canterbury CT2 7NH, UK; R.Barker@kent.ac.uk
4 Jülich Centre for Neutron Science (JCNS) at Heinz Maier-Leibnitz Zentrum (MLZ), Forschungszentrum Jülich, Lichtenbergstr. 1, 85748 Garching bei München, Germany; a.koutsioumpas@fz-juelich.de (A.K.); h.frielinghaus@fz-juelich.de (H.F.)
5 ISIS Neutron and Muon Facility, Rutherford Appleton Laboratory, Science & Technology Facilities Council, Didcot OX11 0QX, UK; mario.campana@stfc.ac.uk
* Correspondence: sebastian.busch@hereon.de (S.B.); jean-francois.moulin@hereon.de (J.-F.M.); Tel.: +49-89-158860-764 (S.B.); +49-89-158860-762 (J.-F.M.)

Citation: Zec, N.; Mangiapia, G.; Hendry, A.C.; Barker, R.; Koutsioubas, A.; Frielinghaus, H.; Campana, M.; Ortega-Roldan, J.L.; Busch, S.; Moulin, J.-F. Mutually Beneficial Combination of Molecular Dynamics Computer Simulations and Scattering Experiments. *Membranes* **2021**, *11*, 507. https://doi.org/10.3390/membranes11070507

Academic Editors: Jordi Marti and Carles Calero

Received: 2 June 2021
Accepted: 29 June 2021
Published: 5 July 2021

Publisher's Note: MDPI stays neutral with regard to jurisdictional claims in published maps and institutional affiliations.

Abstract: We showcase the combination of experimental neutron scattering data and molecular dynamics (MD) simulations for exemplary phospholipid membrane systems. Neutron and X-ray reflectometry and small-angle scattering measurements are determined by the scattering length density profile in real space, but it is not usually possible to retrieve this profile unambiguously from the data alone. MD simulations predict these density profiles, but they require experimental control. Both issues can be addressed simultaneously by cross-validating scattering data and MD results. The strengths and weaknesses of each technique are discussed in detail with the aim of optimizing the opportunities provided by this combination.

Keywords: neutron reflectometry; X-ray reflectometry; small-angle neutron scattering; small-angle X-ray scattering; molecular dynamics simulations; scattering length density profile; phospholipid membrane

1. Introduction

Phospholipid-based bilayers are the main components of biological membranes and represent their basic structural elements [1]. The main role of the cell membrane is to protect the cell from its surroundings, allowing it to have a well-defined environment and accomplish its vital functions [2]. Given the importance of the membrane, structural details for the cell biology, several characterization methods have been used to investigate the structure under different conditions (microscopy [3,4], spectroscopy [5,6], scattering methods [7–10] and simulations [11–14]). Of special interest in this paper are the scattering methods that give access to the structure and dynamics of the system under investigation. These methods are non-invasive, non-destructive over the duration of the data collection and probe a large sample volume, thus providing statistically relevant information [15]. The typical membrane length scales are relatively large compared to atomic dimensions, hence the focus of this work is on scattering by large-scale structures which can be investigated by reflectometry and small-angle scattering methods.

Fundamentally, two main types of probe can be used for these scattering experiments: X-rays and neutrons. Laboratory X-ray sources provide the possibility of performing many useful experiments and high-flux synchrotron sources make very fast and extremely sensitive measurements feasible. Neutron scattering experiments can only be performed at large-scale facilities, but they nevertheless play a fundamental role in the landscape of membrane characterization methods [16–19]. In contrast to X-rays, neutrons interact in a non-destructive fashion with the material under examination. Due to their weak interaction with matter, neutrons have a large penetration depth for most materials, allowing for an elaborate sample environment. Being scattered by the nuclei, and not by the electrons as in the case of X-rays, neutrons offer the possibility to add isotopic sensitivity to the measurements. As a great opportunity for biological systems, the neutron scattering power of hydrogen and deuterium differs widely and neutrons are thus extremely sensitive to the distribution of hydrogen in the sample. The most obvious strategy for taking advantage of this property is to perform measurements on membranes prepared in light or heavy water, but more complex (and more costly) isotopic substitution methods targeting specific molecular sites can also be utilized.

All scattering methods provide a description in reciprocal space, which can be understood as the Fourier transform of the structure of the sample. The data thus tell about periodicity and spatial correlation in the sample [20,21]. The interpretation of such data is by no means intuitive and the eventual aim of all measurements is to describe the actual position of all atoms or molecules in the sample. In the process of inverting the information contained in scattering data from reciprocal to real space, problems arise (in particular, the phase problem [21]), which usually hinders finding an unequivocal solution. To tackle this issue, independent information must be found in order to put constraints on the inversion problem. Unfortunately, there is no available experimental method offering the needed spatial resolution over the required length scales. Nevertheless, computer simulations in the framework of Molecular Dynamics (MD) provide invaluable insights in the real space structure of these complex systems.

MD simulations applied to phospholipid membranes provide an atomic-level description of the system. The positions of individual atoms are followed by numerically solving classical equations of motion. Therefore, MD simulations provide atomic resolution unavailable to the experiments presented here. Combining MD simulations and scattering experiments is beneficial for studying phospholipid membranes but can also be used for the structural analysis of completely unrelated systems [22].

MD and the experimental methods described here (SAS and reflectometry) probe the sample's structure over a limited range of length scales. Those ranges overlap, and, hence, cross-validation is only possible over this restricted domain [23]. Another important aspect to consider is the fact that an MD simulation typically only describes a very brief time interval while the integration times used for data acquisition in NR and SANS are orders of magnitude longer (from seconds to hours). Similarly, simulations typically cover some cube nanometers, while experiments tend to average over cube millimeters. Precautions must thus be taken to ensure that MD simulations do not merely describe transient structures which would be averaged out in the measurements. Conversely, simulations give access to the Ångström scale, which is not directly probed by these experimental techniques.

In classical MD simulations, the interaction potential energy is described in the form of a force field, based on both empirical and quantum chemical data. Validation of the force-field parameters is a tedious and challenging task, so online topology and force-field parameter builders have become popular as a simple solution [24–26]. One has to be very critical of the parameters obtained in this way and ensure that the theoretical model and applied methodology describe the molecule of interest "reasonably well". On the other hand, besides lower spatial resolution compared to the MD and the phase problem, small-angle scattering and reflectometry experiments have additional experimental uncertainties related to the sample and instrumentation. It is therefore theoretically possible that inaccurate experimental data match an incorrect MD simulation perfectly.

To successfully combine these techniques, a certain level of understanding of both scattering experiments and computer simulations is essential in order to fully understand the advantages and limitations of both methods and avoid putting too much (or too little) confidence in the results extracted from one of these methods alone. One can use MD simulation trajectories to extract neutron scattering length density profiles, directly calculate the corresponding reflectivity or small-angle scattering pattern and plan an experiment in order to optimize the use of beam-time at large-scale facilities. One can see the effect of changing different parameters such as instrumental resolution, get a hint as to whether the effect can be experimentally observed and plan an experiment in an effective and efficient manner.

In this article, the study of a single bilayer of DMPC (1,2-dimyristoyl-*sn*-glycero-3-phosphocholine) and multilamellar SoyPC (mainly composed of 1,2-dilinoleoyl-*sn*-glycero-3-phosphocholine) is used to showcase the joint use and cross-validation of MD simulation and scattering experiments. DMPC is a double-saturated phospholipid composed of two myristoyl chains, used in many biophysical studies [27,28] and as an excipient in pharmaceutical formulations [29]. SoyPC is a mixture of phospholipids found in soy and used as a model bilayer in some studies aimed at investigating the interaction of cell membranes and active ingredients [30,31]. The aim of this work is not so much to discuss the properties of the selected phospholipids as to describe the methodology of combining simulation and experiment and the challenges behind it. Since there are many things that can go wrong in both, it is important to establish the methodology and find sources of potential errors before focusing on more complex systems.

The approach to combine simulations with scattering experiments is not new; it was for example used to study peptide self-organization into switchable films at an air–water interface by Xue et al. [32] and by Vanegas et al. [33] to study the insertion of the dengue virus envelope protein into phospholipid bilayers. These techniques were also applied to investigate the contact angles and adsorption energies of nanoparticles at the air-liquid interface [34]. Back in 2005, Benz et al. [23] developed a protocol for comparing MD simulations with X-ray (XRR) and neutron reflectivity (NR) and showed that neither the united-atom GROMACS nor the CHARMM22/27 force fields could reproduce experimental data. More recently, a method for producing continuous scattering length density (SLD) profiles from MD simulations has been presented for interpreting reflectivity data from phospholipid bilayers [35]. Koutsioubas [36] performed coarse-grained MD simulations with the standard MARTINI force field and obtained quantitative and semi-quantitative agreement with neutron reflectivity data for DPPC membranes in the liquid and gel phase, respectively. On the other hand, McCluskey et al. [37] observed that the MARTINI potential model did not accurately describe the 1,2-distearoyl-*sn*-phosphatidylcholine (DSPC) monolayer, while the Berger and Slipid potential models showed better agreement.

Several computer programs for reading MD simulation trajectories, calculating the scattering length density profile and neutron reflectivity, and making direct comparison with the experiment have been developed over the years. SIMtoEXP [38] and NeutronRefTools (as a VMD plug-in) [39] were developed particularly for phospholipid membrane research. The high number of citations shows that these solutions have been accepted and regularly used by the scientific community. Being completely aware of their existence, we employ here a self-written software solution that will be published soon.

In the following, we describe the different techniques, show two examples of phospholipid molecules in two different morphologies and discuss the robustness of experimental features and their constraints on real space structure.

2. Background

In order to provide the tools needed in the discussion, this section introduces some fundamentals of scattering theory and puts them in the context of the present problem. Keeping in mind the typical expectations of the computer simulation community, the strong points as well as the pitfalls of the scattering methods are stressed along the way. Momentum transfer $\vec{Q}$, the natural variable against which scattering intensity is measured in an actual experiment, is introduced first. This variable takes the radiation characteristics (wavelength) and the geometrical details of the experiment into account. The SLD, which describes how strongly a given medium will scatter as a function of its composition, is then introduced and used to express the index of refraction, which in turn is used to predict the propagation of neutrons or photons (both considered as waves) in matter and eventually analyze reflectometry and small-angle experiments.

Neutron and X-ray scattering experiments measure the number of scattered neutrons/photons as a function of the vector $\vec{Q}$, which describes the momentum transfer the wave undergoes upon scattering. $\vec{Q}$ is a function of the experiment geometry, which we symbolically represent here by θ, and of the wavelength of the radiation used, λ:

$$\vec{Q}(\theta, \lambda) = \vec{k_f} - \vec{k_i} \tag{1}$$

where $\vec{k_i}$ and $\vec{k_f}$ are the wave vectors of the incident and scattered radiation, respectively.

$$\left|\vec{Q}\right| = Q \propto \frac{2\pi}{l} \tag{2}$$

signifies that the modulus of each Q vector in reciprocal space is associated with an inter-distance l in direct (real) space, which is characteristic of the size of the scattering structure in the corresponding direction.

The practical problem is to compute the real space sample structure which is compatible with the scattering intensity distribution, measured in the reciprocal or Q space. In tackling this task, which is central to the whole crystallography field, the most fundamental obstacle is the phase problem. What the detectors actually measure is the intensity, i.e., square of the amplitude of the scattered waves. Consequently, all information relative to the phase of those waves is irremediably lost. From the measurement, it is therefore impossible to unequivocally deduce the positions of the scattering particles in an absolute way and a given experimental dataset can correspond to a multitude of real-space structures.

There are several ways to work around this ambiguity: First, one can gain additional experimental data by performing measurements for specifically adjusted scattering contrast of the different constituents without affecting the sample's structure [40]. How this is practically achieved is discussed in detail in the next sections. While this reduces the number of possible real-space structures considerably, a usually unachievable $n(n+1)/2$ contrasts would have to be measured to be able to solve the real-space structure of n components from the data analytically—and even if that many measurements can be performed, experimental imperfections and limited counting statistics limit their usefulness [41].

A second method is to form periodic structures in the system. In the case of membranes, this can for example be achieved by using stacks of bilayers (multilayers) that are periodic in the direction of the membrane normal. This leads to the formation of *Bragg peaks* in the scattered intensity at values of Q where the phase is either 0° or 180°. In a traditional approach, one would then use only the scattered intensity at the Bragg peak positions where the phase can be determined [42]. Although this approach leaves the whole information contained in the rest of the scattering pattern unused, the SLD profile can be reconstructed precisely if many Bragg peaks are measured. In reality, however, it is only possible to measure $\sim$2–5 Bragg peaks due to the disorder inherent in the system and experimental limitations, severely limiting the precision of the extracted information. It is also possible to take the whole scattering pattern into consideration (see [43] and references therein), but the presence of Bragg peaks makes the precise measurement of the specular reflectivity between the Bragg peaks somewhat unreliable, as discussed below.

Complementary to these experimental approaches, one can use a theoretical approach in which a real-space model of the system is built with as many external constraints as possible. The MD method is here the instrument of choice. As shown elsewhere, although one needs to take care of several practical details [22,35,38,39], it is relatively straightforward to compute the scattering pattern corresponding to an MD simulated structure and compare it to the experimental data. This goes beyond the normal fitting procedure in which a set of parameters describing the structure is optimized in order to reproduce the data. While this method cannot prove the accuracy of a given model, it can, however, falsify many models, which must not be underestimated.

2.1. Scattering Length Density

In the following, we briefly describe the scattering processes and introduce the fundamental concept of scattering length density and its influence on the transmission and refraction of the waves in a medium. At the end of this section, the practical possibility of taking advantage of probe type and isotopic composition to control scattering contrast is apparent.

The treatment for X-rays and neutrons is very similar and differs only in the interaction of the corresponding radiation with matter. Neutrons interact with the nuclei (we leave aside the magnetic interactions with electrons, which is generally less relevant for the study of biological materials) while X-rays, being an electromagnetic wave, interact with the electron cloud. We thus here use the general wording "wave" and show probe-specific expressions only where relevant.

The interaction between a wave and a medium is described in quantum mechanical terms by the average potential $\overline{V}$

$$\overline{V} = \frac{2\pi\hbar^2}{m}\rho \tag{3}$$

where m is the neutron mass, $\hbar$ the Planck constant divided by 2π, and

$$\rho = \frac{1}{\text{volume}}\sum_j b_j \tag{4}$$

is the so-called scattering length density of the medium (which we also denote by SLD) and is the result of the superposition of all contributions b_j (scattering lengths) describing the interaction strength of each individual scatterer j.

A general solution of the Schrödinger equation which satisfies the potential $\overline{V}$ and describes the propagation of a wave at every point $\vec{r}$ in the medium is

$$\Psi(\vec{r}) = A\exp\left(in\vec{k_0}\cdot\vec{r}\right) \tag{5}$$

where n is the complex index of refraction relating k_0, the wave momentum in vacuum, and k, the momentum it would have in a material medium,

$$n = \frac{k}{k_0} \quad . \tag{6}$$

The real part of n describes the wave phase velocity in the medium, while the imaginary part describes the absorption phenomena by damping the wave intensity, which is the square of the modulus of Ψ. In the neutron case, the absorption is usually negligibly small and n is a real number.

Similar to what we experience in everyday life while looking at things, scattering methods give us the ability to distinguish different parts of the samples from each other only if their indices of refraction differ, irrespective of their chemical nature.

For X-rays, the scattering length is proportional to the product of the atomic number Z and the classical electron radius or Thomson scattering length $r_0 \approx 2.82$ fm. The energy dependence of the scattering length, which varies abruptly around absorption edges, is described by semi-empirical atomic scattering factors f_1 and f_2, leading to the following expression for the refraction index where N denotes the number concentration of the given atom:

$$n_X = 1 - \frac{1}{2\pi}Nr_0\lambda^2(f_1 + if_2) \tag{7}$$

For neutrons, the energy-independent nuclear scattering length b substitutes the electron radius in the previous expression and one gets

$$n_{n0} = 1 - \frac{1}{2\pi}\lambda^2\rho \quad . \tag{8}$$

As can be intuitively expected from the nature of their interaction, in the case of neutrons the scattering lengths of different isotopes of the same element differ from each other. This isotopic dependency of b is seemingly random [44], but it is very interesting to observe that, in the case of hydrogen and deuterium, the difference is very large, b_H being -3.74 fm for hydrogen and $b_D = 6.67$ fm for deuterium.

From those observations, it is clear that X-rays and neutrons will experience a different index of refraction between the components of the sample, thereby introducing a contrast between those regions. As hinted in the Introduction, one can thus obtain additional independent information about the system under investigation by: (a) combining X-ray and neutron measurements; and/or (b) varying the isotopic composition of the sample used for neutron scattering while keeping its chemical composition and structural details essentially unchanged. In the context of molecular biology, it is clear that advantage can be taken of this method by tuning the isotopic composition of the ubiquitous water molecules. By simply mixing H_2O and D_2O, one can adjust the contrast with precision [45,46]. More complex isotopic substitution schemes, for instance at specific molecular sites, can also be used to achieve more targeted control [47,48].

2.2. Reflectometry

Reflectometry takes advantage of the variation of the index of refraction across planar interfaces in order to investigate structural and compositional profiles.

When a wave impinges on a flat and smooth horizontal surface separating two media (denoted 1 and 2), it can be reflected back into the original medium (reflection into 1) or refracted into medium 2. Since the ideal interface we describe is an SLD fluctuation along the vertical direction only, it cannot affect the in-plane components of the incident wave's momentum. The reflection is purely specular and happens under the same angle as the angle of incidence.

The convenient variable to describe this problem is again the momentum transfer vector $\vec{Q}$, which is here strictly vertical:

$$\vec{Q} = \vec{k_f} - \vec{k_i} = \left|\vec{Q}\right| \cdot \vec{n_z} = Q_z \cdot \vec{n_z} \quad , \tag{9}$$

where $\vec{n_z}$ is the unit vector along the vertical direction, z.

Since we are only considering elastic scattering, the norm of the momentum of the wave is conserved and

$$Q_z = \frac{4\pi \sin(\theta)}{\lambda} \tag{10}$$

where θ is the angle of incidence on the surface.

Regarding the refracted wave, since the index of refraction differs in the two media

$$k_1/k_2 = n_2/n_1, \tag{11}$$

and, following the above argument that the in-plane components of k cannot change, we get

$$n_1 \cos(\theta_1) = n_2 \cos(\theta_2) \tag{12}$$

where θ_1 is the angle of incidence and θ_2 is the refraction angle, both measured between the surface and the corresponding propagation vector. The above relationship is no other than Snell's law of optics, which also holds for neutrons and X-rays.

From this relation, if the index of refraction is smaller than 1 (often the case for neutrons and X-rays), there exists a critical Q below which θ_2 will be zero, i.e., below which the incident wave will undergo total reflection by the interface:

$$Q_c = 4\sqrt{\pi(\rho_1 - \rho_0)} \tag{13}$$

The amplitude reflectivity (r) and amplitude transmitivity (t) of the surface are given by the Fresnel relationships, which can be derived from continuity conditions. Applying the small-angle approximation, which holds in the case of reflectometry measurements, leads to [49]

$$r = \frac{A_{\text{reflected}}}{A_{\text{incident}}} = \frac{\theta_{\text{incident}} - \theta_{\text{refracted}}}{\theta_{\text{incident}} + \theta_{\text{refracted}}} \quad \text{and} \tag{14}$$

$$t = \frac{A_{\text{refracted}}}{A_{\text{incident}}} = \frac{2\theta_{\text{incident}}}{\theta_{\text{incident}} + \theta_{\text{refracted}}} \quad , \tag{15}$$

where A represents the respective amplitudes.

In the case of stratified media on a semi-infinitely thick substrate, a valid description of practical experiments one might perform to study supported thin films, the impinging wave can undergo reflection or refraction at each interface. The wave emerging from the surface is the superposition of all the waves which have traveled paths through the sample that do not end up being transmitted into the semi-infinite substrate.

Similar to the simple case of a single interface, one can express the amplitude reflectivity and the amplitude transmitivity via the Fresnel equations. Starting from the semi-infinite substrate where no multiple reflections are to be considered, one can then recursively reconstruct the reflectivity at the topmost surface. This method, which leads to an exact result, was introduced by Parratt [50]. A computationally convenient method based on the formalism of optical transfer matrices was independently proposed by Abelès [51,52] and leads to the same exact result.

If the interface between regions of different SLD is diffuse rather than sharp as assumed above, two approaches can be used for the evaluation of the reflectivity. The first approximation was proposed by Névot and Croce [53]. It introduces an interfacial roughness factor which damps the reflected waves and which is expressed in a similar way as the Debye–Waller factor describing, in crystallography, the effect of the thermal motion blurring the atomic positions and thereby lowering the diffracted intensities. In this model, the position of the interface is described as normally distributed around its nominal position with a given standard deviation σ. The corresponding SLD profile is a smooth transition from one SLD to the next in a sigmoidal step described by the error function associated to the standard deviation. This approach has the obvious advantage of describing the diffuse interface by a single number. However, it has to be stressed that this approximation is only valid if the roughness is much smaller than the layer thickness.

The second method used to deal with diffuse interfaces, while more computationally demanding, makes it possible to describe arbitrary SLD profiles. In this second approach, the SLD profile is simply discretized into bins thin enough to ensure that they can be considered to be of constant SLD. The reflectivity computation can then follow without further approximation by means of either the Paratt or the Abeles algorithm. Although this approach is potentially able to better "follow" the actual SLD profile and can deal with diffuse areas too broad to be safely described by a Nevot–Croce roughness parameter, it lacks the ability to condense structural information in simple and clear parameters such as layer thickness, width of a transition region, etc. Such a convenient description can of course nevertheless be obtained *a posteriori* by adjusting an analytical model to the binned SLD profile used for the simulations.

It should be kept in mind that, whatever the chosen approach used to describe diffuse SLD transition regions, different lateral distributions of matter could lead to the same SLD profile along the vertical. The in-plane fluctuations, which have been averaged out here, would be the cause of the off-specular scattering. The two types of reflections—specular and off-specular—can be easily understood with the help of an everyday-life analogy: the specular (literally, mirror-like) behavior is what we observe when we contemplate the sunset reflection on the surface of a perfectly still lake, on a windless evening. This image of an undeformed sun tells us that the surface of the water is perfectly flat (and reflective). If we repeat this contemplation while a strong wind is blowing, we will only be able to see a blurry image of the sun on the water surface: the lateral structures on the surface, its roughness, will reflect the light away from the expected ideal trajectory, hence in an off-specular or non-specular way. A detailed analysis of the blurred image could lead to an understanding of the details of the wavy surface. Practical implications of the presence of off-specular scattering are discussed briefly when dealing with actual measurements.

The description of the specular reflectivity evaluation for a given SLD profile given above is exact and can be used for numerical evaluations. It is, however, interesting to keep the results obtained in the framework of the first Born approximation in mind, i.e., in the limit of weak scattering. In this case, one gets the "master-equation of reflectivity" [49]:

$$R = R_{\text{Fresnel}} \left| \int \frac{\mathrm{d}\rho}{\mathrm{d}z} \exp(iQz) \mathrm{d}z \right|^2 \tag{16}$$

which relates the reflectivity of an interface with arbitrary interfaces to the reflectivity of a multilayer with sharp interfaces (R_{Fresnel}) and the spatial rate of change of the SLD, ρ.

The Born approximation clearly does not hold for small values of Q, for which many reflections or even total reflection take place, but it can be used to gain intuition about the reflectivity observed at large Q. This expression of scattering as a Fourier transform of real space makes it clear that the reflectivity curve of a layer of thickness l will display oscillations as a function of Q having a period given by $2\pi/l$. In the case of periodic structures such as those encountered, for instance, in multilayered phospholipids, intensity will build up at specific locations in Q space and appear as the Bragg peaks known in diffraction. Moreover, it is clear from this expression that only regions which display an SLD contrast (i.e., where the derivative of the SLD is not zero) will contribute to the reflectivity. Last but not least, this Fourier-transform approach also helps understand the origins of the spatial resolution limits of the scattering methods: the maximal observed Q value will determine the size of the smallest object which can be resolved by a scattering experiment.

Reflectometry is the method of choice when focusing on planar surfaces or buried interfaces. The sample consists of $\sim$10 cm^2 substrate covered with a sample layer, resulting in very low amounts of sample required for an experiment. The measurement geometry means that one is exclusively sensitive to the direction along the interface normal in specular scattering and can get separate information about in-plane correlations through off-specular scattering.

2.3. Small-Angle Scattering

Differently from reflectometry, which probes the characteristics of planar interfaces, in a small-angle scattering (SAS) experiment, the characteristics of scattering objects (gels, polymer blends, porous structures, micelle aggregates, etc.) are measured in bulk [54]. In SAS geometry, a collimated beam hits a sample, such as an aqueous solution or a solid, and is (elastically) scattered. As the name indicates, only scattering at low angles ($\leq$30 deg) is recorded by a detector. For isotropic samples, the scattering pattern has no azimuthal dependence and depends uniquely on the modulus of the vector $\vec{Q}$, $Q = 4\pi \sin(2\theta/2)/\lambda$, where 2θ is the scattering angle. From reduction of the experimental data, an important quantity, namely the scattering cross section $\mathrm{d}\Sigma/\mathrm{d}\Omega$, is obtained as a function of $\vec{Q}$. This quantity represents the ratio between the number of particles (photons or neutrons) that in the unit of time are scattered in a certain direction reaching a solid angle element $\mathrm{d}\Omega$ and the product between the flux of the incident particles on the sample and the value of the solid angle element itself. $\mathrm{d}\Sigma/\mathrm{d}\Omega$ provides important information about the shape of the scattering structures inside the sample, as well as on the inter-particle interactions [55].

In contrast to reflectometry, the first-order Born approximation is used for the evaluation of SAS data over the whole Q range since multiple scattering effects can usually be neglected. This simplifies data evaluation since, under this approximation, $\mathrm{d}\Sigma/\mathrm{d}\Omega$ may be expressed as the square modulus of the Fourier transform of the SLD profile $\rho(\vec{r})$ [56]:

$$\frac{\mathrm{d}\Sigma}{\mathrm{d}\Omega} = \left|\rho(\vec{r})\exp\left(i\vec{Q}\vec{r}\right)\mathrm{d}^3\vec{r}\right|^2 \tag{17}$$

For the case of scattering from objects with spherical symmetry, integration may be carried out in spherical coordinates and Equation (17) may be simplified to:

$$\frac{\mathrm{d}\Sigma}{\mathrm{d}\Omega} = \left|4\pi \int \rho(r) r^2 \frac{\sin(Qr)}{Qr} \mathrm{d}r\right|^2 \tag{18}$$

which can be used to simulate the cross section starting from the knowledge of the SLD profile obtained from MD.

Compared to reflectometry, where the surface to be probed is suitably prepared on an optically smooth surface, SAS experiments are performed in bulk. The sample is therefore certainly not perturbed by the addition of a substrate. The absence of a substrate also means that it does not have to be described in the model to evaluate the data. Last but not least, the sample preparation is typically easier than the preparation of samples used in reflectometry, where many experimental efforts must be provided to deposit a layer on the substrate.

2.4. Molecular Dynamics Simulations

There are several ways to perform atomistic or coarse-grained computer simulations of phospholipid membranes, in particular using Monte-Carlo (MC) or Molecular Dynamics (MD) approaches. In both cases, the interaction potentials between all atoms in the system have to be defined in a force field. There are two parts to a force field: the functional forms of the potentials (e.g., exponential or polynomial) and the parameters in these functions. The choice of the appropriate force field (all-atom, united-atom or coarse grained) and its parameters is the crucial step in every MD simulation. Among many available force fields (AMBER, GROMOS, OPLS, CHARMM, etc.) and their variations, the one validated against the reliable experimental data for the molecules of interest has to be used [57]. If there is no reliable force-field validation data in the literature or if the simulation does not reproduce experimental data, non-trivial force-field parameterization is required. For generating multi-component lipid membrane configurations for MD simulations, there are the MemGen web server [58] and Packmol package [59].

The simulations necessarily simplify the system enormously; a striking example is the contraction of the atoms' electron clouds into usually fixed point-like partial charges, hereby removing, *inter alia*, polarizability effects. The simulations can therefore not be expected to reproduce all the properties of the membrane at the same time. The art of creating a force field is therefore to tune the functions and parameters such that the quantities of interest are reproduced while others can be incorrect.

MD simulations produce trajectories depicting the motions of atoms over a specified simulation time, usually on the nanosecond to microsecond timescale—depending on the force-field complexity and available computational resources. Some of the most important analyses, technical challenges and existing protocols that can be performed on MD trajectories of the phospholipid membrane were reviewed by Moradi et al. [60]. However, biological processes related to phospholipid membranes are complex and usually challenging either from an experimental or computational aspect. This comprises membrane pore formation, membrane fusion, stalks, domains and curvatures [11,12,61].

3. Materials and Methods

3.1. Materials

For the experiments on a single supported bilayer of DMPC (1,2-dimyristoyl-*sn*-glycero-3-phosphocholine, Avanti polar lipids), lipids were dissolved in chloroform followed by solvent evaporation under a stream of nitrogen gas. The lipids were subsequently dissolved in 50 mM HEPES, 50 mM NaCl pH 7.3 buffer followed by sonication to produce vesicles, before being pumped across the reflectivity cell to form a continuous bilayer in 50 mM HEPES, 50 mM NaCl pH 7.3 buffer.

The substrate for the DMPC bilayer consisted of a highly polished silicon block coated with a natural silicon oxide layer. The reflectivity cell was connected to a system where a HPLC pump was used to run the buffers through the sample. Four buffer contrasts were used in these experiments: H_2O with SLD $= -0.56 \times 10^{-6}$ Å^{-2}, D_2O (SLD $= 6.35 \times 10^{-6}$ Å^{-2}), silicon matched water (SiMW) composed of 38% D_2O and 62% H_2O (SLD $= 2.07 \times 10^{-6}$ Å^{-2}) and 4-matched water (4 MW) composed of 66% D_2O and 34% H_2O, (SLD $= 4.00 \times 10^{-6}$ Å^{-2}). The mass of DMPC in the neutron beam during the reflectometry experiment was on the order of 1 µg.

The experimental procedures used to prepare multilayers of SoyPC have been discussed elsewhere [31] together with the chemicals used. Briefly, the phospholipid mixture was dissolved in pure isopropanol and the resulting solution was poured on top of an ultra-polished silicon mirror. The solvent was then removed by keeping the mirror at first at reduced pressure and then under vacuum for a few hours. The mirror was then mounted into a custom-made sample cell and filled with heavy water. In order to visually inspect the SoyPC layer and check for eventual air bubbles formed after injection of D_2O, the cell was equipped with a glass cover. Samples used for SANS investigations were prepared starting from a stock solution, dissolving a suitable amount of SoyPC in pure chloroform. The dissolution was favored by a slight warming (40 °C) and a very short sonication treatment ($\approx$5 min). A thin film was subsequently obtained through slow evaporation of the chloroform in a stream of argon, in order to prevent phospholipid oxidation. The phospholipid film was hydrated with D_2O, and the resulting suspension was vortexed and then gently sonicated ($\approx$30 min). An aliquot was then repeatedly extruded through a polycarbonate membrane of 100 nm pore size 11 times. The concentration of the hydrogenated SoyPC in D_2O was 5.0 mmol/kg. The mass of SoyPC in the neutron beam during the SANS experiment was on the order of 1 mg. During the SANS experiment, the sample was contained in a closed Hellma 404-QX quartz cell that had a thickness of 2 mm, to prevent solvent evaporation.

3.2. Reflectometry

The neutron reflectivity measurements on DMPC were taken at the ISIS and Muon Source at the Rutherford Appleton Laboratory, Harwell Science and Innovation Centre, using the time-of-flight SURF instrument [62]. The neutron wavelength ranges from 0.5 to 7 Å, a Q range between $\sim$0.01 and 0.3 Å^{-1} was obtained by measuring three different angles $\theta_{\text{incident}} = 0.35°, 0.65°$ and $1.5°$. The slits were chosen to ensure a footprint of 30 mm by 60 mm at the sample stage with an angular resolution of dQ/Q = 3.5%. Vertical slits were scaled linearly with angle. The time-of-flight spectra were recorded with a ^{3}He point detector [63].

Specular and off-specular reflectivities of the SoyPC multilayer were measured at the vertical reflectometer MARIA [64,65] at Heinz Maier-Leibnitz Zentrum (MLZ) in Garching, Germany, as detailed elsewhere [31]. A neutron beam with an average wavelength λ = 10.0 Å and a wavelength spread of $\Delta\lambda/\lambda$ = 0.10 was used. A 4.1 m collimation length with entrance and exit openings of 1.0 mm was used to collimate the incident beam. The sample was mounted on a goniometer and aligned. Reflectivities were measured by varying the incident angle and recording the pattern of the scattered neutrons with a two-dimensional ^{3}He position sensitive detector positioned at 1.9 m from the sample. The experiments were carried out at room temperature.

3.3. Small-Angle Scattering

A SANS measurement on SoyPC liposomes was carried out at the KWS-1 diffractometer [66] installed at the Heinz Maier-Leibnitz Zentrum (MLZ) in Garching, Germany. As detailed elsewhere [31], neutrons with average wavelengths of λ = 5.0 Å and a wavelength spread $\Delta\lambda/\lambda$ = 0.10 were used, by means of a mechanical velocity selector. A two-dimensional 128 × 128 array ^{6}Li scintillation position sensitive detector measured neutrons scattered from the sample. Three collimation (C)/sample-to-detector (D) distances (namely, C_8/D_2, C_8/D_8 and C_{20}/D_{20}, with all distances in meters) allowed collection of data in the scattering vector modulus $Q = 4\pi \sin(2\theta/2)/\lambda$ ranging between 0.0012 and 0.43 Å^{-1}, with 2θ being the scattering angle. The investigated sample was kept under measurement for a period so as to have ≈2 million counts of neutrons. The obtained raw data were corrected for background and empty cell scattering and were then radially averaged. Detector efficiency corrections and transformation to absolute scattering cross sections were executed using a secondary plexiglass standard [67].

3.4. Molecular Dynamics Simulations

MD simulations were carried out using GROMACS 2018.1 package [68]. Initial configurations were generated using Packmol [59].

1,2-dilinoleoyl-*sn*-glycero-3-phosphocholine simulations were carried out in a fully flexible simulation cell containing two phospholipid bilayers consisting of 128 molecules per bilayer (64 molecules per sheet) and 3000 SPC water molecules between the layers was simulated at NPT conditions using Parrinello–Rahman pressure coupling and Nosé–Hoover temperature coupling. The pressure was set to 1 atm through a semi-isotropic coupling with the x/y isothermal compressibility set to 4.5×10^{-5} bar^{-1}, while the phospholipids and water were independently coupled to thermal baths at 300 K with a coupling constant of 0.1 ps. The simulations were run for 100 ns with a time step of 1 fs. The equations of motion were integrated using the Verlet leap-frog algorithm. The long-range electrostatic interactions after a cut-off distance at 0.8 nm were accounted for by the particle-mesh Ewald (PME) algorithm [69]. The 12-6 Lennard–Jones interactions were treated by the conventional shifted force technique with a switch region between 1.2 and 1.4 nm. Cross-interactions between different atom types were derived using the standard Lorentz–Berthelot combination rules. United-atom GROMOS 54A7 force-field parameters were used. The model includes 63 atoms (as opposed to to 134 atoms for the all-atom model) since the hydrogen atoms are integrated into the heavy atoms. Periodic boundary conditions (PBC) were applied in all dimensions. The first step of the simulation was an equilibration process for 5 ns. After that, 100 ns of NPT simulation were performed, saving coordinates every 2 ps for analysis.

DMPC (1,2-dimyristoyl-*sn*-glycero-3-phosphocholine) simulations of 128 DMPC phospholipids and 3655 SPC water molecules were performed with the Berger parameters [70], with the coordinate, force-field and topology files distributed by D. Peter Tieleman (http://wcm.ucalgary.ca/tieleman/downloads, accessed on 1 June 2021). Twenty nanoseconds of NPT simulation were performed with 1 fs time steps, saving coordinates every 2 ps. The resulting area per phospholipid was found to be 60 Å^2, which is the same value as the one obtained by Darré et al. [39] with the CHARMM36 force field and TIP3P water model.

Snapshots were rendered in VMD [71]. The trajectories were either analyzed using TRAVIS-1.14.0 [72,73] and Python scripts written in-house or a new Python program dedicated to this purpose, *Made2Reflect*. This approach allows automatizing analysis of very large trajectories (20–100 ns, i.e., ∼20–30 GB in .pdb format), consequently improving statistics and the calculation of scattering length density profiles on the 10–30 min timescale. Using Python makes the script flexible and easy to adjust to the specific needs of monolayers, multilayers, substrates, etc. In Travis, the density profile function (DProf) was used to calculate the number density distribution of particles along the z axis, i.e., the direction perpendicular to the phospholipid membrane. The result is a histogram that gives the particle density of a selected particle type (either in nm^{-3} or relative to uniform density) in thin slices of the system perpendicular to the chosen vector. The distribution is calculated for every molecule and each atom type. In the next step, the number distribution was multiplied by the atomic scattering length obtaining a scattering length density profile. Since the number distribution is calculated for each atom, it allows for a selective deuteration, i.e., selective isotopic substitution, simply by multiplying the number distribution of selected atoms by the scattering length of D instead of H.

4. Results

4.1. Single Bilayer Neutron Reflectivity of a DMPC Bilayer

To model the scattering of a single DMPC bilayer, a scattering length density profile has to be constructed in real space. This can be achieved either by using an analytic approach where the number densities of different atom types are approximated, e.g., by a Gaussian, or using a numerical approach such as the discretized number density generated from MD simulations. We use here the discretized number density profile of each atom type calculated from the MD simulation so that one can plot and visualize the distribution of the single atom type, specific molecule or its parts. Figure 1 shows the number densities of different elements extracted from an MD simulation, summed up for phospholipid heads and tails separately as an example. Many snapshots along the trajectory were sliced into fine bins (with 0.13 Å thickness) along the z axis, the membrane normal, and the different elements/isotopes were histogrammed in these bins. The time average (using the full length of the trajectory) was taken and the number densities of each atom type were multiplied with their respective neutron scattering lengths. The sum of all contributions, i.e., the total scattering length density profile, is also shown in this figure for different H/D substitutions of the water, i.e., contrast variation—H_2O, D_2O, water with a scattering length density matched to the one of silicon (SiMW, 2.07×10^{-6} Å^{-2}) and water with a scattering length density matched to be 4×10^{-6} Å^{-2} (4 MW).

The first validation step of the calculated SLD profile is to compare the numerical H_2O, D_2O, SiMW and 4MW SLD values with the theoretical bulk SLD values given as dashed lines on the right-hand side of Figure 1. If these were mismatched, either the density obtained from the simulation or the SLD calculation would be incorrect. The next step is to model a semi-infinite silicon substrate with a native SiO_2 layer. The SLD, thickness and roughness of this layer must be obtained through NR measurements and subsequent modeling of Si/SiO_2/D_2O and Si/SiO_2/H_2O. The very same characterized silicon wafer is then used for measuring the NR of the phospholipid bilayer. The modeled substrate is given with dashed lines on the left-hand side of the SLD profile (Figure 1). Merging the simulation SLD with the solid substrate SLD has to be performed with caution since one can produce unwanted artefacts in the reflectivity curve [35]. Particular attention has to be given to the treatment of the substrate roughness, as shown below.

Figure 2 shows the comparison between the measured NR of a single DMPC bilayer and reflectivity calculated directly from an MD simulation. Very good agreement can be observed since the MD curves match all four measured contrasts simultaneously. As the simulations were run without a solid support and the silicon was added by hand while building the SLD profile, the water layer between the substrate and phospholipid head groups also has to be adjusted to fit the experimental data [36]. The layer being about 1 nm thick is in agreement with the literature [74]. The effect of changing this thickness on NR is also shown in Figure 2 (dashed lines). The 5 Å thinner water layer considerably flattens the bump in the reflectivity. As shown below, the influence of this water layer on NR is comparatively minor for multilamellar phospholipid systems.

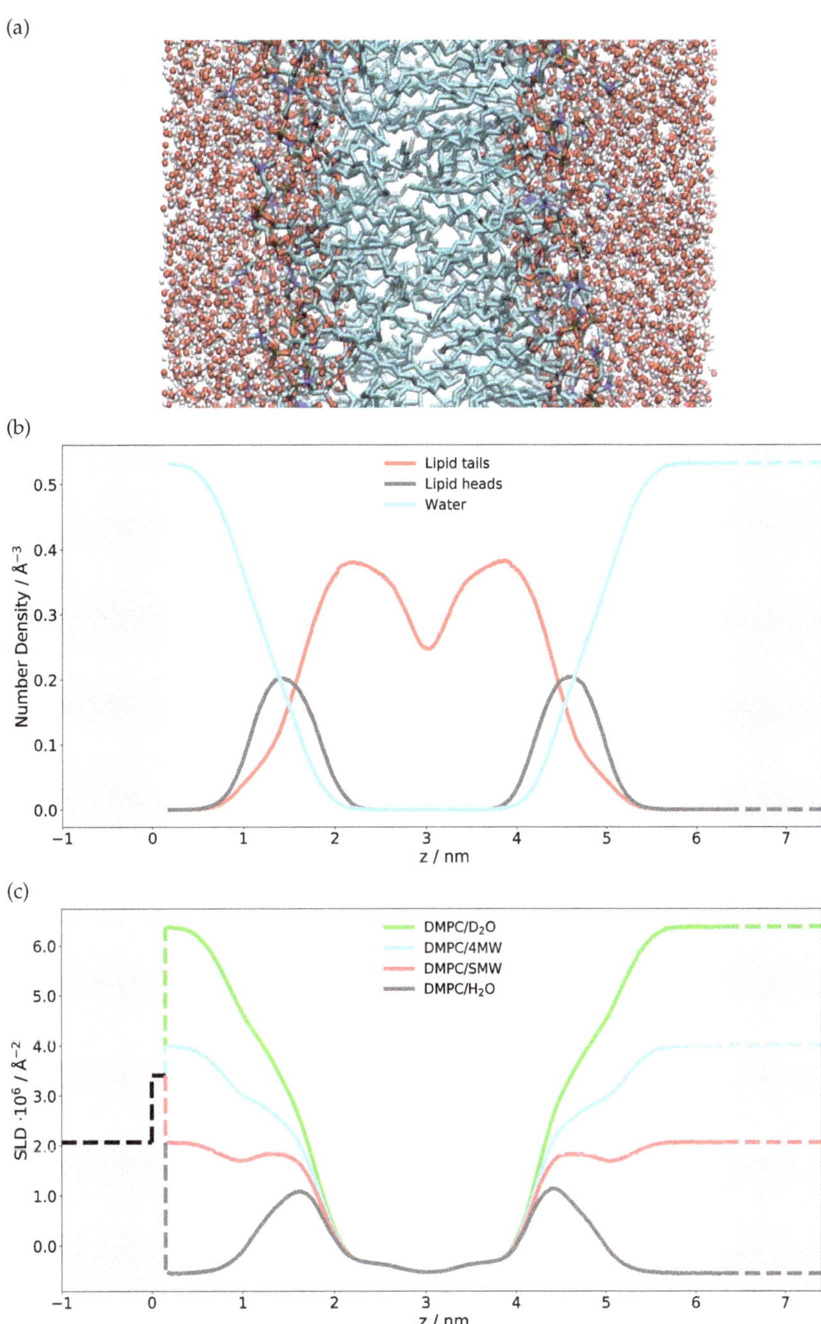

Figure 1. (**a**) Snapshot of the MD simulation of a single free-floating DMPC bilayer in water. The lipid tails can be seen in green, whereas the water molecules are red/white. (**b**) The extracted number density of atoms after averaging over the whole simulation time and summed together based on their presence in a certain group (water/heads/tails). (**c**) Neutron scattering length density (SLD) at four different contrasts calculated from the different atomic number densities. Shaded regions are hand-modeled SLD values for Si/SiO_2 (left) and bulk solvent (right). The SLD can be negative.

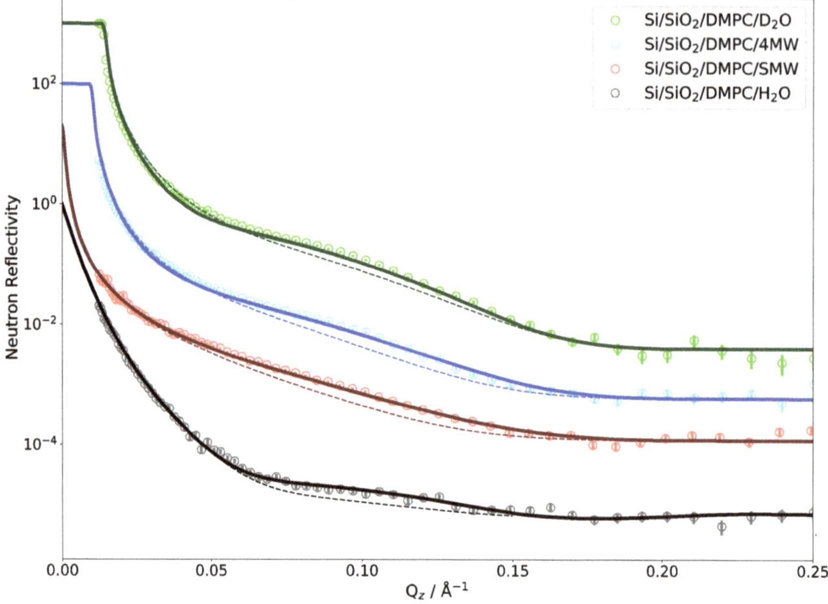

Figure 2. Neutron reflectivity of a DMPC bilayer on a silicon substrate in water for four different H/D contrasts. Points represent experimental data and lines are neutron reflectivity calculated directly from the MD simulations. The curves are shifted along the y axis by a factor 10 each for clarity; they all extrapolate to $R(Q = 0) = 1$. Dashed lines show the effect of reducing the water layer thickness between the substrate and phospholipid head groups from 10 to 5 Å.

There are several parameters related to the experimental setup that have to be taken into consideration when calculating reflectivities from an MD simulation, such as the instrumental resolution, background scattering and substrate roughness. It can also be seen that only two contrasts (D_2O and 4 MW) exhibit a critical edge and that only the D_2O measurement covers it with data points. This means that, for all but the D_2O measurement, one has to rely on the scaling of the measured intensities to absolute values.

4.2. Neutron Reflectivity of a SoyPC Multilayer

The experimented presented in Section 4.1 demonstrated the methodology used on a single phospholipid bilayer. When simulating membrane fusion or stalk formation, of course at least two membranes are required, but, given the low density of stalks, experimentally multilayers (some tens to thousands of bilayers) have to be measured in order to obtain a detectable signal. MD simulations of this many bilayers are neither practically feasible nor useful, since the multilayer can be constructed by repeating the SLD profile of a single bilayer a suitable number of times. This section focuses on the main new features observed and the data evaluation challenges encountered during the study of a multilayer via reflectometry. As an exemplary system, multilamellar SoyPC was chosen since it was hypothesized that, for this phospholipid mixture, the presence of a drug promotes stalks formation [31], and, before MD and scattering methods can be used to look into the details of this question, a good description of the pure and unperturbed system is needed. SoyPC is a mixture of five major lipid components; only the most abundant polyunsaturated 1,2-dilinoleoyl-*sn*-glycero-3-phosphocholine (which we indicate in the following with DLPC, to not be confused with saturated 1,2-dilauroyl-*sn*-glycero-3-phosphocholine) was simulated by MD.

As in the previous case of DMPC, two DLPC bilayers separated by a water layer were simulated and the obtained SLD is presented in Figure 3 for two different contrasts. Once the SLD profile of a single bilayer is extracted from the MD, a multilayer is straightforwardly built by simply repeating the SLD profile n times in z direction. In the case of DLPC, 36 phospholipid bilayer repetitions were used. When applying such a perfect periodicity, this manual merging of two repetitions (usually in the water region) must be performed with caution so that the thickness of each water layer stays the same. Otherwise, an artificial rupture of symmetry can be introduced and the lattice constant would then be doubled. This would introduce a new peak in the calculated NR, at a Q position corresponding to half that of the first Bragg (demonstrated in Figure 4 as a dashed red peak). It is, however, also easily possible to introduce a certain degree of disorder in this step by adding randomness to the water layer thicknesses. In order to simulate the reflectivity in this case, one has to produce a large number of such structures and average the simulations. Such an incoherent addition is valid here since it is expected that the lamellar fluctuations and the corresponding interlamellar distances should be uncorrelated [75].

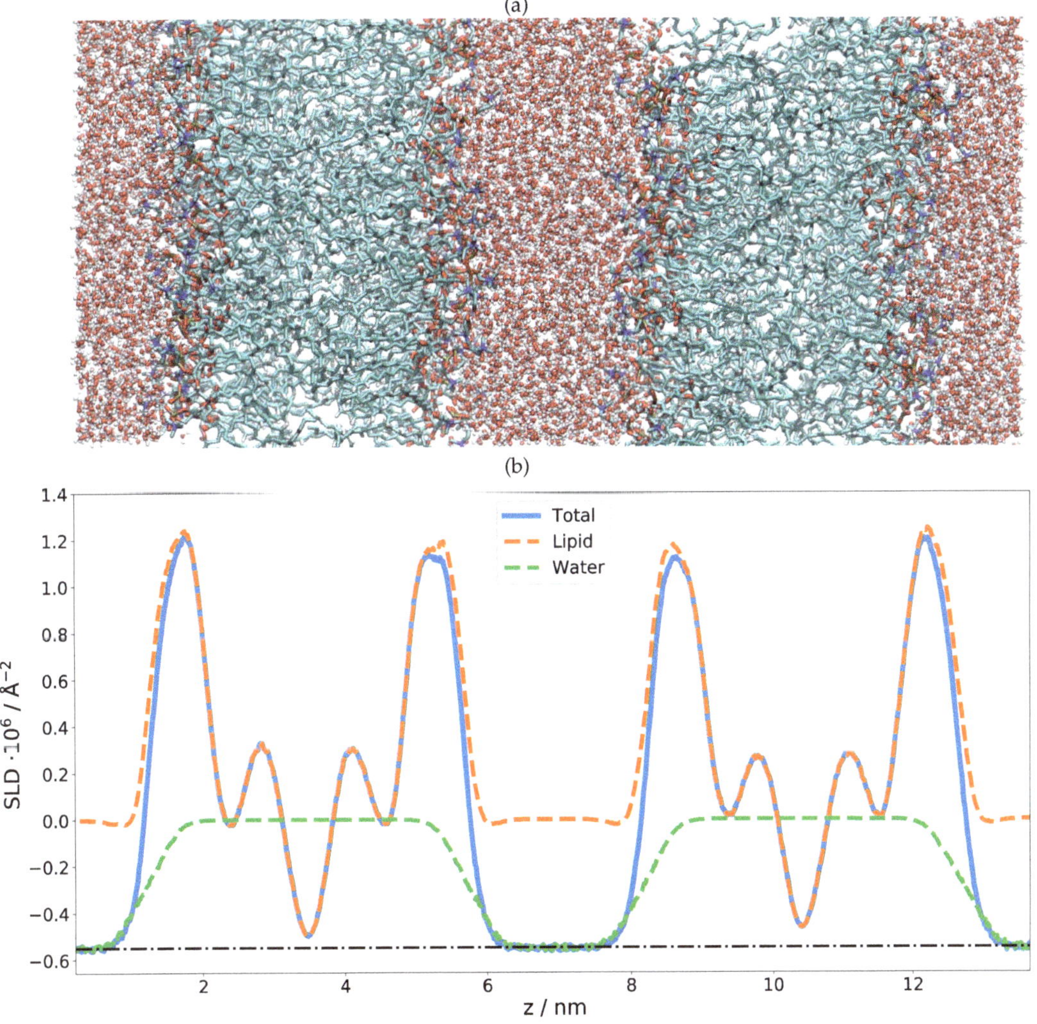

Figure 3. *Cont.*

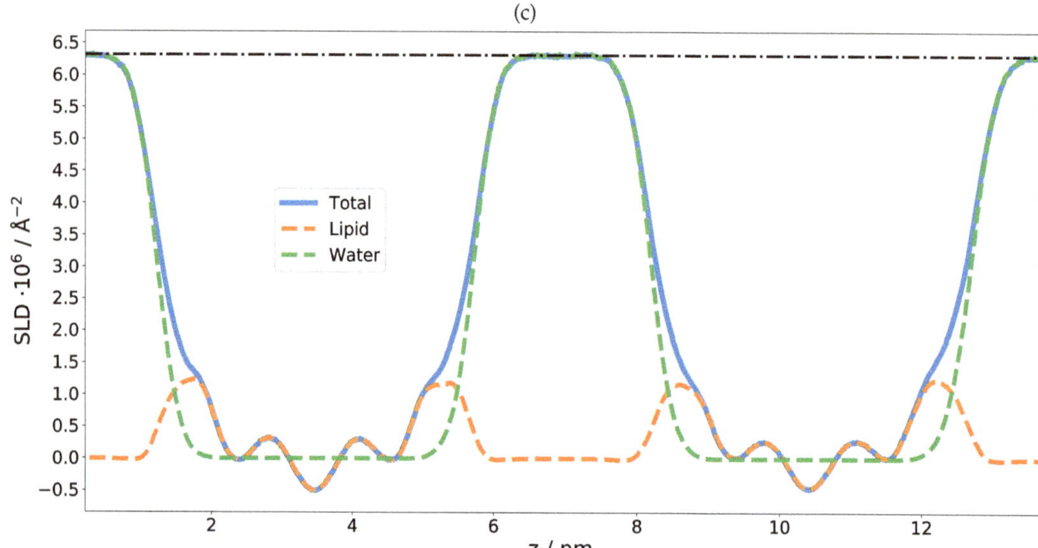

Figure 3. (**a**) MD simulation snapshot of the double phospholipid bilayer. (**b**) SLD profile of the first contrast, H_2O/bilayer/H_2O/bilayer/H_2O. Note that there are negative SLD values. The dash-dotted line indicates the SLD of pure H_2O. (**c**) SLD profile of the second contrast, D_2O/bilayer/D_2O/bilayer/D_2O. The dash-dotted line indicates the SLD of pure D_2O.

The SLD profile obtained for multilamellar DLPC in D_2O was then used to calculate the reflectometry curve. The comparison with the experimental data is given in Figure 4. It is obvious that the simulated reflectivity (blue line) reproduces the first Bragg peak and fits the data well up to $\approx$0.1 Å^{-1}. Between 0.1 and 0.2 Å^{-1}, a discrepancy can be observed.

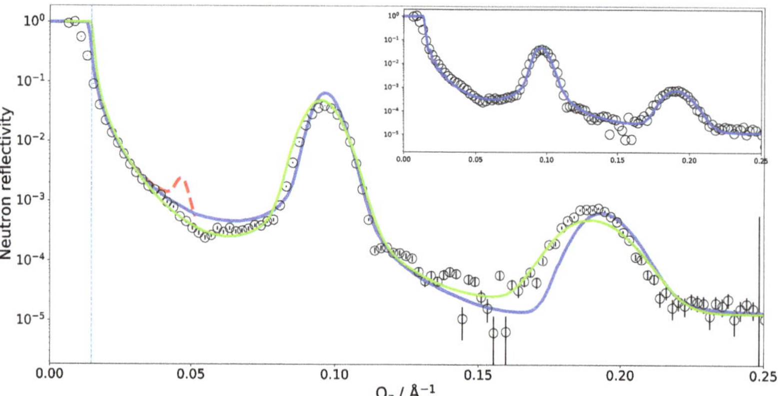

Figure 4. Experimental neutron reflectivity of SoyPC multilayer in D_2O compared with different models. Green line, analytical model published by Mangiapia et al. [31]; blue line, neutron reflectivity calculated directly from the MD simulation. The dashed vertical line marks the theoretical Q_c of Si/D_2O. The dashed red peak reveals the artificial asymmetry in the repartition of the water layer thicknesses (details in the text). The blue line in the inset shows an adjusted MD model (details in the text).

The measurements of SoyPC were in the past evaluated with an analytical model consisting of water, a head group region and a tail group region [31]. The different regions were represented by Gaussian density distributions and repeated without disorder. The algorithm then varied the flexible parameters of the model, e.g., layer thicknesses, etc. The data could be fitted well (green line in Figure 4) and the parameter results were very reproducible and independent of the starting parameters: The phospholipid tail region was fitted to be only 11 Å thick—a surprisingly small value. The water layer was 42 Å thick. The size of the unit cell (the repeat distance of the bilayers) is very well constrained (66 Å) by the positions of the Bragg peaks in the data. The unit cell size was therefore kept constant by the fit, proportionally enlarging the water layer between bilayers when reducing the bilayer thickness.

By comparing the MD SLD profile in Figure 3 with the analytical model published in [31], it is apparent that the analytic fit of the data proposes a smaller membrane thickness than the MD simulation. The water layer thickness in the simulation must be defined a priori by the number of water molecules between the adjacent lipid bilayers in the simulation box. However, this part of the SLD profile can (and must) be adjusted during the SLD profile modeling. Taking the MD model as a starting structure and adjusting the layer thicknesses, it was possible to reproduce the experimental data (see inset in Figure 4). This suggests that the force field and simulation parameters have to be adjusted to increase the density and reduce the thickness of the hydrophobic region. It might also mean that the MD simulation of a pure DLPC system is structurally still different from the mixture present in SoyPC.

Another feature worth noting is shown in Figure 5, which includes a zoom on the critical edge region. The exact position of the critical edge for total reflection is a function of the SLD difference between the semi-infinite medium on which the beam is reflected and the semi-infinite surroundings from which the beam comes. In our case, the beam comes through the side of a thick silicon wafer and is reflected at the interface with D_2O. The exact shape of the reflectivity decay is obviously influenced by the additional layers, but the value of Q_c below which the beam is totally reflected must remain the one of the material combination Si–D_2O. A deviation of the critical edge from the theoretical position leads to the suspicion of a possible contamination of the D_2O by hydrogenated molecules. Two possibilities of hydrogenated molecules come to mind: normal water (H_2O) or phospholipids detached from the multilayer whose hydrogenated tails would significantly lower the overall SLD. In the case of H_2O contamination, one would expect it to be homogeneously distributed across all hydrated parts of the sample. In the case of the phospholipid contamination, however, the contamination would be confined to the bulk water. Figure 5 shows the effect of both scenarios on the NR starting from the MD simulated SLD.

The blue reflectivity curve is obtained by scaling the bulk SLD of D_2O by a factor of 0.8 on account of detached phospholipids with hydrogen-rich tails diffusing to the bulk. In this case, the SLD of D_2O between the bilayers was not scaled. The red dashed curve is obtained by scaling the SLD of all D_2O molecules by a factor of 0.8, simulating D_2O contamination with H_2O during the experiments. Since these corrections did not affect the interlamellar distance, the position of the Bragg peaks is not affected by this change in contrast. Scaling down the D_2O SLD moves Q_c close to the observed value. Adjusting the water SLD inside the lamellar structure, as in the hypothesis of light water contamination, has the effect of reducing the overall contrast of the lamellae and affects the Bragg peak intensity and reflectivity in the 0.1–0.15 $Å^{-1}$ region. One could hope to be able to discriminate two solvent contamination origins on this basis. However, as shown by the green dashed curve in Figure 5, tuning the water SLD has a similar effect on the reflectivity as reducing the number of bilayers in the multilayer.

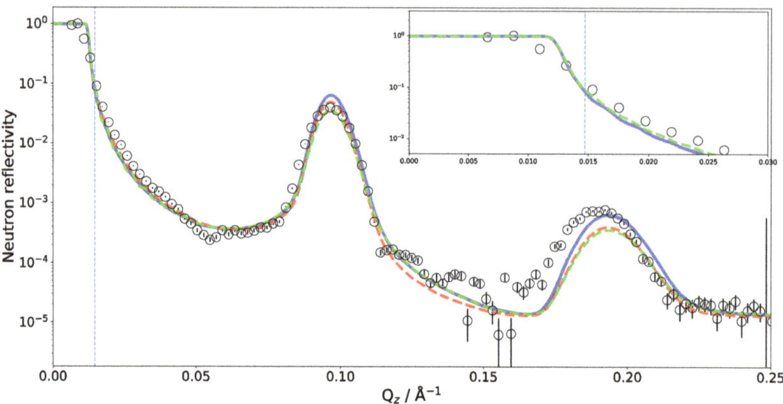

Figure 5. The influence of varying the SLD of D_2O on the critical edge (Q_c) of NR calculated from the MD simulation. Blue line, SLD of only bulk D_2O scaled by a factor of 0.8; dashed red line, SLD of all D_2O in the system scaled by a factor of 0.8; dashed green line, SLD profile after reducing the number of layers in the modeled multilayer from 36 to 20. The dashed vertical line marks the theoretical Q_c of Si/D_2O.

Neutron reflectivity measurements and reflectivity calculated directly from the MD simulations hint at the different structure of the investigated SoyPC multilayer. While the phospholipid bilayer thickness obtained from MD is larger than allowed by the position of the experimental Bragg peaks, the direct unconstrained fitting of the reflectivity using the analytical model suggests an extremely thin hydrophobic region (11 Å). From the MD point of view, such high compression seems hardly achievable since the hydrophobic region in that case has to be thinner than the polar heads. Additional experimental data at different contrasts could help to lift the ambiguity. Since these experimental data were not available, small-angle neutron scattering measurements of a single SoyPC bilayer were compared with the MD simulations.

4.3. Small-Angle Neutron Scattering of SoyPC Bilayers

Scattering cross sections obtained from SANS experiments on SoyPC in heavy water are reported in Figure 6. The data can be fit very well with a model of spherical unilamellar vesicles with polydisperse solvent cores [76,77], which is also expected based on the preparation method. A very careful inspection of the data revealed the presence of a small contribution of multilamellar vesicles [31], which can be neglected in the Q range presented here.

The best fit of the analytical model yields vesicles with a double-layer thickness of (33.0 $\pm$ 1.2) Å. The SLD profile obtained from the MD simulations was used for a comparison to the data. It was inserted into a model of n concentric spherical shells; the inner radius of the vesicle and their polydispersity was optimized by a fitting procedure. The results are displayed with a continuous red curve in Figure 6. There is a clear discrepancy between the experimental data and the description provided by the MD results, which is mainly due to a mismatch of the total membrane thickness. In particular, the oscillation at $Q \approx 0.25$ Å^{-1} is quite sensitive to this parameter. This is illustrated in the inset of Figure 6, where two dashed curves represent adding and subtracting 2.0 Å to the optimized bilayer thickness: a small change shifts the oscillation to higher or lower Q-values. The shoulder at $Q \approx 5 \times 10^{-3}$ Å^{-1} is in contrast not sensitive to the change in bilayer thickness at all and is determined by the total size of the vesicles.

The SANS data on unilamellar vesicles clearly favor a thinner membrane than what is simulated by MD. The associated tail thickness of only 11 Å is, however, so incredibly thin that additional measurements on the pure DLPC system and with a variety of contrasts should be performed before addressing an optimization of the MD force field.

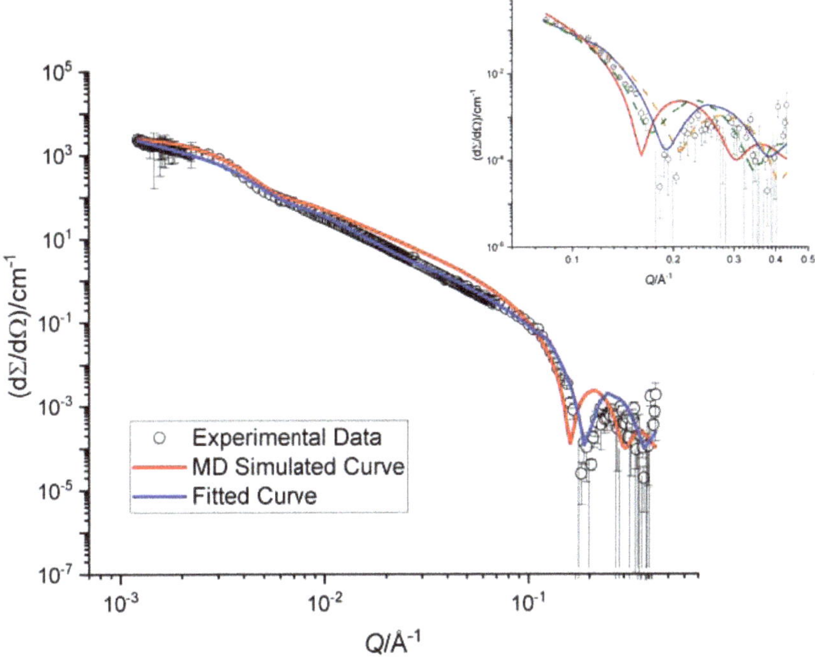

Figure 6. Scattering cross sections obtained for an extruded sample of SoyPC in D_2O. The blue line corresponds to the theoretical cross sections obtained by fitting the model described in the text, whereas the red curve is obtained from the MD SLD profile. The inset shows a zoom on the high-Q region together with two additional dashed curves obtained by adding (in dark green) and subtracting (in orange) 2 Å to the value of the bilayer thickness extracted from the best fit.

5. Discussion

Both reflectometry and small-angle scattering are low-resolution techniques. Their spatial sensitivity is limited to about $d_{\mathrm{min}} \sim 2\pi/Q_{\mathrm{max}} \approx 2\pi/(0.3\,\text{Å}^{-1}) \approx 20\,\text{Å}$. Smaller features can still change the scattering pattern if they change the average SLD of the layer in which they are embedded, but the information content in the data will not confine the shape of this feature. Reflectometry has the advantage over small-angle scattering that the sample is aligned and it is therefore possible to probe the SLD profile along the membrane normal. The scattering signal in small-angle scattering is generally an orientational average.

Despite the limitations of the scattering data, they are some of the very few experimental windows into this nano-world, and it is easy to compare simulations to the data. This combination is particularly powerful since the simulations provide a model that is already heavily constrained by many external inputs via the force field, while the scattering data provide a sensitive indicator of the plausibility of the simulated structures. The ease of comparing simulated and measured scattering curves to each other can, however, lead to an inflated degree of trust—from an experimenter's point of view in the simulations and from a simulator's point of view in the measurements. In the following, we therefore raise the awareness of each of the two communities for the potential problems of the other one—while, and this cannot be stressed enough, unreservedly recommending this combination.

5.1. Reflectometry

For the comparison of the reflectivity curves calculated from analytical or numerical SLD profiles to measured data, one has to take into account instrumental and sample non-idealities.

Effects caused by the instrument vary between different instrument types (e.g., monochromatic or time-of-flight) and even between different instruments of the same type. A non-exhaustive list is given in the following.

- Every instrument will have sources of background which contaminate the intensity with a more or less random noise. These can be independent of the experiment (e.g., the perfectly random detection of cosmic particles) or instrument setting related (e.g., scattering of the probing particles on air or windows in the beam—also random and scattering on slits—a usually more or less strongly peaked effect).
- When the sample is illuminated under a very shallow angle, it might happen that only a fraction of the beam actually illuminates the sample. This geometric effect will be a function of the incident angle and the real intensity distribution in the beam (usually treated as Gaussian). In some configurations, this function could be strongly wavelength dependent, due to the ballistic effect: on their way to the sample, long wavelength neutrons, being slower, fall more than the short wavelength ones under the action of gravitation. They will thus impinge on the sample at a different spot and slightly different incident angle (an effect which also has to be taken into account to properly evaluate Q).
- Both aforementioned effects contribute to a normalization issue: since R is a relative measurement, one must ascertain that the full incident intensity is accurately measured. This can cause practical problems since the primary beam intensity is always orders of magnitude more intense than the reflected one. Gross errors in this step can be detected if enough data points have been taken in the regime of total reflection, but more subtle effects such as the above-mentioned over-illumination are much more difficult to detect if they affect the region where reflectivity intrinsically varies. It should be stressed that it is quite easy to overlook significant systematic errors since the R value is usually plotted on a logarithmic scale with 5–6 orders of magnitude.
- The measured intensity can be described by the convolution of the ideal signal with the instrumental resolution function. This convolution smears the measured curve and limits the possibility to resolve adjacent features (oscillations, peaks) in Q space. In real space, this translates to an upper limit for the measurable layer thickness and sensitivity to long-range correlations. Typically, the instrumental resolution of neutron reflectometers ranges about 1–10% $\Delta Q/Q$ and consists of contributions of the often dominant wavelength uncertainty and the beam divergence.
- A very careful treatment of error propagation during data reduction of the counted intensities is needed in order to preserve the possibility to evaluate the statistical agreement between a simulation and experimental data. Obviously, the error bar validity issue is paramount when dealing with fitting methods, and this is even more so when the fitted data vary over several orders of magnitude, as is the case for both reflectivity and SANS [78].

The sample itself also contributes features to the scattering data that are not reproduced by the computation of the reflectivity from the SLD profile:

- The sample membrane in a reflectometry measurement has to be supported by a substrate, either solid or liquid. This substrate can have an influence on the membrane properties, such as its rigidity. Studies looking at embedding larger proteins into the membrane might even experience collisions between the proteins and the substrate [46]. Further, the surface of the substrate can be ill-defined. While a reasonably thin silicon oxide layer usually does not influence the scattering data too much, the surface roughness of the substrate has an immediate effect on the data and can render the data useless if the roughness is not controlled to be below ~5 Å. Besides influencing the quality of the data, it is clear that large surface irregularities will also affect the membrane morphology. The other half-infinite side also adds possibilities for imperfections that might not be mapped into the simulation: the solvent (especially when deuterated) might be contaminated with another isotope—either from the experimental setup of channels leading to the sample chamber or by hydrogen or hydrogen-containing groups escaping from the sample layer. This might have happened in the DLPC multilayer presented here where an amount of bilayers could have detached from the multilayer and float in some form through the solvent, lowering its SLD.

- A more subtle point concerns the water contrast variation. In order to make use of the different contrasts that can be achieved by isotopic substitution, one has to assume that the exchange between hydrogen and deuterium changes only the scattering lengths and not the actual structure. This is generally a justifiable approximation. The density [79,80] and many of the molecular interactions do change between H_2O and D_2O [80,81], but mostly very slightly. These—usually small—changes that happen in the real sample will of course not be reproduced by only one simulation where the isotopic exchange is performed a posteriori by assigning different scattering lengths to the atoms.
- The sample layer itself can also deviate from the modeled version in several aspects: concerning the SLD, it is basically impossible for an experimenter to ascertain the deuteration degree of the purchased phospholipids. Further, the deposition of phospholipids on the substrate might not have produced the structure that was intended (e.g., a single bilayer)—either on the complete sample or as inhomogeneities within the membrane plane. Neutron reflectometry measurements probe a surface on the order of 10–100 cm^2: it is rather unrealistic that a phospholipid layer would coat such a large area homogeneously. Last, inhomogeneities can of course also occur in the direction of the membrane normal, such as a disorder of the water layer thickness between neighboring membranes in a multilayer.
- The sample will not only scatter neutrons/X-rays into the specular spot, but will also itself contribute an isotropic background which will add up to the extrinsic background sources discussed above. In the case of neutron scattering, this sample-related background level is dominated by the incoherent scattering from hydrogen atoms in the sample and will therefore vary between different contrasts of a given system. In the case of X-rays, the diffuse background is generated by the inelastic Compton scattering. It is measured and subtracted from the signal together with the off-specular scattering (see below). A remaining Q-independent background has to be accounted for in the modeling.
- Most importantly, in reflectometry, the sample will also generate *off-specular* scattering. This scattering intensity is caused by fluctuations of the SLD profile parallel to the membrane due, for instance, to membrane fluctuations. In the current context, this additional intensity overlaps the purely specular signal and needs to be subtracted from the experimental values before R can be evaluated. The length scale of the fluctuations responsible for off-specular scattering is up to the micrometer regime [20], which renders, as hinted above, an evaluation from the computer simulations impossible. The usual approach is therefore to measure the scattered intensity on both sides of the specular condition near to it in order to then interpolate the background intensity. On modern instruments using bidimensional detectors, one does not need to perform any additional experiment since a whole range of reflected angles is being covered around the specular direction. Figure 7 shows the intensity distribution as a function of incident angle (θ_{incident}) and reflection angle ($\theta_{\text{reflected}}$). The specular line is seen along the main diagonal ($\theta_{\text{incident}} = \theta_{\text{reflected}}$), and it shows the total reflection region at the smallest angles. Along this line, the intensity maxima correspond to the Bragg peaks. The most prominent feature of this intensity map is, however, the broad off-specular band which follows the condition $\theta_{\text{incident}} + \theta_{\text{reflected}} = \theta_{\text{Bragg}}$,

which in Q space translates to $Q_z = Q_{\text{Bragg}}$. As hinted above, this intensity band is thus characteristic of the in-plane correlations of the structures responsible for the Bragg peak. The broad width of the Bragg peak in the reciprocal space shows that the bilayers are only coherent over very small length scales. Subtraction of the underlying off-specular signal is clearly a challenging task, especially in regions where the overall reflectivity is low, leading to poor statistics. One needs to be aware of the risk of introducing systematic deviations from the true specular reflectivity during this data reduction step.

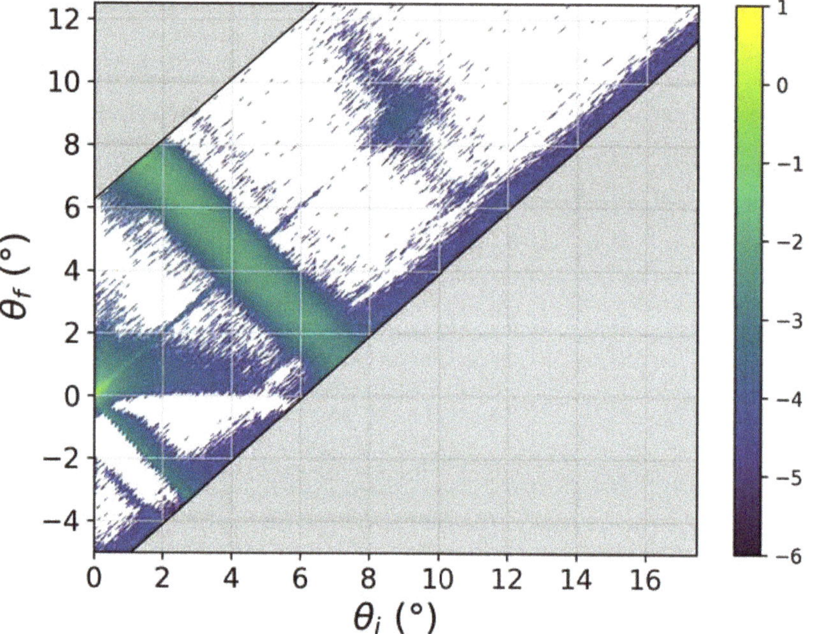

Figure 7. Off-specular reflectivity map (log scale) of the multilayer sample as a function of the angle of incidence (θ_{incident}) and of the reflection angle ($\theta_{\text{reflected}}$). The specular reflected beam contributes only at $\theta_{\text{incident}} = \theta_{\text{reflected}}$. The width of the recorded band around the specular line is defined by the detector size.

5.2. Small-Angle Scattering

While there is no substrate in small-angle scattering experiments, the other issues mentioned for reflectometry also exist for this technique. For example, unilamellar vesicles might be neither as spherical nor as homogeneously unilamellar as assumed. If they were produced by extensive sonication, the phospholipid molecules might even have been damaged and lyso-phospholipids can be present in the sample [82].

Factors that do not play a role for the samples used in reflectometry but have to be considered in SAS include the vesicle size polydispersity, which is basically always modeled with rather simple assumptions, such as a the Schulz–Zimm distribution. It is absolutely possible that the real size distribution is far more complex and does not smear the features in the scattering curve in exactly the way that is modeled. In addition, the positioning of different vesicles with respect to each other has an—often subtle—influence on the scattering data. If the interaction between the vesicles is known, it can (and should) be taken into consideration in the model.

The effects of an isotropic background and instrumental resolution are very similar to those observed in reflectometry. The background due to the sample itself is usually accounted for in the modeling, while all other background contributions are subtracted from the data using measurements of the empty sample container and the intrinsic noise of the detector. The instrumental resolution has to be taken into consideration much in the same way as for reflectometry.

Instrumental challenges that are more specific to small-angle scattering than reflectometry are related to the calibration of the detector efficiency [83], especially when attempting to obtain absolute units for the data in order to measure concentrations. Additionally, a full small-angle scattering curve is often measured in several steps with varying collimation lengths and the individual curves have to be stitched together before modeling, mostly manually.

As discussed in the Introduction, the theoretical treatment of SANS curves relies on the applicability of the first Born approximation. Care must be taken to avoid multiple scattering as this is not incorporated in the theoretical evaluation of the data—unfortunately, the absence or presence of multiple scattering is usually not apparent in the data. A practical rule of thumb is to lower the concentration until the sample transmits at least roughly 85% of the beam without interaction.

5.3. Molecular Dynamics Simulations

As in the case of the experiments, there are several aspects of computer simulations that might not be immediately clear to the non specialist, which might lead to misinterpretation of the results. In contrast to the DMPC phospholipid, containing saturated fatty acid chains, where simulations match experimental data very well, there is a disagreement in the case of polyunsaturated DLPC (the main component of SoyPC). The DLPC lipid bilayer thickness and area per lipid obtained in the MD simulation are not in agreement with the NR and SANS measurements. The experiments suggest a larger area per lipid (75 $Å^2$) than the MD simulations (60 $Å^2$) using the GROMOS 54A7 force field. The same problem was observed for polyunsaturated 1-stearoyl-2-docosahexaenoyl-*sn*-glycerco-3-phosphocholine (SDPC) bilayers [84] where high-level quantum mechanical calculations are used to improve the force fields' (CHARMM36) dihedral potential of neighboring double bonds. An approach was proposed by Marquardt et al. [8] constraining the average area per lipid while allowing the z axis to expand and contract. This issue, including the force field reparameterization, is out of the scope of this article and will be addressed in our future work on SoyPC, including the measurements and comparison of pure mono- and multilamellar DLPC with the simulations.

It is important to say that simulated SLD profiles cannot be used "as received" and several adjustments have to be made. A first crucial point is related to the thickness of water layers: since the number of water molecules is fixed at the start of the simulation, the water layer thickness will also be artificially defined by the MD simulation. However, this issue can easily be solved and the water layer thickness can be adjusted as a parameter during the SLD profile calculation while keeping all the other parameters untouched. There are other practical reasons why the simulated system does not perfectly describe the actual sample used during the neutron/X-ray experiments. The MD simulation will not include the substrate and even less its interface imperfections. Due to limited computation power, the simulated volume usually represents a small fraction of the actual sample and cannot reproduce large-scale features such as the radius of curvature or fluctuations. The same argument justifies why the simplest subelements of quasi-periodic systems are simulated and then artificially reproduced, as for instance a bilayer is simulated in details rather than a true multilayer system.

There are some more obvious reasons a simulation might not be a true representation of reality: in a complex lipid mixture, the composition of the system under study has to be drastically simplified. In the current example of SoyPC, it was approximated by the most abundant phospholipid. In addition, the effects of pH are hard to reproduce: while the pH clearly plays a major role in reality, a simulation box will contain only very few OH^-/H_3O^+ ions if they are included at all—and the most common force fields will not allow the phospholipids to change their protonation state. Besides these factors that limit the realism of the simulations, two more craftsmanship-related issues have to be considered since MD is nothing more than a means of sampling the phase space. First, the simulation has to be equilibrated for a long enough time. Second, the production run has to cover a long enough time for the system to sample the configurations in the vicinity of the energy minimum with accurate statistics. Transitions over energy barriers or phase transitions can pose significant obstacles against sufficient sampling. Further, there are severe limitations in simulations when it comes to temperature and pressure: both of these can typically not be expected to have a 1:1 relation to the real physical quantities. Just to name one example, the freezing point of commonly used water models varies between 213 and 271 K [85]. The last, even more general comment is that all available force fields represent only a part of the interactions that happen in reality. Moreover, even for the interactions taken into account, the force fields are always not more than mere simplifying parameterization. One might be tempted to discard classical MD and make use of a priori more realistic methods based on first principles. Apart from the fact that computing power severely limits the applicability of those methods to a smaller number of atoms, it should be noted that even the ab initio molecular dynamics simulations, which explicitly deal with many more effects than the classical simulations, still contain a number of adjustable parameters that have to be chosen by the simulation operator.

One aspect which has only been discussed briefly here and deserves some more comments is related to lateral correlations in the plane of the sample (such as lamellar fluctuations, undulations and fusion sites, e.g., stalks). We mention this point in the discussion of data reduction and specifically of background subtraction. Obviously, this scattering transports very important information about the exact nature of the in-plane structure and treating it as mere background is wasting information. However, given the very large length scales involved, atomic MD cannot produce relevant data and other methods, such as coarse-grained simulations, need to be used. Once a model of the structure has been constructed, however, well-documented software packages already exist which could be used to compute the corresponding off-specular scattering patterns from the reconstructed SLD distribution [86].

6. Conclusions

In this study, we showed, on the basis of a set of actual examples, how the structures obtained from MD simulations can be used to compute the corresponding scattering patterns in SANS and NR. It appeared along the way that several oversimplifications and assumptions have to be carefully dealt with, notably in producing a reliable description of the sample involving some "details" which are not simulated by MD (the substrate in reflectometry, the multilayer, D_2O/H_2O contamination, substrate roughness, etc.).

Clearly, potential imperfections and intrinsic limitations of all the techniques have to be kept in mind and overconfidence in a single observation to draw conclusions is at best risky. In our experience, however, confronting the experiment to the simulation and *vice versa* is a beneficial process for both sides as it opens opportunities for further understanding of the systems under study and, on a more mundane level, it helps to detect and/or understand inconsistencies (such as solvent contamination, the importance of fluctuations, etc.).

The complementarity of scattering methods and MD simulations is striking, not only because it bridges the divide between direct and reciprocal space and helps to solve the age old phase problem, but also because each method sheds light in the blind spot of the other, for instance in terms of accessible length scales and timescales.

Apart from suggesting new experiments to be performed on this very system (e.g., monolayers and increasingly thick multilayer systems, more contrasts and eventually moving to more complex/interesting systems such as those including drugs), this work hints at possible methodological developments such as the systematic use of MD models for the preparation and analysis of scattering experiments.

The calculation of expected, reasonable scattering patterns can assist in the preparation and the optimization of experiments to be performed at large-scale facilities. Subtle instrumental effects could be simulated in the framework of virtual experiments such as those performed using Monte-Carlo simulation packages [87]. Knowing where to expect important features or how these would differ between competing models would allow measurements to be tailored to concentrate on these regions.

One can further imagine a range of tools that would allow the investigator to alter the MD simulation to optimize the agreement between calculated scattering curves and experimental data: first, one could tweak the SLD profile without touching the simulation itself, simply using it as a suitable starting point. Second, the deviation between calculated and measured scattering curves can be employed as an additional contribution to the potential, driving the simulation into a compatible configuration [88]. Third, it might be possible to adjust individual parameters in the force field, possibly via big data/machine learning approaches. Scattering methods are possibly the only class of experiments that probe directly the very thing MD simulates, giving a unique angle in this ambitious endeavor.

To facilitate these ideas, a new software for calculating neutron and X-ray small-angle scattering and reflectivity patterns directly from the MD simulation trajectory, *Made2Reflect*, will be published soon. This standalone Python program allows the fast and simple analysis of large trajectories and is applicable not only for phospholipid membranes but also for electrochemistry, corrosion and batteries, i.e., solid–liquid and liquid–liquid interfaces in general.

Author Contributions: Conceptualization, S.B. and J.-F.M.; methodology, N.Z. and G.M.; software, N.Z., G.M. and J.-F.M.; validation, N.Z. and G.M.; formal analysis, N.Z. and G.M.; investigation, N.Z., G.M., A.C.H., R.B. and J.L.O.-R.; resources, A.K., H.F. and M.C.; data curation, N.Z. and G.M.; writing—original draft preparation, N.Z., G.M., S.B. and J.-F.M.; writing—review and editing, N.Z., G.M., A.C.H., A.K., H.F., M.C., J.L.O.-R., S.B. and J.-F.M.; visualization, N.Z., G.M. and J.-F.M.; supervision, S.B. and J.-F.M.; project administration, S.B. and J.-F.M.; and funding acquisition, J.L.O.-R., S.B. and J.-F.M. All authors have read and agreed to the published version of the manuscript.

Funding: This research was partially funded by the Wellcome Trust [207743/Z/17/Z]; the Royal Society [RGS\R1\191414].

Data Availability Statement: The data presented in this study are openly available: DMPC specular reflectivity (10.5286/ISIS.E.RB1820565 and 10.5281/zenodo.4882721). SoyPC specular and off-specular neutron reflectivity, SoyPC SANS data, DLPC force field parameters (Gromos54a7) and DLPC.pdb structure (10.5281/zenodo.4882721). Publicly available DMPC force field parameters were used in this study. This data can be found here: http://wcm.ucalgary.ca/tieleman/downloads, accessed on 1 June 2021.

Acknowledgments: This work was based on experiments performed at the MARIA and KWS-1 instruments operated by JCNS at the Heinz Maier-Leibnitz Zentrum (MLZ), Garching, Germany. We gratefully acknowledge the Science and Technology Facilities Council (STFC) for access to neutron beam time at ISIS and the instrument SURF.

Conflicts of Interest: The authors declare no conflict of interest. The funders had no role in the design of the study; in the collection, analyses, or interpretation of data; in the writing of the manuscript, or in the decision to publish the results.

References

1. Cooper, G. *The Cell: A Molecular Approach*; Sinauer Associates, Inc.: Sunderland, MA, USA, 2016.
2. Alberts, B.; Johnson, A.; Lewis, J.; Morgan, D.; Raff, M.; Roberts, K.; Walter, P. *Molecular Biology of the Cell*; Garland Science, Taylor and Francis Group: New York, NY, USA, 2015.
3. Stone, M.B.; Shelby, S.A.; Veatch, S.L. Super-Resolution Microscopy: Shedding Light on the Cellular Plasma Membrane. *Chem. Rev.* **2017**, *117*, 7457–7477. [CrossRef]
4. Connell, S.D.; Smith, D.A. The atomic force microscope as a tool for studying phase separation in lipid membranes (Review). *Mol. Membr. Biol.* **2006**, *23*, 17–28. [CrossRef] [PubMed]
5. Macháň, R.; Hof, M. Recent Developments in Fluorescence Correlation Spectroscopy for Diffusion Measurements in Planar Lipid Membranes. *Int. J. Mol. Sci.* **2010**, *11*, 427–457. [CrossRef]

6. Binder, H. The Molecular Architecture of Lipid Membranes—New Insights from Hydration-Tuning Infrared Linear Dichroism Spectroscopy. *Appl. Spectrosc. Rev.* **2003**, *38*, 15–69. [CrossRef]
7. Salditt, T.; Münster, C.; Mennicke, U.; Ollinger, C.; Fragneto, G. Thermal Fluctuations of Oriented Lipid Membranes by Nonspecular Neutron Reflectometry. *Langmuir* **2003**, *19*, 7703–7711. [CrossRef]
8. Marquardt, D.; Heberle, F.A.; Pan, J.; Cheng, X.; Pabst, G.; Harroun, T.A.; Kučerka, N.; Katsaras, J. The structures of polyunsaturated lipid bilayers by joint refinement of neutron and X-ray scattering data. *Chem. Phys. Lipids* **2020**, *229*, 104892. [CrossRef] [PubMed]
9. Mills, T.T.; Toombes, G.E.; Tristram-Nagle, S.; Smilgies, D.M.; Feigenson, G.W.; Nagle, J.F. Order Parameters and Areas in Fluid-Phase Oriented Lipid Membranes Using Wide Angle X-Ray Scattering. *Biophys. J.* **2008**, *95*, 669–681. [CrossRef] [PubMed]
10. Kučerka, N.; Nagle, J.F.; Sachs, J.N.; Feller, S.E.; Pencer, J.; Jackson, A.; Katsaras, J. Lipid Bilayer Structure Determined by the Simultaneous Analysis of Neutron and X-Ray Scattering Data. *Biophys. J.* **2008**, *95*, 2356–2367. [CrossRef] [PubMed]
11. Marrink, S.J.; de Vries, A.H.; Tieleman, D.P. Lipids on the move: Simulations of membrane pores, domains, stalks and curves. *Biochim. Biophys. Acta BBA Biomembr.* **2009**, *1788*, 149–168. [CrossRef] [PubMed]
12. Kirsch, S.A.; Böckmann, R.A. Membrane pore formation in atomistic and coarse-grained simulations. *Biochim. Biophys. Acta BBA Biomembr.* **2016**, *1858*, 2266–2277. [CrossRef] [PubMed]
13. Yang, J.; Martí, J.; Calero, C. Pair interactions among ternary DPPC/POPC/cholesterol mixtures in liquid-ordered and liquid-disordered phases. *Soft Matter* **2016**, *12*, 4557–4561. [CrossRef] [PubMed]
14. Yang, J.; Calero, C.; Martí, J. Diffusion and spectroscopy of water and lipids in fully hydrated dimyristoylphosphatidylcholine bilayer membranes. *J. Chem. Phys.* **2014**, *140*, 104901. [CrossRef] [PubMed]
15. Harroun, T.A.; Wignall, G.D.; Katsaras, J. Neutron Scattering for Biology. In *Neutron Scattering in Biology: Techniques and Applications*; Springer: Berlin/Heidelberg, Germany, 2006; pp. 1–18.
16. Krueger, S.; Meuse, C.W.; Majkrzak, C.F.; Dura, J.A.; Berk, N.F.; Tarek, M.; Plant, A.L. Investigation of Hybrid Bilayer Membranes with Neutron Reflectometry: Probing the Interactions of Melittin. *Langmuir* **2001**, *17*, 511–521. [CrossRef]
17. Fragneto, G. Neutrons and model membranes. *Eur. Phys. J. Spec. Top.* **2012**, *213*, 327–342. [CrossRef]
18. Campbell, R.A. Recent advances in resolving kinetic and dynamic processes at the air/water interface using specular neutron reflectometry. *Curr. Opin. Colloid Interface Sci.* **2018**, *37*, 49–60. [CrossRef]
19. Luchini, A.; Gerelli, Y.; Fragneto, G.; Nylander, T.; Pálsson, G.K.; Appavou, M.S.; Paduano, L. Neutron Reflectometry reveals the interaction between functionalized SPIONs and the surface of lipid bilayers. *Colloids Surf. B Biointerfaces* **2017**, *151*, 76–87. [CrossRef]
20. Pynn, R.; Baker, S.M.; Smith, G.; Fitzsimmons, M. Off-specular scattering in neutron reflectometry. *J. Neutron Res.* **1999**, *7*, 139–158. [CrossRef]
21. Sivia, D.S. *Elementary Scattering Theory: For X-ray and Neutron Users*; Oxford University Press: Oxford, UK; New York, NY, USA, 2011.
22. Zec, N.; Mangiapia, G.; Zheludkevich, M.L.; Busch, S.; Moulin, J.F. Revealing the interfacial nanostructure of a deep eutectic solvent at a solid electrode. *Phys. Chem. Chem. Phys.* **2020**, *22*, 12104–12112. [CrossRef]
23. Benz, R.W.; Castro-Roman, F.; Tobias, D.J.; White, S.H. Experimental validation of molecular dynamics simulations of lipid bilayers: A new approach. *Biophys. J.* **2005**, *88*, 805–817. [CrossRef]
24. Dodda, L.S.; de Vaca, I.C.; Tirado-Rives, J.; Jorgensen, W.L. LigParGen web server: An automatic OPLS-AA parameter generator for organic ligands. *Nucleic Acids Res.* **2017**, *45*, W331–W336. [CrossRef]
25. Malde, A.K.; Zuo, L.; Breeze, M.; Stroet, M.; Poger, D.; Nair, P.C.; Oostenbrink, C.; Mark, A.E. An Automated Force Field Topology Builder (ATB) and Repository: Version 1.0. *J. Chem. Theory Comput.* **2011**, *7*, 4026–4037. [CrossRef] [PubMed]
26. Zoete, V.; Cuendet, M.A.; Grosdidier, A.; Michielin, O. SwissParam: A fast force field generation tool for small organic molecules. *J. Comput. Chem.* **2011**, *32*, 2359–2368. [CrossRef]
27. Siligardi, G.; Hussain, R.; Patching, S.G.; Phillips-Jones, M.K. Ligand- and drug-binding studies of membrane proteins revealed through circular dichroism spectroscopy. *Biochim. Biophys. Acta BBA Biomembr.* **2014**, *1838*, 34–42. [CrossRef] [PubMed]
28. Upert, G.; Luther, A.; Obrecht, D.; Ermert, P. Emerging peptide antibiotics with therapeutic potential. *Med. Drug Discov.* **2021**, *9*, 100078. [CrossRef]
29. van Hoogevest, P.; Wendel, A. The use of natural and synthetic phospholipids as pharmaceutical excipients. *Eur. J. Lipid Sci. Technol.* **2014**, *116*, 1088–1107. [CrossRef]
30. Jaksch, S.; Lipfert, F.; Koutsioubas, A.; Mattauch, S.; Holderer, O.; Ivanova, O.; Frielinghaus, H.; Hertrich, S.; Fischer, S.F.; Nickel, B. Influence of ibuprofen on phospholipid membranes. *Phys. Rev. E* **2015**, *91*, 022716. [CrossRef] [PubMed]
31. Mangiapia, G.; Gvaramia, M.; Kuhrts, L.; Teixeira, J.; Koutsioubas, A.; Soltwedel, O.; Frielinghaus, H. Effect of benzocaine and propranolol on phospholipid-based bilayers. *Phys. Chem. Chem. Phys.* **2017**, *19*, 32057–32071. [CrossRef]
32. Xue, Y.; He, L.; Middelberg, A.P.J.; Mark, A.E.; Poger, D. Determining the Structure of Interfacial Peptide Films: Comparing Neutron Reflectometry and Molecular Dynamics Simulations. *Langmuir* **2014**, *30*, 10080–10089. [CrossRef]
33. Vanegas, J.M.; Heinrich, F.; Rogers, D.M.; Carson, B.D.; La Bauve, S.; Vernon, B.C.; Akgun, B.; Satija, S.; Zheng, A.; Kielian, M.; et al. Insertion of Dengue E into lipid bilayers studied by neutron reflectivity and molecular dynamics simulations. *Biochim. Biophys. Acta BBA Biomembr.* **2018**, *1860*, 1216–1230. [CrossRef]

34. Reguera, J.; Ponomarev, E.; Geue, T.; Stellacci, F.; Bresme, F.; Moglianetti, M. Contact angle and adsorption energies of nanoparticles at the air–liquid interface determined by neutron reflectivity and molecular dynamics. *Nanoscale* **2015**, *7*, 5665–5673. [CrossRef]
35. Hughes, A.V.; Ciesielski, F.; Kalli, A.C.; Clifton, L.A.; Charlton, T.R.; Sansom, M.S.P.; Webster, J.R.P. On the interpretation of reflectivity data from lipid bilayers in terms of molecular-dynamics models. *Acta Crystallogr. Sect. Struct. Biol.* **2016**, *72*, 1227–1240. [CrossRef]
36. Koutsioubas, A. Combined Coarse-Grained Molecular Dynamics and Neutron Reflectivity Characterization of Supported Lipid Membranes. *J. Phys. Chem. B* **2016**, *120*, 11474–11483. [CrossRef]
37. McCluskey, A.R.; Grant, J.; Smith, A.J.; Rawle, J.L.; Barlow, D.J.; Lawrence, M.J.; Parker, S.C.; Edler, K.J. Assessing molecular simulation for the analysis of lipid monolayer reflectometry. *J. Phys. Commun.* **2019**, *3*, 075001. [CrossRef]
38. Kučerka, N.; Katsaras, J.; Nagle, J.F. Comparing Membrane Simulations to Scattering Experiments: Introducing the SIMtoEXP Software. *J. Membr. Biol.* **2010**, *235*, 43–50. [CrossRef] [PubMed]
39. Darré, L.; Iglesias-Fernandez, J.; Kohlmeyer, A.; Wacklin, H.; Domene, C. Molecular Dynamics Simulations and Neutron Reflectivity as an Effective Approach To Characterize Biological Membranes and Related Macromolecular Assemblies. *J. Chem. Theory Comput.* **2015**, *11*, 4875–4884. [CrossRef]
40. Grillo, I. Small-Angle Neutron Scattering and Applications in Soft Condensed Matter. In *Soft Matter Characterization*; Springer: Dordrecht, The Netherlands , 2008; pp. 723–782.
41. Fischer, H.E.; Barnes, A.C.; Salmon, P.S. Neutron and X-ray diffraction studies of liquids and glasses. *Rep. Prog. Phys.* **2005**, *69*, 233–299. [CrossRef]
42. Tristram-Nagle, S.; Liu, Y.; Legleiter, J.; Nagle, J.F. Structure of Gel Phase DMPC Determined by X-ray Diffraction. *Biophys. J.* **2002**, *83*, 3324–3335. [CrossRef]
43. Salditt, T.; Li, C.; Spaar, A.; Mennicke, U. X-ray reflectivity of solid-supported, multilamellar membranes. *Eur. Phys. J. E* **2002**, *7*, 105–116. [CrossRef]
44. Sears, V.F. Neutron scattering lengths and cross sections. *Neutron News* **1992**, *3*, 26–37. [CrossRef]
45. Heinrich, F.; Kienzle, P.A.; Hoogerheide, D.P.; Lösche, M. Information gain from isotopic contrast variation in neutron reflectometry on protein–membrane complex structures. *J. Appl. Crystallogr.* **2020**, *53*, 800–810. [CrossRef]
46. Böhm, P.; Koutsioubas, A.; Moulin, J.F.; Rädler, J.O.; Sackmann, E.; Nickel, B. Probing the Interface Structure of Adhering Cells by Contrast Variation Neutron Reflectometry. *Langmuir* **2018**, *35*, 513–521. [CrossRef]
47. Junghans, A.; Watkins, E.B.; Barker, R.D.; Singh, S.; Waltman, M.J.; Smith, H.L.; Pocivavsek, L.; Majewski, J. Analysis of biosurfaces by neutron reflectometry: From simple to complex interfaces. *Biointerphases* **2015**, *10*, 019014. [CrossRef] [PubMed]
48. Mushtaq, A.U.; Ådén, J.; Clifton, L.A.; Wacklin-Knecht, H.; Campana, M.; Dingeldein, A.P.G.; Persson, C.; Sparrman, T.; Gröbner, G. Neutron reflectometry and NMR spectroscopy of full-length Bcl-2 protein reveal its membrane localization and conformation. *Commun. Biol.* **2021**, *4*, 507. [CrossRef]
49. Nielsen, J. *Elements of Modern X-ray Physics*; Wiley: Hoboken, NJ, USA, 2011.
50. Parratt, L.G. Surface Studies of Solids by Total Reflection of X-Rays. *Phys. Rev.* **1954**, *95*, 359–369. [CrossRef]
51. Abelès, F. La théorie générale des couches minces. *J. Phys. Radium* **1950**, *11*, 307–309. [CrossRef]
52. Born, M. *Principles of Optics: Electromagnetic Theory of Propagation, Interference, and Diffraction of Light*; Cambridge University Press: Cambridge, UK, 2019.
53. Névot, L.; Croce, P. Caractérisation des surfaces par réflexion rasante de rayons X. Application à l'étude du polissage de quelques verres silicates. *Rev. Phys. Appl.* **1980**, *15*, 761–779. [CrossRef]
54. Penfold, J.; Thomas, R.K. Neutron reflectivity and small angle neutron scattering: An introduction and perspective on recent progress. *Curr. Opin. Colloid Interface Sci.* **2014**, *19*, 198–206. [CrossRef]
55. Brumberger, H. (Ed.) *Modern Aspects of Small-Angle Scattering*; Springer: Dordrecht, The Netherlands , 1995.
56. Svergun, D.I. *Structure Analysis by Small-Angle X-ray and Neutron Scattering*; Plenum Press: New York, NY, USA, 1987.
57. Poger, D.; Caron, B.; Mark, A.E. Validating lipid force fields against experimental data: Progress, challenges and perspectives. *Biochim. Biophys. Acta BBA Biomembr.* **2016**, *1858*, 1556–1565. [CrossRef]
58. Knight, C.J.; Hub, J.S. MemGen: A general web server for the setup of lipid membrane simulation systems: Fig. 1. *Bioinformatics* **2015**, *31*, 2897–2899. [CrossRef] [PubMed]
59. Martínez, L.; Andrade, R.; Birgin, E.G.; Martínez, J.M. PACKMOL: A package for building initial configurations for molecular dynamics simulations. *J. Comput. Chem.* **2009**, *30*, 2157–2164. [CrossRef]
60. Moradi, S.; Nowroozi, A.; Shahlaei, M. Shedding light on the structural properties of lipid bilayers using molecular dynamics simulation: A review study. *RSC Adv.* **2019**, *9*, 4644–4658. [CrossRef]
61. Khattari, Z.; Köhler, S.; Xu, Y.; Aeffner, S.; Salditt, T. Stalk formation as a function of lipid composition studied by X-ray reflectivity. *Biochim. Biophys. Acta BBA Biomembr.* **2015**, *1848*, 41–50. [CrossRef] [PubMed]
62. Penfold, J.; Richardson, R.M.; Zarbakhsh, A.; Webster, J.R.P.; Bucknall, D.G.; Rennie, A.R.; Jones, R.A.L.; Cosgrove, T.; Thomas, R.K.; Higgins, J.S.; et al. Recent advances in the study of chemical surfaces and interfaces by specular neutron reflection. *J. Chem. Soc. Faraday Trans.* **1997**, *93*, 3899–3917. [CrossRef]

63. Roldan, J.L.O.; Campana, M.; Barker, R.; Yoldi, I.; Hendry, A. *Elucidating the Membrane Interaction Mechanism of Chloride Intracellular Channel Proteins*; STFC ISIS Neutron and Muon Source: 2018. Available online: https://doi.org/10.5286/ISIS.E.RB1820565 (accessed on 1 June 2021).
64. Mattauch, S.; Koutsioubas, A.; Pütter, S. MARIA: Magnetic reflectometer with high incident angle. *J. Large Scale Res. Facil.* **2015**, *1*, 8. [CrossRef]
65. Mattauch, S.; Koutsioubas, A.; Rücker, U.; Korolkov, D.; Fracassi, V.; Daemen, J.; Schmitz, R.; Bussmann, K.; Suxdorf, F.; Wagener, M.; et al. The high-intensity reflectometer of the Jülich Centre for Neutron Science: MARIA. *J. Appl. Crystallogr.* **2018**, *51*, 646–654. [CrossRef] [PubMed]
66. Frielinghaus, H.; Feoktystov, A.; Berts, I.; Mangiapia, G. KWS-1: Small-angle scattering diffractometer. *J. Large Scale Res. Facil. JLSRF* **2015**, *1*, 28. [CrossRef]
67. Wignall, G.D.; Bates, F.S. Absolute calibration of small-angle neutron scattering data. *J. Appl. Crystallogr.* **1987**, *20*, 28–40. [CrossRef]
68. Hess, B.; Kutzner, C.; van der Spoel, D.; Lindahl, E. GROMACS 4: Algorithms for Highly Efficient, Load-Balanced, and Scalable Molecular Simulation. *J. Chem. Theory Comput.* **2008**, *4*, 435–447. [CrossRef]
69. Darden, T.; York, D.; Pedersen, L. Particle mesh Ewald: An $N \cdot \log(N)$ method for Ewald sums in large systems. *J. Chem. Phys.* **1993**, *98*, 10089–10092. [CrossRef]
70. Berger, O.; Edholm, O.; Jähnig, F. Molecular dynamics simulations of a fluid bilayer of dipalmitoylphosphatidylcholine at full hydration, constant pressure, and constant temperature. *Biophys. J.* **1997**, *72*, 2002–2013. [CrossRef]
71. Humphrey, W.; Dalke, A.; Schulten, K. VMD: Visual molecular dynamics. *J. Mol. Graph.* **1996**, *14*, 33–38. [CrossRef]
72. Brehm, M.; Thomas, M.; Gehrke, S.; Kirchner, B. TRAVIS—A free analyzer for trajectories from molecular simulation. *J. Chem. Phys.* **2020**, *152*, 164105. [CrossRef]
73. Brehm, M.; Kirchner, B. TRAVIS—A Free Analyzer and Visualizer for Monte Carlo and Molecular Dynamics Trajectories. *J. Chem. Inf. Model.* **2011**, *51*, 2007–2023. [CrossRef]
74. Krueger, S.; Koenig, B.W.; Orts, W.J.; Berk, N.F.; Majkrzak, C.F.; Gawrisch, K. Neutron Reflectivity Studies of Single Lipid Bilayers Supported on Planar Substrates. In *Neutrons in Biology*; Springer: Boston, MA, USA, 1996; pp. 205–213.
75. Nouhi, S.; Koutsioubas, A.; Kapaklis, V.; Rennie, A.R. Distortion of surfactant lamellar phases induced by surface roughness. *Eur. Phys. J. Spec. Top.* **2020**, *229*, 2807–2823. [CrossRef]
76. Degiorgio, V. *Physics of Amphiphiles—Micelles, Vesicles, and Microemulsions: Varenna on Lake Como, Villa Monastero, 19–29 July 1983*; Elsevier Science Publishers: New York, NY, USA, 1983.
77. Kotlarchyk, M.; Chen, S.H. Analysis of small angle neutron scattering spectra from polydisperse interacting colloids. *J. Chem. Phys.* **1983**, *79*, 2461–2469. [CrossRef]
78. Pardo, L.C.; Rovira-Esteva, M.; Busch, S.; Moulin, J.F.; Tamarit, J.L. Fitting in a complex χ^2 landscape using an optimized hypersurface sampling. *Phys. Rev. E* **2011**, *84*, 046711. [CrossRef] [PubMed]
79. Soper, A.K. The Radial Distribution Functions of Water as Derived from Radiation Total Scattering Experiments: Is There Anything We Can Say for Sure? *ISRN Phys. Chem.* **2013**, *2013*, 1–67. [CrossRef]
80. Jancsó, G. Isotope Effects. In *Handbook of Nuclear Chemistry*; Springer: Boston, MA, USA, 2011; pp. 699–725.
81. Jelińska-Kazimierczuk, M.; Szydłowski, J. Isotope effect on the solubility of amino acids in water. *J. Solut. Chem.* **1996**, *25*, 1175–1184. [CrossRef]
82. Silva, R.; Ferreira, H.; Cavaco-Paulo, A. Sonoproduction of Liposomes and Protein Particles as Templates for Delivery Purposes. *Biomacromolecules* **2011**, *12*, 3353–3368. [CrossRef]
83. Karge, L.; Gilles, R.; Busch, S. Calibrating SANS data for instrument geometry and pixel sensitivity effects: Access to an extendedQrange. *J. Appl. Crystallogr.* **2017**, *50*, 1382–1394. [CrossRef]
84. Klauda, J.B.; Monje, V.; Kim, T.; Im, W. Improving the CHARMM Force Field for Polyunsaturated Fatty Acid Chains. *J. Phys. Chem. B* **2012**, *116*, 9424–9431. [CrossRef] [PubMed]
85. Fernández, R.G.; Abascal, J.L.F.; Vega, C. The melting point of ice Ih for common water models calculated from direct coexistence of the solid-liquid interface. *J. Chem. Phys.* **2006**, *124*, 144506. [CrossRef] [PubMed]
86. Pospelov, G.; Van Herck, W.; Burle, J.; Carmona Loaiza, J.M.; Durniak, C.; Fisher, J.M.; Ganeva, M.; Yurov, D.; Wuttke, J. *BornAgain*: Software for simulating and fitting grazing-incidence small-angle scattering. *J. Appl. Crystallogr.* **2020**, *53*, 262–276. [CrossRef]
87. Udby, L.; Willendrup, P.; Knudsen, E.; Niedermayer, C.; Filges, U.; Christensen, N.; Farhi, E.; Wells, B.; Lefmann, K. Analysing neutron scattering data using McStas virtual experiments. *Nucl. Instrum. Methods Phys. Res. Sect. A Accel. Spectrom. Detect. Assoc. Equip.* **2011**, *634*, S138–S143. [CrossRef]
88. Chen, P.C.; Shevchuk, R.; Strnad, F.M.; Lorenz, C.; Karge, L.; Gilles, R.; Stadler, A.M.; Hennig, J.; Hub, J.S. Combined Small-Angle X-ray and Neutron Scattering Restraints in Molecular Dynamics Simulations. *J. Chem. Theory Comput.* **2019**, *15*, 4687–4698. [CrossRef] [PubMed]

Article

Study of the Interaction of a Novel Semi-Synthetic Peptide with Model Lipid Membranes

Lucia Sessa [1,2,*], Simona Concilio [1,2,*], Peter Walde [3], Tom Robinson [4], Petra S. Dittrich [5], Amalia Porta [1,2], Barbara Panunzi [6], Ugo Caruso [7] and Stefano Piotto [1,2]

1 Department of Pharmacy, University of Salerno, 84084 Fisciano (SA), Italy; aporta@unisa.it (A.P.); piotto@unisa.it (S.P.)
2 Research Centre for Biomaterials BIONAM, University of Salerno, Via Giovanni Paolo II 132, 84084 Fisciano (SA), Italy
3 Department of Materials, ETH Zürich, 8093 Zürich, Switzerland; peter.walde@mat.ethz.ch
4 Department of Theory and Bio-Systems, Max Planck Institute of Colloids and Interfaces, D-14424 Potsdam, Germany; tom.robinson@mpikg.mpg.de
5 Department of Biosystems Science and Engineering, ETH Zurich, 4058 Basel, Switzerland; petra.dittrich@bsse.ethz.ch
6 Department of Agriculture, University of Napoli Federico II, 80055 Portici (NA), Italy; barbara.panunzi@unina.it
7 Department of Chemical Sciences, University of Napoli Federico II, 80126 Napoli, Italy; ugo.caruso@unina.it
* Correspondence: lucsessa@unisa.it (L.S.); sconcilio@unisa.it (S.C.)

Received: 30 September 2020; Accepted: 16 October 2020; Published: 19 October 2020

Abstract: Most linear peptides directly interact with membranes, but the mechanisms of interaction are far from being completely understood. Here, we present an investigation of the membrane interactions of a designed peptide containing a non-natural, synthetic amino acid. We selected a nonapeptide that is reported to interact with phospholipid membranes, ALYLAIRKR, abbreviated as ALY. We designed a modified peptide (azoALY) by substituting the tyrosine residue of ALY with an antimicrobial azobenzene-bearing amino acid. Both of the peptides were examined for their ability to interact with model membranes, assessing the penetration of phospholipid monolayers, and leakage across the bilayer of large unilamellar vesicles (LUVs) and giant unilamellar vesicles (GUVs). The latter was performed in a microfluidic device in order to study the kinetics of leakage of entrapped calcein from the vesicles at the single vesicle level. Both types of vesicles were prepared from a 9:1 (mol/mol) mixture of POPC (1-palmitoyl-2-oleoyl-*sn*-glycero-3-phosphocholine) and POPG (1-palmitoyl-2-oleoyl-*sn*-glycero-3-phospho(1′-*rac*-glycerol). Calcein leakage from the vesicles was more pronounced at a low concentration in the case of azoALY than for ALY. Increased vesicle membrane disturbance in the presence of azoALY was also evident from an enzymatic assay with LUVs and entrapped horseradish peroxidase. Molecular dynamics simulations of ALY and azoALY in an anionic POPC/POPG model bilayer showed that ALY peptide only interacts with the lipid head groups. In contrast, azoALY penetrates the hydrophobic core of the bilayers causing a stronger membrane perturbation as compared to ALY, in qualitative agreement with the experimental results from the leakage assays.

Keywords: peptide; MD; GUV; LUV; azo-amino acid

---

1. Introduction

Membrane interacting peptides are an exciting topic of research, because they cover different classes of peptides with several biological activities. The interactions between peptides and lipid membranes are involved in many critical biological processes [1,2]. Depending on their structural characteristics, different peptides employ different mechanisms of interaction with the membrane, causing membrane alteration or permeation [3,4]. Moreover, during their interactions, peptides and membranes may undergo a sequence of structural changes. Even for short and linear peptides, the molecular details of the process are often not completely understood. Numerous studies have been carried out to clarify the interactions of peptides with lipid bilayers, when considering the position, orientation, structure, and the effects on the surrounding lipids [5–10]. A better understanding of peptide-membrane interactions at a molecular level is not only essential in the study of various biological processes, but it could also help in designing peptides with specific functionalities that may be exploited for therapeutic applications. We performed theoretical and experimental studies using model lipid membranes to develop a novel semi-synthetic peptide with a direct effect on membrane perturbation [11]. Here, we report the results of such investigations highlighting the interactions of a designed peptide with lipid membranes. We selected a membrane interacting nonapeptide, ALYLAIRKR (abbreviated as ALY) [12], after a screening of peptides in the database YADAMP [13]. This peptide is known to interact with liposomal membranes and cause an increase in membrane permeability proportional to the peptide concentration. It is known that linear peptides and membrane proteins affect membrane structure [14–17], but little is known regarding peptides with azo-modified amino acids.

We used this peptide as a reference peptide to develop a novel modified peptide (azoALY) replacing the tyrosine ("Y") in the ALY sequence with an unnatural amino acid bearing an azo group, which may affect the membrane permeability and the antimicrobial activity. We chose to modify the tyrosine residue of ALY by replacing one of the hydrogen atoms of the phenyl ring in the side chain with an azobenzene group (via a diazo coupling reaction).

The novel azo-amino acid (azoTyr) was then employed in order to synthesize the peptide chain without further modifications. The choice of azobenzene in the peptide chain was determined by its hydrophobicity and intrinsic antimicrobial properties. In our previous works [18,19], we have already studied the antibacterial and antifungal activity of the azobenzene group. Therefore, a modified peptide might have antimicrobial activity due to the presence of azobenzene, and it might act as a prodrug, releasing the antimicrobial azo compound in vivo. However, this work is not focused on the antimicrobial aspects, rather on the membrane perturbation effect. With this purpose, molecular dynamics (MD) simulations are a useful method for studying peptide-membrane interactions. They provide a detailed description of the processes at a molecular level, in which all of the components can be studied, as well as being able to visualize their organization and dynamics [20].

In the first part of the work, we used MD simulations to understand the different modes of interactions between the two peptides (ALY and azoALY) and a model bilayer. We employed a mixture of POPC/POPG at a molar ratio of 9:1 to study the interaction of the peptides with the surface of lipid bilayers, according to previous studies [21,22]. POPC/POPG lipids are frequently used to build a model of bacterial membranes (see, e.g., [23–25]). It is relevant to evidence that, although POPC is widely used in both molecular dynamics simulations and vesicle preparations, it is not common to find it in bacterial membranes. Still, the preparation of POPC/POPG membranes offers two essential advantages. First, the results can be easily compared with similar settings in the scientific literature. This kind of vesicle is prepared in a standard and easy way due to low phase transition temperature of the lipids. Second, it allows for the building of a negatively charged membrane without introducing curvature stress. The MD trajectories allowed for us to analyse the dynamics of the peptides in the lipid bilayer and their effects on the lipid molecules. Membrane-active peptides can induce an immediate and complete release of entrapped solutes from the aqueous interior of lipid vesicles (liposomes) into the bulk solution by forming membrane pores [26,27]. In the second part of the work, we prepared vesicles with the same composition (POPC/POPG, 9:1) to assess the amount of membrane perturbation that is caused by the interaction with the peptides. Two different approaches were used. First, we estimated the release of entrapped calcein that is induced by the two peptides from giant unilamellar vesicles

(GUVs). Secondly, we performed measurements of the leakage of entrapped horseradish peroxidase isoenzyme C enzyme (HRPC) from large unilamellar vesicles (LUVs) after peptide addition.

2. Materials and Methods

2.1. Selection of the Reference Peptide ALY

The database YADAMP was used to select a membrane-interacting peptide [13]. The criteria for the selection were: (i) a tyrosine in the amino acid sequence; (ii) a sequence length of ≤10 residues; (iii) a helicity value higher than 4.5; and, (iv) a cell-penetrating potential value higher than 0.5.

The helicity value is estimated on a scale range between 0 and 9, whereby the value 9 corresponds to the highest probability of α-helix secondary structure formation [28]. The cell-penetrating value is estimated on a scale range between 0 and 1, whereby 1 corresponds to the highest probability of a peptide in order to penetrate a membrane and 0 indicates the impossibility to enter a membrane [29].

We selected the peptide ALYLAIRKR (H-Ala-Leu-Tyr-Leu-Ala-Ile-Arg-Lys-Arg-NH_2) [12]. The peptide is abbreviated as ALY, whereas the modified peptide was named azoALY.

2.2. Molecular Dynamics (MD) Simulations

The three-dimensional structures of the peptides were built by MD simulations. In MD simulations, Newton's equations of motion are solved at each step of the atom movement, which is probably the most reliable method for investigating protein interactions. MD simulations were performed while using the script protein_folding_by_MD.js, accessible in the Abalone software 2.1.4.2 Version [30], setting the AMBER94 force field, the temperature to 350 K, and the implicit water model (continuum solvent). After force field assigning, we performed structure optimization to avoid overlapping. The simulation time of 10 ns was enough to reach the native protein conformation. The interactions of peptides with membranes were evaluated on the model membrane of POPC/POPG. The membrane was generated while using the web tool CHARMM-GUI Membrane Builder [31] using a relative composition POPC/POPG of 9 to 1. Each monolayer of the membrane was made of 110 lipids. A periodic simulation cell (X = 85.58 Å, Y = 100.00 Å, Z = 84.44 Å) was built around the entire complex. The charges were assigned at physiological conditions (pH 7.4). The MD simulations were performed using the software YASARA Structure 17.3.30 [32]. We used AMBER14 as a force field with long-ranged PME potential and a cutoff of 8.0 Å. The simulation box was filled with TIP3P water, choosing a density of 0.997 g/mL. The system was neutralized with NaCl at a concentration of 0.9%. The membranes were equilibrated during 200 ps. After equilibration, both peptides ALY and azoALY were embedded in the membrane, placing the transmembrane regions of the peptides perpendicular to the membrane. Short energy minimization was performed in order to optimize the membrane geometry and fill the membrane pores. The simulation was then initiated at 298 K and integration time steps for intramolecular forces every 1.25 fs. The simulation snapshots were saved at regular time intervals of 100 ps. The total simulation time was 50 ns.

The bilayer thickness is defined as the average distance between the lipid phosphorus atoms of the opposing leaflets. It was calculated averaging the 50 snapshots of the last 5 ns of simulations. In addition, in order to provide an overview of the local membrane deformation along with the simulation, we used MembPlugIn in VMD [33] to interpolate the lipid head's positions and compute a 2D thickness map. The fluidity was indirectly measured by calculating the deuterium order parameter (SCD) with the VMD MembPlugin software version 1.1.

The mass density profile indicates the atom distribution of the membrane component along the bilayer. This analysis offers useful information regarding the structural changes in membranes [11]. Profiles are determined by dividing the simulation box along the perpendicular to the z-axis (normal to the bilayer) into several thin portions 1 Å thick, and by finding the mass density of the atoms that are positioned in each portion [34].

2.3. *Synthesis of Azobenzene-Modified Tyr (azoTyr) and the Peptides ALY and azoALY*

All of the reagents and solvents were purchased from Sigma–Aldrich (Milan, Italy) and used without further purification. The ^{1}H NMR spectra were recorded with a Bruker DRX/400 spectrometer (Bruker, Billerica, MA, USA). Chemical shifts are reported relative to the residual solvent peak (dimethylsulfoxide-d$_6$: $\delta = 2.50$ ppm). Mass spectrometry measurements were performed while using a Q-TOF premier instrument (Waters, Milford, MA, USA) that was equipped with an electrospray ion source and hybrid quadrupole-time of flight analyser. The mass spectra were acquired in positive ion mode, in 50% CH_3CN solution, over the 200–800 *m*/*z* range. Instrument mass calibration was achieved by a separate injection of 1 mM NaI in 50% CH_3CN. The data were processed using MassLynx software (Waters, Milford, MA, USA).

The azo-amino acid 3-(*p*-tolyldiazenyl)-*N*-Fmoc-L-tyrosine was synthesized according to a diazo coupling reaction, as illustrated in the following Scheme 1:

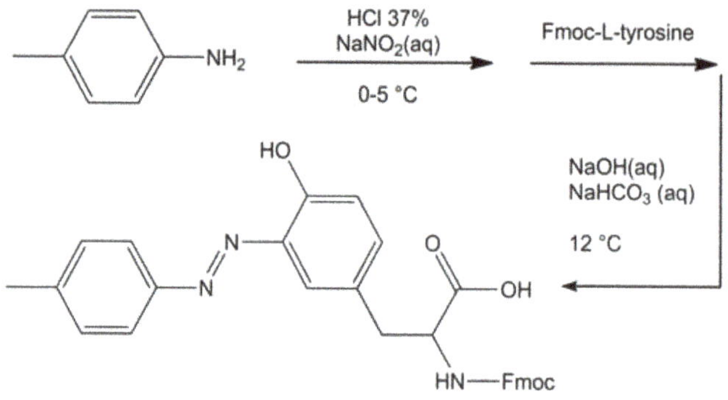

Scheme 1. Synthetic path of amino acid azoTyr.

0.011 mol of p-toluidine were suspended in a solution containing 19.2 mL of water and 4.8 mL of HCl 37% (*w*/*w*). The solution was cooled at 0–5 °C and a solution of 0.85 g of sodium nitrite (0.012 mol) dissolved in 2.4 mL of water was added dropwise, obtaining a suspension of the diazonium salt (suspension A). Separately, a solution containing 0.018 g of NaOH and 0.068 g of $NaHCO_3$ in 20 mL of water with 2.00 g of Fmoc-L-tyrosine (0.00496 mol) was prepared (solution B). Suspension A was added dropwise to solution B, under stirring at 15 °C, adjusting the pH at 9–10, with the addition of NaOH, if necessary. The system was left reacting for 20 min. A reddish precipitate of the azo compound formed. The crude precipitate was filtered, dried under vacuum, and then crystallized from chloroform. The product was recovered as a dark red microcrystalline powder, with a final yield of 85%. ^{1}H NMR (DMSO-d$_6$): (δ, ppm) = 11.25 (s, OH); 7.87 (d, 2H); 7.80 (d, 2H); 7.63 (dd, 2H); 7.60 (d, 2H); 7.38 (m, 1H); 7.34 (m, 2H); 7.28 (m, 2H); 7.19 (d, 1H); 6.89 (d, 1H); 4.34 (m, 1H); 4.18 (m, 2H); 3.10, 2.98 (m, 2H); and, 2.40 (s, 3H). HRMS (ESI): *m*/*z*: 522.20 [M^+H^+].

Peptides ALY (ALYLAIRKR; Mw = 1103.361 g mol^{-1}) and azoALY (ALXLAIRKR; Mw = 1221.497 g mol^{-1}) were synthesized by Zhejiang Ontores Biotechnologies Co., Ltd (Shanghai, China). Purity for ALY peptide was 96.44% and for azoALY peptide was 96.27%.

A stock solution of ALY was prepared by dissolving a few µg of the peptide in 1 mL of MilliporeQ water. The exact peptide concentration was calculated from the absorbance at 275 nm while using $\varepsilon_{275} = 1400$ M^{-1} cm^{-1} for the single tyrosine chromophore [35,36]. The solution was filtered through a 0.22 µm pore size syringe filter.

Similarly, a stock solution of azoALY was prepared and the concentration was calculated from the absorbance at 334 nm using $\varepsilon_{334} = 36805$ M^{-1} cm^{-1} for the single modified tyrosine amino acid. The molar extinction coefficient was determined by measuring the UV-Vis spectrum of a solution of azoTyr at a determined concentration. The solution was filtered through a 0.22 µm pore size syringe filter.

2.4. Permeability Tests with POPC/POPG (9:1) Vesicles

2.4.1. Materials

POPC (1-palmitoyl-2-oleoyl-*sn*-glycero-3-phosphocholine) and POPG (1-palmitoyl-2-oleoyl-*sn*-glycero-3-phospho-(1′-*rac*-glycerol) (sodium salt)) (≥99%) were from Sigma–Aldrich Chemie GmbH (Buchs, Switzerland). The membrane indocarbocyanine dye DiI and phosphate-buffered saline (PBS, pH = 7.2) were obtained from Invitrogen Thermo Fisher Scientific (Thermo-Fisher, Waltham, MA, USA)). Calcein was purchased from Fisher Scientific AG (Wohlen, Switzerland). HRPC (Horseradish peroxidase isoenzyme C, EC 1.11.1.7, RZ A_{403}/A_{280} > 3.1) was from Toyobo Enzymes (Osaka, Japan). The diammonium salt of $ABTS^{2-}$ (2,2′-azinobis(3-ethylbenzothiazoline-6-sulfonate)) and Triton X-100 were purchased from Sigma–Aldrich Chemie GmbH (Buchs, Switzerland). Hydrogen peroxide (H_2O_2, 30%) was from Acros Organics Fisher Scientific AG (Wohlen, Switzerland), 4-morpholineethanesulfonic acid (MES ≥ 99%) was purchased from Fluka Fisher Scientific AG (Reinach, Switzerland). Sepharose 4B and chloroform (stabilized with ethanol, 99.8%) were purchased from Sigma–Aldrich Chemie GmbH (Buchs, Switzerland).

2.4.2. Calcein Release Test with Giant Unilamellar Vesicles (GUVs)

Wide-field microscopy was performed with an inverted microscope (IX70, Olympus America, Melville, NY, USA) equipped with a mercury lamp and a 40×/0.65 NA air objective lens. The images were recorded with an EMCCD camera (iXon DV887, Andor Technology, South Windsor, CT)). The fluorescence intensity of the calcein within three separate vesicles was monitored while using a confocal laser-scanning microscope (Axiovert 200 M, Zeiss, Oberkochen, Germany), and an appropriate optical filter sets for DiI and calcein. The GUVs were prepared by electroformation following the procedure that was originally described by Angelova et al. [37], but conducted in a modified chamber [38]. Briefly, the setup used for the preparation of the vesicles consisted of two conductive indium tin oxide (ITO) coated glasses separated by a 1.5 mm thick silicone rubber spacer to maximize the yield of vesicles. POPC/POPG lipids (9:1 mol/mol) were dissolved in chloroform/methanol (9:1 *v*/*v*) at a concentration of 1 mM. The orange-red fluorescent dye DiI was added at a concentration of 1 μM. A drop of 2.5 μL of the mixture was deposited on one of the conductively coated glasses, and this was repeated in 12 locations. The lipid film was then dried under vacuum overnight and hydrated with MilliQ water containing 10 μM calcein. The chambers were sealed by a second ITO slide and held at 60 °C within a custom-built heating device. GUVs were formed by applying 0.7 V at a frequency of 10 Hz for 4 h while using a function generator. After applying 1 V at 4 Hz for 30 min. to detach the vesicles from the surface, harvesting was achieved by careful pipetting. GUVs were stored at room temperature and used within 48 h. For the analysis of calcein release from the GUVs, we used a microfluidic platform that was able to trap single GUVs in an array of chambers [39] (see Figure S1 in Supp. Mat.). By exchanging the solution inside the chip and subsequently opening the valves, it is possible to perform fast kinetic studies from the seconds to minutes timescale. GUVs containing fluorescent calcein were trapped in order to investigate the membrane perturbation and subsequent leakage of calcein across the membrane. Because of the hydrophilic nature of calcein, it does not permeate the membrane and remains encapsulated. Once trapped, the GUVs remain stable for long times (at least 12 h), and no significant deformation or rupture of the vesicles was observed. The solution was then exchanged with MilliQ water while using a syringe-pump (neMESYS, Cetoni, Germany) until only calcein fluorescence inside GUV was visible and no more calcein was detected in the surrounding solution (See Figure S2 in Supp. Mat.). The same pump was used to exchange the aqueous solution outside the donuts with an aqueous solution of the peptide, with a total flow rate of 5 μL min^{-1}. The release measured in the absence of the peptide was used as the control, and the amount of calcein within the GUV remained constant, indicating that photobleaching did not occur. The error bars were calculated from the standard error, estimated by population standard deviation divided by the square root of the sample number. The number of individual experiments at a given peptide concentration was n = 3.

2.4.3. Enzymatic Permeability Assay with Large Unilamellar Vesicles (LUVs)

LUVs that were composed of POPC/POPG (9:1 mol/mol) and loaded with HRPC were prepared by lipid film hydration, followed by mechanical extrusion using for final extrusions 200 nm polycarbonate membranes, in a similar way described before for POPC vesicles containing HRPC [40]. Weighted amounts of lipids (20 mM) were first dissolved in chloroform that was stabilized with ethanol and then mixed in a 100 mL round bottom flask. The solvent was removed by rotatory evaporation at 35 °C and dried in high vacuum overnight. The obtained thin lipid film was hydrated with 3.5 mL of 20 μM HRPC solution in 10 mM (MES) buffer (pH = 5.0) at room temperature. The hydration was made by gentle agitation of the flask to exclude enzyme denaturation. Ten freezing-thawing cycles were carried out in order to homogenize and equilibrate the enzyme-containing multilamellar vesicles suspension (MLV), cooling the flask in liquid nitrogen for 20 seconds followed by placing the flask in a warm (50 °C) water bath for 30 seconds. The MLV suspension was first extruded 10 times through track-etch polycarbonate membranes with cylindrical pores of 400 nm size under moderate pressure (three-bar) at room temperature, followed by extrusion with a 200 nm membrane increasing the pressure to five-bar at the same temperature conditions. The non-entrapped enzyme molecules were separated from the enzyme-containing vesicles by size-exclusion chromatography by using a 2 × 20 cm glass column that was filled with Sepharose 4B equilibrated with MES buffer (pH = 5), the flow rate of 0.5 mL/min.

We performed turbidity measurements of each eluted fraction to select the most concentrated fraction of HRPC-containing vesicles with the lowest amount of free enzyme (pI (HRPC) ≈ 10) possibly bound to the outer membrane of the anionic vesicles (see Figure S3 in Supp. Mat. for details). The fraction with the highest optical density at 403 nm was used in order to establish the release of HRPC from enzyme containing POPC/POPG vesicles.

The activity of HRPC was measured with $ABTS^{2-}$. $ABTS^{2-}$ is frequently used as a sensitive chromogenic substrate for the quantification of horseradish peroxidase [40,41]. $ABTS^{2-}$ has an absorption maximum at 340 nm. Depending on the experimental conditions, HRPC catalyses the oxidation of $ABTS^{2-}$ forming a stable nitrogen-centered radical cation, so that the obtained product is overall negatively charged, $ABTS^{\bullet-}$. $ABTS^{\bullet-}$ has an absorption maximum at 414 nm and three other bands centered around 650, 735, and 814 nm. Therefore, UV/Vis-spectroscopy is most convenient for determining the catalytic activity of HRPC by following the rate of reaction product formation [42,43]. The rate of $ABTS^{\bullet-}$ formation (monitored at λ_{max} = 414 nm) during the first 5 min. of the reaction was followed spectrophotometrically and found to linearly depend on the HRPC concentration between 50 and 350 pM under the conditions used (see the curve in Figure S4, Supp. Mat.). The following reaction conditions for the spectrophotometric quantification of the HRPC activity were found to be appropriate: $[ABTS^{2-}]_0$ = 0.25 mM, [peptide] = 4 μM, $[H_2O_2]_0$ = 80 μM, 100 μL of pooled fraction diluted at 25 °C in MES buffer pH 5, for a total assay volume of 1 mL and reaction time of 5 min. The activity measurements were carried out, as follows: buffer to reach the assay volume of 1 mL, $ABTS^{2-}$ (final concentration of 0.25 mM), and HRPC stock solution was mixed in a reaction tube. Immediately before the spectrophotometric analysis, H_2O_2 was added (final concentration of 80 μM). The mixture was transferred into a polystyrene cuvette (path length = 1 cm) and the development of the absorption spectrum of the reaction mixture was monitored as a function of time (see Figure S4 in Supp. Mat.) while using a diode array spectrophotometer (Specord 5600 Analytik Jena, Jena Germany). The spectra were recorded every 10 sec immediately after the initiation of the reaction (up to 5 min.). All of the measurements were carried out three times. Triton X-100 was used as the control. For the control assay, we used $[ABTS^{2-}]_0$ = 0.25 mM, $[H_2O_2]_0$ = 80 μM, 100 μL of pooled fraction 100× diluted and 20 μL 5% *v*/*v* Triton X-100 at 25 °C in MES buffer, pH 5. Because of the delay of inactivation in the course of substrate reaction [44], we built a calibration curve in the presence of Triton X-100 (see Figure S5 in Supp. Mat.). For the assay, we used a surfactant concentration of 0.1% in the assay mixture containing HRPC and $ABTS^{2-}$ solutions. After 5 min. from Triton X-100 addition, the reaction was started adding hydrogen peroxide, and the enzymatic reaction was followed by UV/Vis spectrophotometry.

3. Results and Discussion

3.1. Peptide Localization within Lipid Bilayers: a MD Simulation Analysis

In this study, we analysed two peptides, a natural peptide, named ALY, and a modified peptide, azoALY. They both have a short sequence of nine amino acids that fold into an alpha helix structure. The azo amino acid slightly deforms the helix structure adopted by azoALY. In the MD simulations, the hydrophobic regions of the peptides ALY and azoALY were embedded perpendicularly to the membrane bilayer of POPC/POPG (9:1) at the starting point. We used an online tool to predict the position of membrane proteins (https://opm.phar.umich.edu). The model suggested the insertion for both peptides in membrane with a tilt angle of 48° and with the hydrophobic portion (Ala-Leu-Tyr-Leu-Ala) immersed in the core membrane. The charged portion of the peptide (Arg-Lys-Arg) is external to the membrane for both peptides (see Figure S6 in Supp. Mat.). After the first 200 ps of the simulation, the ALY peptide comes out of the bilayer and it is located between the head groups and the water phase in the outer part of the membrane. In contrast, the azoALY peptide is anchored to the membrane for the whole simulation time (see Figure S6 in Supp. Mat.). The density profile indicates the position of the different atoms of the system along the normal phospholipid bilayer in the last 5 ns of a simulation of 50 ns. In Figures 1 and 2, the density profiles of ALY and azoALY (normalized to 1 by water mass) in the POPC/POPG membrane is reported. The cyan line represents the water continuum; it is outside the bilayer, ensuring membrane hardness. The phosphorus and nitrogen atoms (continuous blue and dashed blue line. respectively) represent the head group of the phospholipids; the dashed black line indicates the carbon chains of the phospholipid, while the black line represents the terminal carbons. After a simulation time of 45 ns, the ALY peptide (red line in Figure 1) is placed below the phosphorus atoms between the head group and the water phase, as seen in Figure 1. In contrast, the peptide azoALY (red line in Figure 2) is anchored to the core membrane for the same simulation time (from 45 to 50 ns).

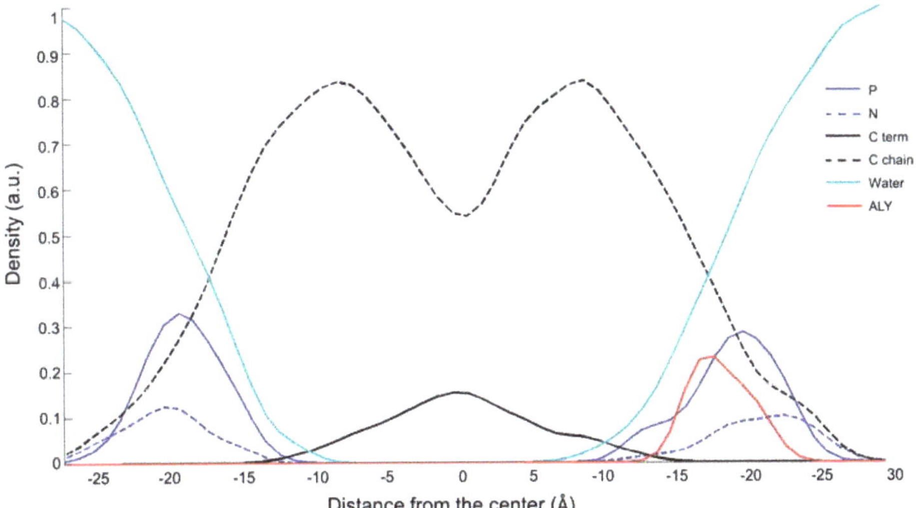

Figure 1. Density profile of the system ALY-POPC/POPG (9:1) after 45 ns of molecular dynamic (MD) simulation. Cyan line: water molecules; blue line: phosphorous atoms; dashed blue line: nitrogen atoms; dashed black line: carbon chains of the phospholipid, black line: terminal carbons of the phospholipid; red line ALY peptide.

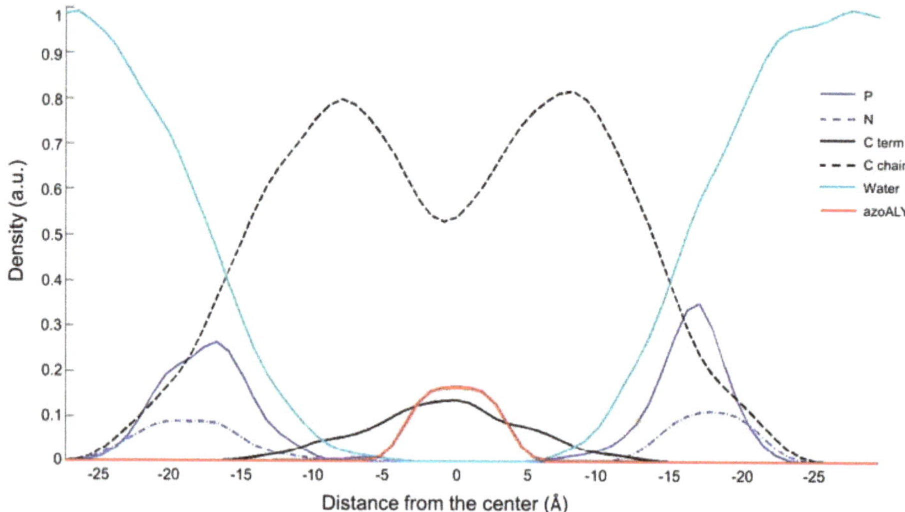

Figure 2. Density profile of the system azoALY-POPC/POPG (9:1) after 45 ns of MD simulation. Cyan line: water molecules; blue line: phosphorous atoms; dashed blue line: nitrogen atoms; dashed black line: carbon chains of the phospholipid, black line: terminal carbons of the phospholipid; red line azoALY peptide.

For both peptides, the membrane thickness of the peptide/membrane systems was measured in order to estimate the perturbation of the membrane packing induced by the peptide, when only considering the lipids surrounding the peptide with a distance of 4 Å. The membrane thickness was measured as the average distance between the center of mass of the phosphate atoms of the inner layer and the outer layer. The thickness of the pure POPC/POPG bilayer did not change during the 50 ns simulation, suggesting that the system was equilibrated with an initial value of 38.5 ± 0.3 Å. In the presence of the two peptides, in the last 5 ns of the simulations, the membranes thickness values were 38.7 ± 0.3 Å with ALY and 36.4 ± 0.5 with azoALY. The modified peptide causes a reduction of the membrane thickness. This different perturbation is due to different peptide-membrane interactions. The peptide ALY is placed between the water molecules and head groups, and it does not cause perturbation of the membrane packing. On the other hand, azoALY is anchored to the membrane core, which affects the invagination of the lipid surrounding and produces a significant reduction of the membrane thickness.

From Figure 2, in fact, the peptide azoALY (red line) is in the same position as the terminal carbon atoms of the lipid tails. This simulation suggests that the peptide is anchored to the core of the membrane in the last 5 ns of the simulation time. The azo group has a significant influence on the peptide-membrane interaction: due to its rigid and planar structure, the azo amino acid permeates the membrane, causing disorder in the lipid leaflet. We performed a deuterium order parameter (SCD) analysis. SCD studies provide comprehensive information on membrane fluidity of lipophilic phospholipid chains near peptide molecules to assess whether the modified peptide causes a change in membrane fluidity. The analysis, performed on the surrounding lipids with a distance of 5 Å from the peptides, confirmed a different behavior of the two peptides. azoALY increases the mobility in the lipid head more than peptide ALY. The results are shown in Figure S7 in the Supplementary Materials section.

The thickness maps in Figure 3 show the local membrane deformation along with the simulation. The ALY peptide has a minimal effect on membrane thickness with a constant fluctuation of lipids around a value of 38 Å (Figure 3a). The azo-modified peptide causes the invagination of the membrane, reducing the thickness of the surrounding lipids that are represented by the blue hole in top leaflet of the membrane (Figure 3b).

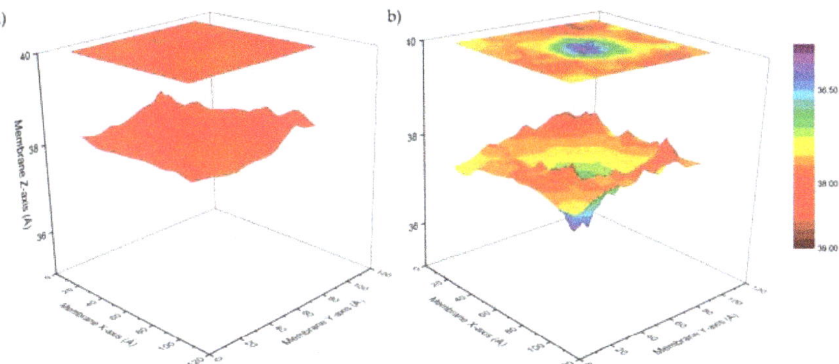

Figure 3. Thickness maps of POPC/POPG/ALY (**a**) and POPC/POPG/azoALY (**b**) during the last 5 ns of the MD simulation (total time 50 ns). The top panel is the deformation profile as a color map projected onto the surface defined by fitting a grid (spacing 2 Å) to the positions of the phosphate atoms in the top leaflet during the trajectory, followed by time averaging and spatial smoothing.

3.2. *Experimental Membrane Permeability Measurements*

3.2.1. Calcein Leakage from GUVs

The membrane perturbing effects of the modified peptide azoALY were also investigated while using calcein leakage experiments in a suspension of giant unilamellar vesicles (GUVs) formed by POPC/POPG (9:1, mol/mol). In this assay, a fluorescent probe (calcein) was employed, and its release from the GUVs was followed by measuring its fluorescence intensity inside the vesicles. Excessive leakage upon adding peptides or other compounds to the GUVs indicates instability in the structure of the vesicle membrane (alteration of the packing of the lipids). This, in turn, suggests that the peptide added to the GUVs strongly interacts with the lipid membrane. GUVs represent an excellent model representing the lipid membrane of biological cell membranes and they have become an essential tool in biophysical research [45,46]. The experiments were carried out using two different concentrations of peptides: 1 µM and 50 µM of ALY and azoALY. Figure 4 shows a representative confocal image series after the addition of 50 µM azoALY, where calcein diffuses out of the GUVs via a decrease in fluorescence intensity.

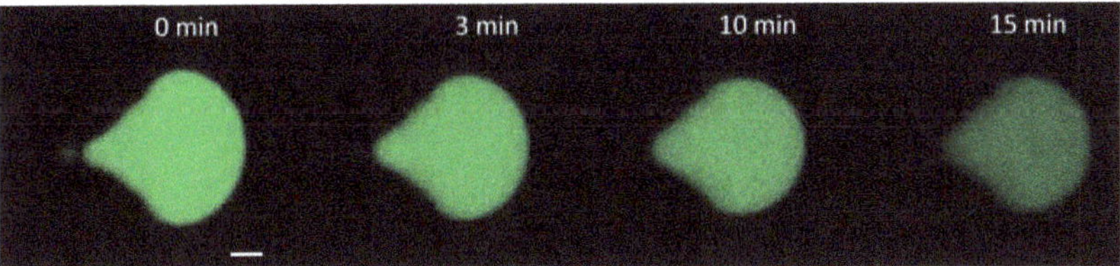

Figure 4. Confocal fluorescence images of calcein leakage from an exemplary POPC/POPG (9:1) GUV with azoALY added externally (50 µM at time 0 min. A single GUV was held in one spatial location using a microfluidic platform; for details, see reference [39]. Scale bar: 5 µm.

After 15 min. of incubation with the peptides, the GUVs preserved their membrane structure (i.e., no visible defects), so we exclude the destruction of the membrane or micron-sized pore formation. However, a considerable amount of calcein remained entrapped inside the vesicles after 15 min. under the used conditions. The analysed GUVs were selected to have the same diameter of approximately 12 μm to maintain the same surface area for comparable results. The mean pixel intensities within each GUV were normalized and plotted against time (Figure 5).

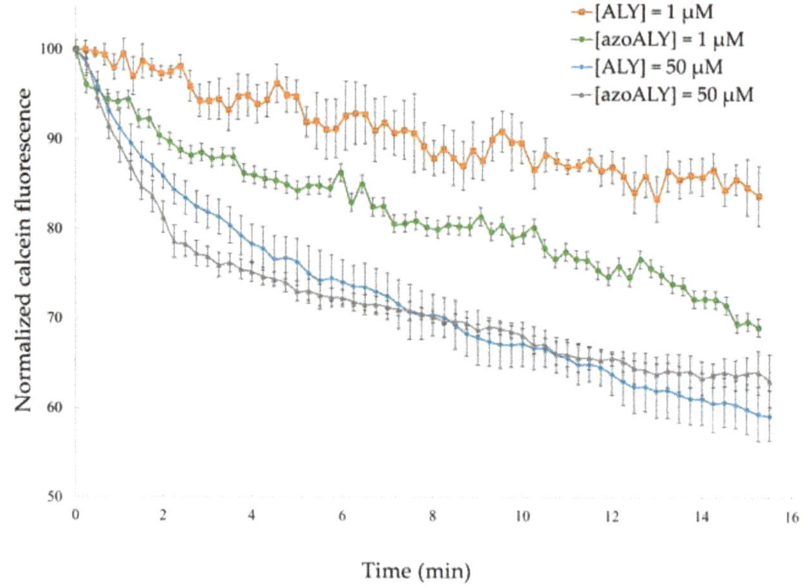

Figure 5. Kinetics of calcein release from POPC/POPG (9:1) GUVs for externally added ALY or azoALY (1 and 50 μM) using 3 separate vesicles. Control measurements without the addition of peptide showed no calcein release (data not shown).

At low concentrations of ALY peptide (1 μM), during the first 15 min. of exposure, the calcein fluorescence undergoes a decrease of 17%. Nevertheless, while using the same concentration of the modified azoALY peptide, a decrease of 31% in intensity was observed at the same time. In contrast, when high concentrations of peptides were used (50 μM), it was possible to observe that the percentage of calcein release is similar for both of the peptides. In that case, the addition of azoALY resulted in a faster calcein release than the reference peptide ALY. In fact, during the first three minutes of the experiment, 25% of calcein leakage was observed, though, after 15 min., both of the peptides reached the same calcein leakage.

3.2.2. Permeabilization of LUVs

We performed an enzymatic assay for detecting possible changes in the permeability of LUVs to further evaluate the membrane permeability increase upon peptide addition to the calcein-containing GUVs. For this purpose, LUVs containing entrapped horseradish peroxidase isoenzyme C (HRPC) were prepared. Afterward, the chromogenic substrate $ABTS^{2-}$ and hydrogen peroxide (H_2O_2) were added to the enzyme-containing vesicles. An unperturbed phospholipid membrane is impermeable for HRPC and $ABTS^{2-}$. The HRPC molecules cannot cross the membrane due to their large size, and $ABTS^{2-}$ molecules cannot move from the external bulk solution into the interior of the vesicles due to their negative charge. On the contrary, H_2O_2, being small and uncharged, can easily permeate across fluid phospholipid bilayers [47]. The release of HRPC from the LUVs and/or the uptake of $ABTS^{2-}$ by the LUVs after peptide addition can be conveniently monitored by UV-vis spectrophotometry as the HRPC-catalysed oxidation of $ABTS^{2-}$ to $ABTS^{\bullet-}$ results in increased absorbance (see Figure 6 for a schematic representation of the permeability assay).

HRPC was entrapped in LUVs with a diameter of about 200 nm. We used the same formulation for the vesicles, as in the case of the experiments with GUVs, i.e., POPC/POPG (9:1). The amount of entrapped HRPC was quantified by measuring the HRPC activity after adding Triton X-100 (0.1 vol%) that causes immediate vesicles destruction and the release of the entrapped enzyme into the assay solution (dilution factor 2000×). The increase of membrane permeability (without added Triton X-100, but with added peptides) was estimated by measuring the enzyme activity with $ABTS^{2-}$ and H_2O_2 as HRPC substrates, monitoring the formation of $ABTS^{\bullet -}$.

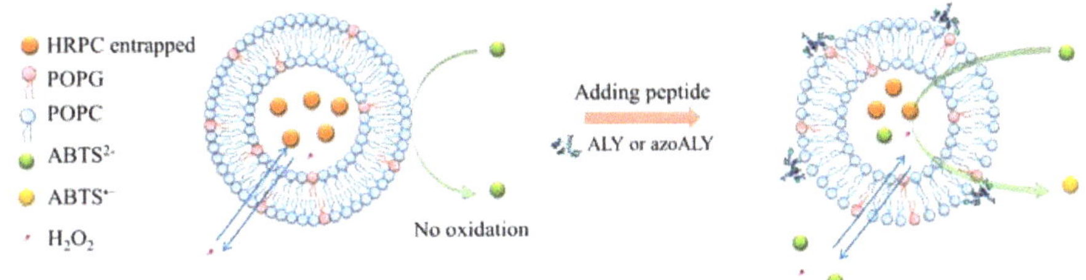

Figure 6. Horseradish peroxidase isoenzyme C enzyme (HRPC) enzyme leakage from large unilamellar vesicles (LUVs), measured with H_2O_2 and $ABTS^{2-}$, the latter being membrane impermeable.

We analysed the interactions of ALY and azoALY with POPC/POPG (9:1) LUVs containing entrapped HRPC (dilution factor 100×). We used as blank the system POPC/POPG (9:1) LUVs with entrapped HRPC, without peptides (dilution factor 100×), in order to exclude the presence of enzyme molecules bound to the outer membrane surface of the vesicles. Measurements were performed after different incubation times (0, 5, 10 min.) to estimate a possible time dependence of the interactions and understand if and how much the azo-amino acid inserted in the amino acidic sequence could affect the membrane permeability. The preparation of the vesicles with entrapped enzyme was repeated three times and, for each mixture, we evaluated the membrane perturbation due to the action of ALY and azoALY. In all cases, we obtained analogous enzyme leakage. Therefore, we excluded the possibility of false-positive results by the oxidation of the phenol moiety (in tyrosine) and azo-group in the peptides. Moreover, we excluded a possible shift of the characteristic peaks of the oxidized $ABTS^{2-}$ form (414 nm, 650, 735, and 814 nm) due to the presence of peptides. We also checked if a mixture of "empty" vesicles and peptide influences the enzymatic reaction and we exclude enzyme inhibition by the assay mixture.

Linear regression of $ABTS^{\bullet -}$ absorbance at λ_{max} = 414 nm as a function of reaction time allowed for the determination of the HRPC activity, read as the slope of the linear fit (dA_{414nm}/dt), see Figure 7.

The enzymatic activity for the blank was very low, and the corresponding HRPC concentration in the assay solution was lower than 50 pM (for the calibration curve see Figure S5 in Supp. Mat.). Besides, the activity for this system was not time-dependent: the slope value did not change for the three different incubation times. This means that the amount of HRPC that was possibly bound to the outer membrane was quite low; additionally, there was no leakage of the enzyme or uptake of the substrate. *Vice versa*, a significant increase in enzyme activity was observed after 5 min. of incubation in enzyme-containing vesicles plus the surfactant Triton X-100 (0.1%). Analysing the calibration curve in the presence of Triton X-100 (see Figure S5 in Supp. Mat.), the concentration of the entrapped enzyme was estimated to be around 400 nM (when considering a dilution factor 2000×). The ALY peptide produced a small increase in enzyme activity. At the beginning (incubation 0 min.), the effect on the membrane permeability was comparable to the blank. After 10 minutes of incubation, we recorded an enzymatic activity corresponding to an enzyme concentration lower than 5 nM (dilution factor 100×). The situation was completely different while using the modified peptide azoALY: an immediate significant increase in HRPC activity was registered, and the activity increase was time dependent. When considering a dilution factor 100×, the amount of enzyme released for 0 min. of incubation corresponds to an enzyme concentration of HRPC around 10 nM, for 5 min. was 25 nM and for 10 min.

was 33 nM. We could estimate an increase of enzyme release or $ABTS^{2-}$ uptake time dependent with an enzyme concentration in the pooled fraction of 10 nM, 25 nM, and 32.5 nM, respectively.

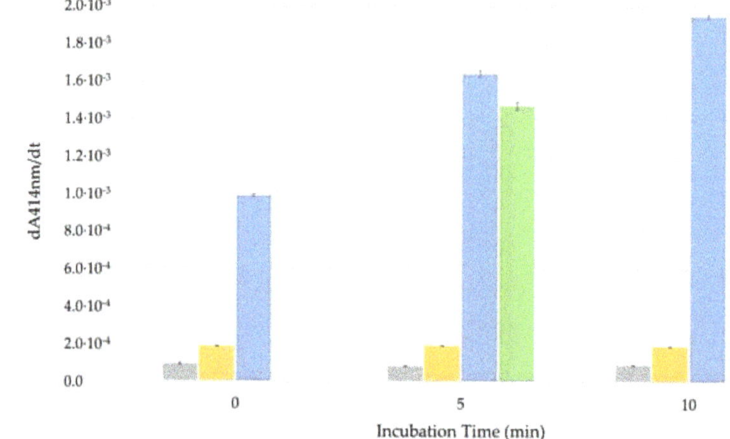

Figure 7. Enzymatic activity time dependence in four different systems: (grey bar) Blank vesicles with HRPC entrapped without peptides (dilution factor 100×); (green bar) control vesicles with HRPC entrapped, adding Triton X-100 (dilution factor 2000×); (yellow bar) vesicles with HRPC entrapped, adding a solution of ALY at 4 μM (dilution factor 100×); and (blue bar) vesicles with HRPC entrapped, adding a solution of azoALY at 4 μM (dilution factor 100×). Each data point shown is the average from three measurements and the small standard deviation is indicated with a bar.

The HRPC/LUV experiments indicate a time-dependent increased membrane permeability upon the addition of azoALY. However, it remains to be clarified whether the higher enzymatic activity is ascribed to $ABTS^{2-}$ uptake or HRPC release from the vesicles.

4. Conclusions

The combined use of MD simulations, together with experimental tests, allowed for us to investigate the interaction of a natural peptide and its azobenzene modified analog with artificial membrane models. We have designed a novel peptide (azoALY) by modifying a natural membrane-active peptide (ALY) with a tyrosine containing an azo-group in the side chain. Computational and experimental permeability studies with model membrane systems (vesicles of POPC and POPG at a molar ration of 9:1) suggest that the insertion of the modified tyrosine residue in the peptide sequence increases the peptide affinity to the vesicle membrane. A different membrane permeability for ALY and azoALY was observed in experimental membrane permeability assays while using GUVs and LUVs: increased calcein release form GUVs and increased HRPC leakage from the LUVs or increased $ABTS^{2-}$ uptake by the LUVs. The MD simulations on model membranes support these results. The all-atom molecular dynamics showed a higher membrane perturbation by the modified peptide (azoALY) than the unmodified peptide (ALY) under the same conditions. The azoALY peptide penetrates the hydrophobic core of the bilayers, while the ALY structure only interacts with the lipid headgroups, causing a lower extent of membrane alteration. As a more general consideration, our investigations showed that the results obtained from physical-chemical in vitro investigations with simple membrane model systems might not directly correlate with the in vivo behavior of biological membranes of living cells. The vesicle and the bacterial membrane compositions were very different, and the actual concentrations in terms of peptides and membrane components were difficult to compare. Similar conclusions were drawn before from other studies [48,49].

Still, based on the obtained results, the presence of a modified amino acid residue appears to be essential in determining the perturbation effect of the azobenzene peptide membrane. Under such conditions, modified peptides could increase peptide-membrane interaction and complete their biological task (for example, antimicrobial agents). The novel semi-synthetic peptide could act as a prodrug, generating in vivo the antimicrobial azo compound. The preliminary results of membrane permeability underlined a higher membrane perturbation with time-dependence leakage from lipid vesicles after interaction with modified peptide as compared to the unmodified peptide under the same conditions. In conclusion, we evidenced a different behavior of the natural and the azo-modified peptide, and we suggested the use of azo-modified amino acid moieties to increase the membrane modification. Although only one peptide pair and one membrane model were studied, the different effects on membrane perturbation suggest that these effects are potentially of broader significance. Additional work is needed in order to clarify the possible generalization of the results that were observed in this study.

Supplementary Materials: The following are available online at http://www.mdpi.com/2077-0375/10/10/294/s1. Figure S1. Scheme showing a trapped GUV isolated by a donut valve; Figure S2. Wide-field fluorescence image of a single GUV trapped hydrodynamically by the posts; Figure S3. OD_{403} originating from light scattering (turbidity) of the vesicles present in the fractions eluting up to an elution volume of about 12 mL, and HRPC activity of the different fractions, measured with $ABTS^{2-}/H_2O_2$ as substrates. Figure S4. Changes of the absorption spectrum of the reaction solution as a function of reaction time; Figure S5. HRPC concentration dependency of the absorbance of the reaction solution at 414 nm; Figure S6. Snapshot of the starting configuration of both peptide/POPC-POPG systems and the last frame at 50 ns of simulation time. Figure S7. Order parameter SCD for (a) the unsaturated oleic and (b) the saturated palmitoyl acyl chains of phospholipids in POPC/POPG/peptide.

Author Contributions: Conceptualization, L.S., P.W. and S.P.; Data curation, L.S., S.C., P.W., A.P. and B.P.; Investigation, L.S., S.C. and A.P.; Methodology, L.S., T.R. and U.C.; Resources, S.C., P.W., P.S.D. and S.P.; Writing–original draft, L.S.; Writing–review & editing, L.S., S.C., P.W., T.R., P.S.D. and S.P. All authors have read and agreed to the published version of the manuscript.

Funding: This research was funded by the Italian Ministry of Education, University and Research (MIUR) [grant number 300395FRB17] and by the European Grant CLINGLIO, Research & Innovation H2020-SC1-2017-Two-Stage-RTD Call H2020-SC1-2016-2017.

Conflicts of Interest: The authors declare no conflict of interest.

References

1. Guha, S.; Ghimire, J.; Wu, E.; Wimley, W.C. Mechanistic Landscape of Membrane-Permeabilizing Peptides. *Chem. Rev.* **2019**, *119*, 6040–6085. [CrossRef]
2. Hasan, M.; Yamazaki, M. Elementary Processes and Mechanisms of Interactions of Antimicrobial Peptides with Membranes—Single Giant Unilamellar Vesicle Studies. In *Antimicrobial Peptides. Advances in Experimental Medicine and Biology*; Springer: Berlin/Heidelberg, Germany, 2019; Volume 1117, pp. 17–32.
3. Shai, Y. Mechanism of the binding, insertion and destabilization of phospholipid bilayer membranes by α-helical antimicrobial and cell non-selective membrane-lytic peptides. *Biochim. Biophys. Acta (BBA) Biomembr.* **1999**, *1462*, 55–70. [CrossRef]
4. Ergene, C.; Yasuhara, K.; Palermo, E.F. Biomimetic antimicrobial polymers: Recent advances in molecular design. *Polym. Chem.* **2018**, *9*, 2407–2427. [CrossRef]
5. Galdiero, S.; Falanga, A.; Cantisani, M.; Vitiello, M.; Morelli, G.; Galdiero, M. Peptide-Lipid Interactions: Experiments and Applications. *Int. J. Mol. Sci.* **2013**, *14*, 18758–18789. [CrossRef]
6. Hall, B.A.; Chetwynd, A.P.; Sansom, M.S.P. Exploring Peptide-Membrane Interactions with Coarse-Grained MD Simulations. *Biophys. J.* **2011**, *100*, 1940–1948. [CrossRef]
7. Sato, H.; Feix, J.B. Peptide–membrane interactions and mechanisms of membrane destruction by amphipathic α-helical antimicrobial peptides. *Biochim. Biophys. Acta (BBA) Biomembr.* **2006**, *1758*, 1245–1256. [CrossRef]
8. Hollmann, A.; Martinez, M.; Maturana, P.; Semorile, L.C.; Maffia, P.C. Antimicrobial Peptides: Interaction With Model and Biological Membranes and Synergism With Chemical Antibiotics. *Front. Chem.* **2018**, *6*. [CrossRef]
9. Travkova, O.G.; Moehwald, H.; Brezesinski, G. The interaction of antimicrobial peptides with membranes. *Adv. Colloid Interface Sci.* **2017**, *247*, 521–532. [CrossRef]

10. Pillong, M.; Hiss, J.A.; Schneider, P.; Lin, Y.-C.; Posselt, G.; Pfeiffer, B.; Blatter, M.; Müller, A.T.; Bachler, S.; Neuhaus, C.S.; et al. Rational Design of Membrane-Pore-Forming Peptides. *Small* **2017**, *13*, 1701316. [CrossRef]
11. Friedman, R.; Khalid, S.; Aponte-Santamaría, C.; Arutyunova, E.; Becker, M.; Boyd, K.J.; Christensen, M.; Coimbra, J.T.S.; Concilio, S.; Daday, C.; et al. Understanding Conformational Dynamics of Complex Lipid Mixtures Relevant to Biology. *J. Membr. Biol.* **2018**, *251*, 609–631. [CrossRef]
12. Ishibashi, J.; Saido-Sakanaka, H.; Yang, J.; Sagisaka, A.; Yamakawa, M. Purification, cDNA cloning and modification of a defensin from the coconut rhinoceros beetle, Oryctes rhinoceros. *Eur. J. Biochem.* **1999**, *266*, 616–623. [CrossRef] [PubMed]
13. Piotto, S.P.; Sessa, L.; Concilio, S.; Iannelli, P. YADAMP: Yet another database of antimicrobial peptides. *Int. J. Antimicrob. Agents* **2012**, *39*, 346–351. [CrossRef] [PubMed]
14. Piotto, S.; Di Biasi, L.; Sessa, L.; Concilio, S. Transmembrane Peptides as Sensors of the Membrane Physical State. *Front. Phys.* **2018**, *6*. [CrossRef]
15. Scrima, M.; Di Marino, S.; Grimaldi, M.; Campana, F.; Vitiello, G.; Piotto, S.P.; D'Errico, G.; D'Ursi, A.M. Structural features of the C8 antiviral peptide in a membrane-mimicking environment. *Biochim. Biophys. Acta (BBA) Biomembr.* **2014**, *1838*, 1010–1018. [CrossRef] [PubMed]
16. Chen, C.H.; Melo, M.C.; Berglund, N.; Khan, A.; de la Fuente-Nunez, C.; Ulmschneider, J.P.; Ulmschneider, M.B. Understanding and modelling the interactions of peptides with membranes: From partitioning to self-assembly. *Curr. Opin. Struct. Biol.* **2020**, *61*, 160–166. [CrossRef] [PubMed]
17. Mondal, S.; Khelashvili, G.; Shan, J.; Andersen, O.S.; Weinstein, H. Quantitative modeling of membrane deformations by multihelical membrane proteins: Application to G-protein coupled receptors. *Biophys. J.* **2011**, *101*, 2092–2101. [CrossRef]
18. Piotto, S.; Concilio, S.; Sessa, L.; Diana, R.; Torrens, G.; Juan, C.; Caruso, U.; Iannelli, P. Synthesis and antimicrobial studies of new antibacterial azo-compounds active against staphylococcus aureus and listeria monocytogenes. *Molecules* **2017**, *22*, 1372. [CrossRef]
19. Concilio, S.; Sessa, L.; Petrone, A.M.; Porta, A.; Diana, R.; Iannelli, P.; Piotto, S. Structure modification of an active azo-compound as a route to new antimicrobial compounds. *Molecules* **2017**, *22*, 875. [CrossRef]
20. Chan, Y.H.M.; Boxer, S.G. Model membrane systems and their applications. *Curr. Opin. Chem. Biol.* **2007**, *11*, 581–587. [CrossRef]
21. Kumagai, P.S.; Sousa, V.K.; Donato, M.; Itri, R.; Beltramini, L.M.; Araujo, A.P.; Buerck, J.; Wallace, B.; Lopes, J.L. Unveiling the binding and orientation of the antimicrobial peptide Plantaricin 149 in zwitterionic and negatively charged membranes. *Eur. Biophys. J.* **2019**, *48*, 621–633. [CrossRef]
22. Ning, L.; Mu, Y. Aggregation of PrP106–126 on surfaces of neutral and negatively charged membranes studied by molecular dynamics simulations. *Biochim. Biophys. Acta (BBA) Biomembr.* **2018**, *1860*, 1936–1948. [CrossRef]
23. Leber, R.; Pachler, M.; Kabelka, I.; Svoboda, I.; Enkoller, D.; Vácha, R.; Lohner, K.; Pabst, G. Synergism of antimicrobial frog peptides couples to membrane intrinsic curvature strain. *Biophys. J.* **2018**, *114*, 1945–1954. [CrossRef]
24. Salnikov, E.S.; Bechinger, B. Lipid-controlled peptide topology and interactions in bilayers: Structural insights into the synergistic enhancement of the antimicrobial activities of PGLa and magainin 2. *Biophys. J.* **2011**, *100*, 1473–1480. [CrossRef]
25. Tzong-Hsien, L.; Kristopher, N.H.; Marie-Isabel, A. Antimicrobial Peptide Structure and Mechanism of Action: A Focus on the Role of Membrane Structure. *Curr. Top. Med. Chem.* **2016**, *16*, 25–39. [CrossRef]
26. Matsuzaki, K.; Harada, M.; Handa, T.; Funakoshi, S.; Fujii, N.; Yajima, H.; Miyajima, K. Magainin 1-Induced Leakage of Entrapped Calcein out of Negatively-Charged Lipid Vesicles. *Biochim. Biophys. Acta* **1989**, *981*, 130–134. [CrossRef]
27. Matsuzaki, K.; Fukui, M.; Fujii, N.; Miyajima, K. Interactions of an antimicrobial peptide, tachyplesin I, with lipid membranes. *Biochim. Biophys. Acta* **1991**, *1070*, 259–264. [CrossRef]
28. King, R.D.; Sternberg, M.J. Identification and application of the concepts important for accurate and reliable protein secondary structure prediction. *Protein Sci.* **1996**, *5*, 2298–2310. [CrossRef] [PubMed]
29. Holton, T.A.; Pollastri, G.; Shields, D.C.; Mooney, C. CPPpred: Prediction of cell penetrating peptides. *Bioinformatics* **2013**, *29*, 3094–3096. [CrossRef]

30. Altomare, A.; Burla, M.C.; Camalli, M.; Cascarano, G.L.; Giacovazzo, C.; Guagliardi, A.; Moliterni, A.G.; Polidori, G.; Spagna, R. SIR97: A new tool for crystal structure determination and refinement. *J. Appl. Crystallogr.* **1999**, *32*, 115–119. [CrossRef]
31. Jo, S.; Lim, J.B.; Klauda, J.B.; Im, W. CHARMM-GUI Membrane Builder for Mixed Bilayers and Its Application to Yeast Membranes. *Biophys. J.* **2009**, *97*, 50–58. [CrossRef]
32. Krieger, E.; Vriend, G. YASARA View-molecular graphics for all devices-from smartphones to workstations. *Bioinformatics* **2014**, *30*, 2981–2982. [CrossRef] [PubMed]
33. Guixà-González, R.; Rodriguez-Espigares, I.; Ramírez-Anguita, J.M.; Carrió-Gaspar, P.; Martinez-Seara, H.; Giorgino, T.; Selent, J. MEMBPLUGIN: Studying membrane complexity in VMD. *Bioinformatics* **2014**, *30*, 1478–1480. [CrossRef]
34. Concilio, S.; Ferrentino, I.; Sessa, L.; Massa, A.; Iannelli, P.; Diana, R.; Panunzi, B.; Rella, A.; Piotto, S. A novel fluorescent solvatochromic probe for lipid bilayers. *Supramol. Chem.* **2017**, *29*, 887–895. [CrossRef]
35. Maget-Dana, R.; Bonmatin, J.M.; Hetru, C.; Ptak, M.; Maurizot, J.C. The secondary structure of the insect defensin A depends on its environment. A circular dichroism study. *Biochimie* **1995**, *77*, 240–244. [CrossRef]
36. Pieri, E.; Ledentu, V.; Huix-Rotllant, M.; Ferré, N. Sampling the protonation states: The pH-dependent UV absorption spectrum of a polypeptide dyad. *Phys. Chem. Chem. Phys.* **2018**, *20*, 23252–23261. [CrossRef]
37. Angelova, M.I.; Soleau, S.; Meleard, P.; Faucon, J.F.; Bothorel, P. Preparation of Giant Vesicles by External Ac Electric-Fields-Kinetics and Applications. *Prog. Coll. Pol. Sci.* **1992**, *89*, 127–131. [CrossRef]
38. Robinson, T.; Dittrich, P.S. Observations of membrane domain reorganization in mechanically compressed artificial cells. *ChemBioChem* **2019**, *20*, 2666–2673. [CrossRef] [PubMed]
39. Robinson, T.; Kuhn, P.; Eyer, K.; Dittrich, P.S. Microfluidic trapping of giant unilamellar vesicles to study transport through a membrane pore. *Biomicrofluidics* **2013**, *7*, 044105. [CrossRef]
40. Zhang, Y.; Schmid, Y.R.F.; Luginbühl, S.; Wang, Q.; Dittrich, P.S.; Walde, P. Spectrophotometric Quantification of Peroxidase with p-Phenylene-diamine for Analyzing Peroxidase-Encapsulating Lipid Vesicles. *Anal. Chem.* **2017**, *89*, 5484–5493. [CrossRef] [PubMed]
41. Childs, R.E.; Bardsley, W.G. The steady-state kinetics of peroxidase with 2,2′-azino-di-(3-ethyl-benzthiazoline-6-sulphonic acid) as chromogen. *Biochem. J.* **1975**, *145*, 93–103. [CrossRef]
42. Welinder, K.G.; Smillie, L.B. Amino acid sequence studies of horseradish peroxidase. II. Thermolytic peptides. *Can. J. Biochem.* **1972**, *50*, 63–90. [CrossRef] [PubMed]
43. Veitch, N.C. Horseradish peroxidase: A modern view of a classic enzyme. *Phytochemistry* **2004**, *65*, 249–259. [CrossRef] [PubMed]
44. Porstmann, B.; Porstmann, T.; Gaede, D.; Nugel, E.; Egger, E. Temperature dependent rise in activity of horseradish peroxidase caused by non-ionic detergents and its use in enzyme-immunoassay. *Clin. Chim. Acta* **1981**, *109*, 175–181. [CrossRef]
45. Walde, P.; Cosentino, K.; Engel, H.; Stano, P. Giant vesicles: Preparations and applications. *ChemBioChem* **2010**, *11*, 848–865. [CrossRef] [PubMed]
46. Dimova, R. Giant vesicles and their use in assays for assessing membrane phase state, curvature, mechanics, and electrical properties. *Annu. Rev. Biophys.* **2019**, *48*, 93–119. [CrossRef] [PubMed]
47. Yoshimoto, M.; Higa, M. A kinetic analysis of catalytic production of oxygen in catalase-containing liposome dispersions for controlled transfer of oxygen in a bioreactor. *J. Chem. Technol. Biotechnol.* **2014**, *89*, 1388–1395. [CrossRef]
48. Shimanouchi, T.; Walde, P.; Gardiner, J.; Mahajan, Y.R.; Seebach, D.; Thomae, A.; Krämer, S.D.; Voser, M.; Kuboi, R. Permeation of a β-heptapeptide derivative across phospholipid bilayers. *Biochim. Biophys. Acta (BBA) Biomembr.* **2007**, *1768*, 2726–2736. [CrossRef]
49. Zepik, H.H.; Walde, P.; Kostoryz, E.L.; Code, J.; Yourtee, D.M. Lipid vesicles as membrane models for toxicological assessment of xenobiotics. *Crit. Rev. Toxicol.* **2008**, *38*, 1–11. [CrossRef] [PubMed]

 *membranes*

Article

Influence of Cholesterol on the Orientation of the Farnesylated GTP-Bound KRas-4B Binding with Anionic Model Membranes

Huixia Lu and Jordi Martí *

Department of Physics, Technical University of Catalonia-Barcelona Tech, 08034 Barcelona, Spain; huixia.lu@upc.edu

* Correspondence: jordi.marti@upc.edu

Received: 6 October 2020; Accepted: 18 November 2020; Published: 22 November 2020

Abstract: The Ras family of proteins is tethered to the inner leaflet of the cell membranes which plays an essential role in signal transduction pathways that promote cellular proliferation, survival, growth, and differentiation. KRas-4B, the most mutated Ras isoform in different cancers, has been under extensive study for more than two decades. Here we have focused our interest on the influence of cholesterol on the orientations that KRas-4B adopts with respect to the plane of the anionic model membranes. How cholesterol in the bilayer might modulate preferences for specific orientation states is far from clear. Herein, after analyzing data from in total 4000 ns-long molecular dynamics (MD) simulations for four KRas-4B systems, properties such as the area per lipid and thickness of the membrane as well as selected radial distribution functions, penetration of different moieties of KRas-4B, and internal conformational fluctuations of flexible moieties in KRas-4B have been calculated. It has been shown that high cholesterol content in the plasma membrane (PM) favors one orientation state (OS_1), exposing the effector-binding loop for signal transduction in the cell from the atomic level. We confirm that high cholesterol in the PM helps KRas-4B mutant stay in its constitutively active state, which suggests that high cholesterol intake can increase mortality and may promote cancer progression for cancer patients. We propose that during the treatment of KRas-4B-related cancers, reducing the cholesterol level in the PM and sustaining cancer progression by controlling the plasma cholesterol intake might be taken into account in anti-cancer therapies.

Keywords: KRas-4B; mutation; post-translational modification; HVR; anionic plasma membrane; signaling; cholesterol

---

1. Introduction

The cell membrane plays an important role in controlling the passing of nutrients, wastes, drugs, and heat between the inner and outer parts of a cell. The principal components of human cellular membranes are phospholipids, cholesterol, and proteins, etc. Moreover, the concentration of each constituent differs for different types of cells. Phospholipids provide the framework to biomembranes and they consist of two leaflets of amphiphilic lipids with a hydrophilic head and one or two hydrophobic tails which self-assemble due to the hydrophobic effect [1,2]. For instance, 1,2-dioleoyl-sn-glycero-3-phosphocholine (DOPC) belongs to the unsaturated phospholipids which is a typical constituent of real biological membranes. Furthermore, 1,2-dioleoyl-sn-glycero-3-phospho-L-serine (DOPS) is the most common anionic lipid in the plasma membrane (PM) of mammalian cells, which is preferentially targeted by the PM intracellular surface protein KRas-4B [3] for signal transduction.

PM systems have been extensively studied over several decades [4–8] on their association with proteins. Recent studies have shown that the role of proteins and their interactions with components of

PM is extremely important to understand the mechanisms of protein anchoring into the membrane that can lead to oncogenesis [9]. GTPases are a large family of hydrolase enzymes that bind to the nucleotide guanosine triphosphate (GTP) and hydrolyze it to guanosine diphosphate (GDP). GDP/GTP cycling is controlled by two main classes of regulatory proteins. Guanine-nucleotide-exchange factors (GEFs) promote the formation of the active, GTP-bound form, while GTPase-activating proteins (GAPs) inactivate Ras by enhancing the intrinsic GTPase activity to promote the formation of the inactive GDP-bound form [10–12]. Ras proteins are small molecular weight GTPases and function as GDP/GTP-regulated molecular switches controlling pathways involved in critical cellular functions, like cell proliferation, signaling, cell growth, and anti-apoptosis pathways [13]. The three Ras genes give rise to three base protein sequences: KRas, HRas, and NRas. Over 30% of cancers are driven by mutant Ras proteins, thereinto, one method called catalog of somatic mutations in cancer (COSMIC) [14] confirms that HRas (3%) are the least frequently mutated Ras isoforms in human cancers, where KRas (86%) are the predominantly mutated isoforms followed by NRas (11%) [15].

KRas can be found as two splice variants designated KRas-4A and KRas-4B. They both have polybasic sequences that facilitate membrane-association in acidic membrane regions [16], however, for KRas-4A, it is covalently modified by a single palmitic acid. KRas-4B is distinguished from KRas-4A isoform in the residue 181 that serves as a phosphorylation site within its flexible hypervariable region (HVR, residues 167–185) that contains the farnesyl group (FAR) serving as the lipid anchor. The HVR of KRas-4B contains multiple amino-acid lysines that act as an electrostatic farnesylated switch which guarantees KRas-4B's association with the negatively charged phospholipids in the inner PM leaflet. It has been reported that the KRas-4B activation level in diseased cells is linked to phosphatidylserine contents [17]. Anionic lipids could influence the membrane potential which in turn regulates the orientation, location, and signaling ability of KRas-4B [18,19].

The catalytic domain (CD, residues 1–166), which contains the catalytic lobe (lobe 1, residues 1–86) and the allosteric lobe (lobe 2, residues 87–166), highly homologous, conserved, and the structure is shared and identical for KRas-4A and KRas-4B. According to P. Prakash et al. [20–22], three distinct orientation states of the oncogenic G12V-KRas-4B mutant on the membrane have been reported, namely, OS_1, OS_2, and OS_0. OS_1, with an accessible effector-binding loop, and OS_2, with the effector-binding loop occluded by the membrane, has been reported. They differ in the accessibility of functionally critical switch loops to the downstream effectors, suggesting that membrane reorientation of KRas-4B on the inner cell leaflet may modulate its signaling [21]. The idea of the more flexible in the structure of proteins, the larger the number of their populated states have been pointed out [23]. We and other researchers have recently shown that despite the HVR and FAR anchor, the CD of KRas-4B could interact with anionic model membranes by forming steady salt bridges and hydrogen bonds to help organize its orientations in cells [18,24,25].

All Ras proteins' signaling strongly depends on their correct localization in the cell membrane and it is essential for activating downstream signaling pathways. KRas-4B function, membrane association and interaction with other proteins are regulated by post-translational modifications (PTMs) [26–28], including ubiquitination, acetylation, prenylation, phosphorylation, and carboxymethylation, see Figure 1. Firstly, the prenylation reaction, catalyzed by cytosolic farnesyltrasferase (FTase) or geranylgeranyltransferase (GGTase), proceeds through the addition of an isoprenyl group to the Cys-185 side chain. Then, farnesylated KRas-4B is ready for further processing: hydrolysis, catalyzed by the endopeptidase enzyme called Ras-converting enzyme 1 (RCE1); during the process, the VIM motif (HVR tail is composed of three amino acids: valine-isoleucine-methionine) of the C-terminal Cys-185 is lost in step 2.

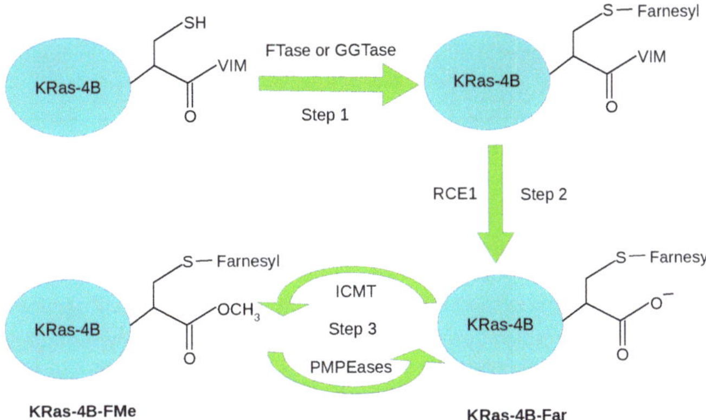

Figure 1. Post-translational modifications (PTMs) steps of KRas-4B: prenylation, hydrolysis, carboxymethylation and decarboxymethylation.

In step 3, KRas-4B is transferred to the endoplasmic reticulum for carboxymethylation at the carboxyl terminus of Cys-185 catalyzed by isoprenylcysteine carboxyl methyltransferase (ICMT), forming a reversible ester bond. The reversible ester bond can go through decarboxymethylation, catalyzed by prenylated/polyisoprenylated methylated protein methyl esterases (PMPEases) giving rise to a farnesylated and demethylated KRas-4B (KRas-4B-Far). Carboxymethylation is one of the best known reversible PTMs in HVR [29]. This reversible reaction can modulate the equilibrium of methylated/demethylated KRas-4B population (KRas-4B-FMe/KRas-4B-Far) in tumors and consequently can impact downstream signaling, protein–protein interactions, or protein–lipid interactions [30]. Another well-known reversible PTM in the HVR is phosphorylation [28,31,32]. There are two sites (Ser-171 and Ser-181) within HVR that could be phosphorylated. Phosphorylation involves the addition of phosphate (PO_4^{3-}) group to the side chain of the amino acid serine, then the phosphorylated serine is obtained. In this work, we have only applied the phosphorylation at Ser-181 (PHOS) for the oncogenic KRas-4B. Phosphorylation at Ser-181 operates a farnesyl-electrostatic switch that reduces but does not completely inhibit membrane association and clustering of KRas-4B, leading to the redistribution of the cytoplasm and endomembranes [27,33,34]. Functionally, the phosphorylation of KRas-4B can have either a negative [35,36] or positive [34,37] regulatory effect on tumor cell growth, depending on the conditions [30]. For instance, from a molecular dynamics (MD) simulation of the HVR peptide with the FAR at Cys-185 of KRas-4B in two types of model membranes, it has been observed that phosphorylation at Ser-181 prohibits spontaneous FAR membrane insertion [38]. According to Agell et al., KRas-4B binding with calmodulin leads to different behaviors: short or prolonged signaling whether KRas-4B is at its phosphorylated state on residue Ser-181 [34,39]. Moreover, according to Barcelo et al. [37], phosphorylation at Ser-181 of oncogenic KRas is required for tumor growth. In summary, the phosphorylation of the HVR of KRas-4B can affect its function, membrane association, and reacting with downstream effectors [30].

Phosphodiest-eraseδ (PDEδ) has been revealed to promote effective KRas-4B signaling by sequestering KRas-4B-FMe from the cytosol by binding the prenylated HVR and help to enhance its diffusion to the PM throughout the cell, where it is released to activate various signaling pathways required for the initiation and maintenance of cancer [40–43], hoping to identify a panel of novel PDEδ inhibitors. As described in our earlier work [24], despite KRas-4B-Far's poor affinity for PDEδ [40], it can still be transferred to the PM through trapping and vesicular transport without the help of

PDEδ [44]. Moreover, according to Ntai et al., 91% of the mutant KRas-4B and 51% of wild-type KRas-4B proteins in certain colorectal tumor samples have been found to exist in its KRas-4B-Far form. While there is a relatively high abundance of KRas-4B-Far (wild-type and mutant) lacking the methyl group of Cys-185 in tumors, the effects of demethylated KRas-4B-Far on downstream signaling have yet to be determined [45]. While extensive research has been focused on methylated KRas-4B-FMe, we believe that demethylated KRas-4B-Far could play a big role in the signaling pathway that happens on the inner leaflet of the membrane bilayers.

Cholesterol plays an important role in maintaining the structure of different membranes and regulating their functions [46,47], and cancer development as well [48]. In some types of cells, however, the distribution of cholesterol is different in the inner and outer leaflets of the membrane, ranging from 0.1% to 50% of total membrane lipids depending on the cell type [49]. The average value of cholesterol concentration in the PM fell in the range of the reported value of 19–40 % [50–52]. For example, in red cells, the percentage of cholesterol differs in the outer leaflet (51%) and the inner leaflet (49%) [53]. In other types of cells, cholesterol constitutes about 33.3% of the outer leaflet in healthy colorectal cells [50]. Previous simulations with percentages of 10%, 20%, and 40% for DPPC lipid bilayers showed no further relevant physical changes compared to the cholesterol percentages of 0%, 30% and 50% adopted in our previous work [54].

The fluidity of the membrane is mainly regulated by the amount of cholesterol, in such a way that membranes with high cholesterol contents are stiffer than those with low amounts but keeping the appropriate fluidity for allowing normal membrane functions. Extensive research has been done on the influence of cholesterol on the mechanism of membrane structures [55], the 18-kDa translocator protein (TSPO) binding in the brain [56], etc. Pancreatic ductal adenocarcinoma (PDAC) is one of the most lethal cancers with the lowest survival rate (five-year survival of only 8%) among the cancers reported by the American Cancer Society [57]. There is evidence that shows that high cholesterol increases PDAC cancer risk. According to Chen et al. [58], a linear dose–response relationship has attested that the risk of pancreatic cancer rises by 8% with 100 mg/day of cholesterol intake through the dose-response analysis. In addition, cholesterol does not influence the mortality among patients with PDAC cancer for both statin users and nonusers measured at different time windows and analyzed as continuous, dichotomous, and categorical variables [59].

The mechanisms underlying the cholesterol-cancer correlation have not been fully elucidated. In the present work, we used molecular dynamics (MD) simulations, a very successful tool to describe a wide variety of molecular setups at the all-atom level, such as complex biological and aqueous systems [60–62]. We have investigated whether cholesterol in membranes affects the signaling of Ras proteins by interfering with their orientations when the oncogenic and wild-type KRas binding with the membrane. Moreover, two percentages of cholesterol (0% and 30%) have been considered. Gaining a precise understanding of the influence of cholesterol on the reorientation of mutant and wild-type KRas-4B-Far binding at the anionic model membranes is the goal of the current work.

2. Results and Discussion

2.1. Area Per Lipid

Area per lipid is often used as the key parameter when assessing the validity of MD simulations of cell membranes. It has been proposed that a good test for such validation is the comparison of the area per lipid and thickness of the membrane with experimental data obtained from scattering density profiles [63]. The area per lipid and thickness along the simulation time of the last 500 ns have been computed (see Figure A1 of Appendix A.1) and the average values are reported in Table 1.

Table 1. The average area per lipid (A) and thickness (Δz) of the anionic membrane for four KRas-4B-Far systems studied in this work. The thickness of the membrane Δz by computing the mean distance between phosphorus atoms of DOPC head groups from both leaflets. Estimated errors in parenthesis.

System	A (nm^2)	Δz (nm)
wt. chol.-0%	0.679 (0.008)	3.84 (0.04)
wt. chol.-30% [24]	0.523 (0.007)	4.35 (0.05)
onc. chol.-0%	0.679 (0.008)	3.89 (0.04)
onc. chol.-30% [24]	0.525 (0.006)	4.23 (0.04)

The experimental value of 0.71 nm^2 for area per lipid of DOPC/DOPS (4:1) at 297 K was reported in Ref. [64] and the experimental value of thickness to be of 3.94 nm at 303 K was reported by Novakova et al. [65]. As temperature increases, atoms in the lipid structure oscillate more perpendicularly to their bonds. So, increasing the temperature of the system leads to an increased area per lipid of certain phosphatidylcholine lipids, as was observed for temperatures below 420 K [66]. It was previously reported that KRas-4B can interact with head groups of DOPC and DOPS lipid molecules through long-lived salt bridges and hydrogen bonds [24]. Accordingly, when the system temperature rises, for the same model membrane, its thickness decreases. Our results of the area per lipid (0.03 nm^2) and thickness ($\sim$0.1 nm) are smaller than the experiment values for pure lipid systems. The main reason is the contribution of the joint effect of raising the system temperature and the appearance of KRas-4B-Far, showing a slight condensing effect on the membrane. According to earlier research [67], in the case of relatively high cholesterol concentration, 10 $\sim$ 20% smaller area per lipid will be considered to be reasonable and close to equilibrium ones. From another work [68], compared to pure DMPC bilayer, the area per lipid of DMPC with cholesterol (30%) has been decreased by 32% from 0.62 to 0.42 nm^2. In the regime with chol. $\leq$ 30%, the area per lipid has been reported to decrease sharply as cholesterol is added into the system [69]. In Table 1, the area per lipid for high cholesterol cases (chol.-30%) has been decreased by 23% when compared with the cholesterol-free cases (chol.-0%). The results make much sense when compared with the experimental values confirming cholesterol's condensation effect on DOPC/DOPS membrane bilayers. The results of the area per lipid and thickness of membrane bilayers we have investigated are in good agreement with experimental values. Hence, the validity of MD simulations reported here, regarding the structural characteristics of the membrane, has been established.

2.2. Preferential Localization of Kras-4b-Far on Membranes

Ras proteins are activated following an incoming signal from their upstream regulators and interact with their downstream effectors only when they are anchored into the membrane and being at the GTP-bound state. Tracking the movement of the FAR of KRas-4B-Far and GTP along the membrane normal could give us direct information on how the KRas-4B proteins and GTP molecule regulate each other. We report in Figure 2, the Z-axis positions of the centers of FAR and GTP from the center of lipids (i.e., $Z = 0$) using the second half of 1000 ns simulation for all cases.

As is described in Ref. [24], the FAR of the wild-type KRas-4B-Far is revealed to be able to anchor into and depart from the membrane without difficulty in the chol.-30% case when GTP favors bind with the interface of the membrane through salt bridges and hydrogen bonds, located at around 2.39 nm from the membrane center. The FAR can have two preferred localisations: (1) 3.90 nm when FAR wanders in the water region, and (2) 1.73 nm when FAR anchors inside of the PM. However, in the chol.-free case, the FAR of the wild-type KRas-4B-Far is found to be anchoring constantly into the anionic membrane for the entire duration of the simulations. FAR keeps locating around 1.30 nm, while GTP keeps binding to the CD, staying around 4.38 nm.

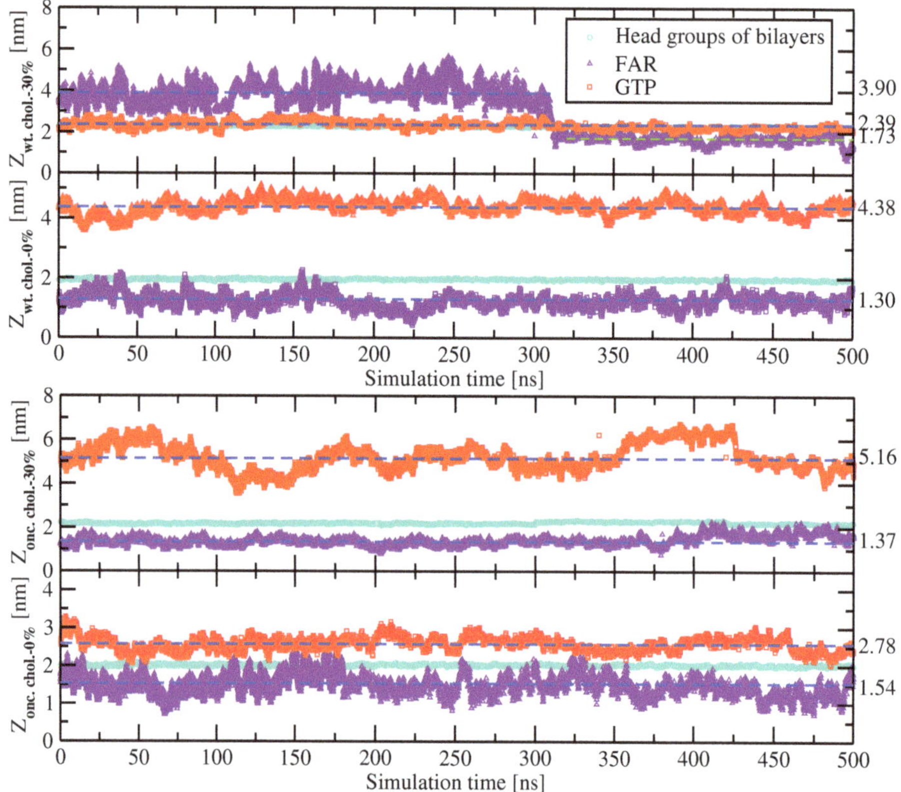

Figure 2. The localizations of the farnesyl group (FAR) and GTP of four KRas-4B systems studied in this work with respect to the center of the membrane along with the membrane normal as a function of simulation time. Geometric centers of the FAR, GTP, and phosphorus atoms of DOPC lipids from both leaflets are indicated as triangle up in violet and circle in turquoise, respectively. Data for chol.-30% shown here are adopted from our previous work [24] for the convenience of the audience. The average values of the FAR and GTP are indicated in dashed lines.

For the mutant KRas-4B-Far, when diffusing in the DOPC/DOPS (4:1) bilayer, GTP tends to wander around 2.78 nm away from the membrane center along with the membrane normal direction. When 30% of cholesterol was considered, GTP favors binding with the CD instead of wandering near the interface region of the membrane. For both oncogenic cases, FAR is revealed to be anchoring constantly into the anionic membrane as a function of the simulation time, as might be expected.

By comparing the four systems we studied, we propose that adding cholesterol into the system has less influence on the behavior of FAR of the oncogenic KRas-4B-Far anchoring to the anionic membrane. Moreover, for a different type of KRas-4B-Far, GTP's localization cannot be predicted according to different types of mutations in the KRas-4B's structure and the constitution of the cell membrane we are studying. Remarkably, the existence of cholesterol helps FAR of the mutant KRas-4B-Far anchor 0.17 nm deeper into the anionic membrane than the chol.-0% case.

2.3. Conformation of the 5-Aa-Sequence in The Hvr

As suggested by Dharmaiah et al. [40], a 5-amino-acid-long sequence motif in its HVR (K-S-K-T-K, residues 180-184), which is shared by KRas-4B-Far and KRas-4B-FMe, may enable PDEδ to bind prenylated KRas-4B.

The root mean square deviation (RMSD) of certain atoms in a molecule with respect to a reference structure, defined as Equation (1), is the most commonly used quantitative measure of the similarity between two superimposed atomic coordinates [70].

$$RMSD = \sqrt{\frac{1}{n}\sum_{i=1}^{n} d_i^2} \tag{1}$$

where d_i is the distance between the two atoms in the i-th pair and the averaging is performed over the n pairs of atoms.

A closer investigation of RMSD of this 5-aa-sequence of the HVR of two KRas-4B-Far (wt. and onc.) binding to two different anionic model membranes has been done.

Figure 3 presents the results of adding cholesterol into the system to their respective reference structures. Obviously, cholesterol doesn't have as much impact as two mutations (G12D and PHOS) in its sequence for the same type of KRas-4B-Far, highlighting the significant influence of the mutations on the conformational change of the 5-aa-sequence in the HVR. It also demonstrates that for the wild-type KRas-4B-Far protein, the RMSD of the 5-aa-sequence ranges from 0.24 to 0.3 nm, and for the oncogenic one, the value ranges from 0.35 to 0.42, due to inherent structural flexibility.

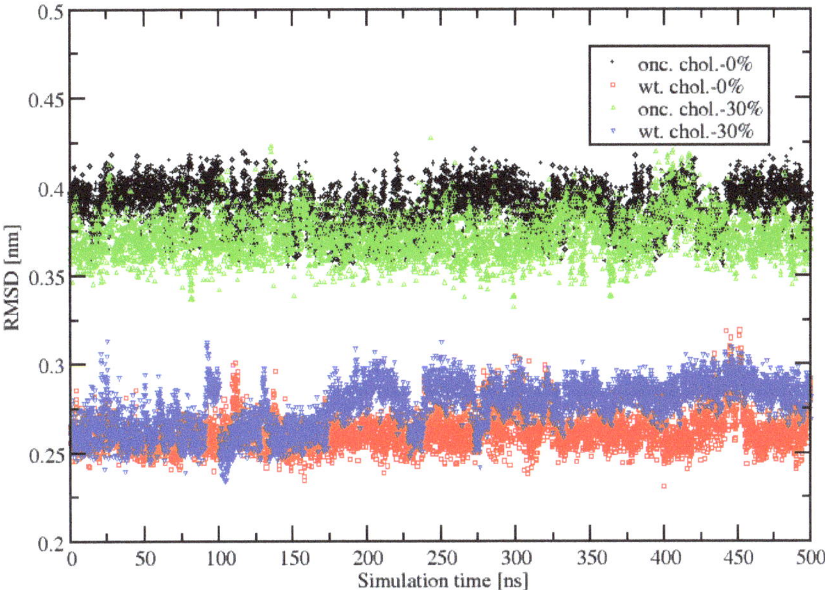

Figure 3. The RMSD of the 5-aa-sequence of the HVR during the last 500 ns of the 1 μs time span for four systems.

2.4. Orientational Distributions of Kras-4b-Far on Different Anionic Membranes

Several previous studies have shown that the orientations of Ras proteins on membranes significantly impact their function in cell [20,21,71–73]. Cell membranes are platforms for cellular signal transduction. Their structure and function depend on the composition of cholesterol and related phospholipids [74]. Furthermore, both clinical and experimental studies have found that hypercholesterolemia and a high-fat high-cholesterol diet can affect cancer development [48,75]. Increased cholesterol levels in the human body are associated with a higher cancer incidence, and reducing its level through drugs (for instance: statins) could reduce the risk and mortality of some cancers, such as prostate, colorectal, and breast cancer [76–78]. Increased serum cholesterol levels could be used as an indicator for developing cancers, such as colon, prostatic, testicular, and rectal cancer [79,80].

To explore the influence of cholesterol on the orientation of KRas, we employed the definition of the orientation of KRas-4B-FMe described by Prakash et al. [20] to compare with the results from this work and propose a new method to define the orientation of KRas binding to the PM. In general, two order parameters have been adopted: (1) the distance (z) between C_α atoms of the residue 132 on the lobe 2 and the residue 183 on the HVR, and (2) the angle Θ between the membrane normal direction and a vector running the C_α atoms of the residue 5 and the residue 9 which belong to the first strand $\beta1$ in the structure of KRas-4B. The results of the density distribution of conformations defined by the order parameter z and cosΘ during the last 500 ns simulation time for four systems studied in this work have been presented in Figure 4.

Two distinct orientations of KRas were proposed in their work: OS_1, in which the loop is solvent-accessible, and OS_2, in which the effector-binding loop is occluded by the membrane. The remaining conformations are categorized into the intermediate state OS_0.

Moreover, we define a new parameter, the angle Φ that runs the membrane normal and a vector running the C_α atoms of residues 163 and 156 on the last helix $\alpha5$ of lobe 2. According to data from references [81–83], it is known that dimerization of KRas-4B is a requirement for KRas signaling activity and tumor growth. The helix $\alpha5$ has been reported to be involved in its dimerization interface. However, despite this being a relevant and interesting topic, dimerization of KRas-4B is outside of the scope of the MD study reported here, since the classical force field we have employed in the present work (CHARMM36) does not allow us to simulate the breaking and formation of chemical bonds. Due to its highly conserved structure for the CD, providing the information of the angle Θ along with the Φ could provide a new way for researchers to define the movement and orientation for KRas when binding to the membranes. We have calculated the angle and distance as described above. Moreover, the parameter we newly introduce here will be discussed further later.

Through the two-dimensional histogram, (z, cosΘ), three orientation states OS_1 , OS_2, and OS_0 were reported to be centered around (1.86, −0.5), (4.97, 1), and (3.33, 0.9), respectively, according to Prakash et al. [20]. In OS_1, KRas-4B can interact with other proteins in cells, confirming that cholesterol has an important impact on the signaling activity for KRas-4B, especially for mutant ones, by increasing the flexibility and fluctuation in its CD with the exposed effector-binding loop. In Figure 4, we can observe that only when oncogenic KRas-4B-Far is bound to the chol.-30% membrane, can OS_1 be shown for KRas-4B in a time span of 500 ns. As expected, when binding to the chol.-30% membrane, the wild-type Kras-4B-Far has been observed to stay in its inactive state (OS_2). From the last 500 ns simulation time analyzed in this work, when binding to the anionic cholesterol-rich membrane, the wild-type and mutant KRas-4B-Far proteins can reach all three conformational regions, indicating more flexibility for the CD in the membrane normal direction and less affinity to the cholesterol-rich PM.

However, using these two coordinates (z, cosΘ) defined by Prakash et al. makes it difficult for us to categorize the conformational states of (wild-type and oncogenic) KRas-4B-Far for the cholesterol-free systems, and no clear OS_1, OS_2, and OS_0 have been observed, see the upper panels in Figure 4. However, OS_1 for the oncogenic KRas-4B-Far in Figure A3 and OS_2 for the wild-type KRas-4B-Far in Figure A4 have been observed for the chol.-0% membrane systems. So, using two well-defined angles to describe the orientation of KRas-4B on the anionic membrane could be a good idea.

We have also investigated the time evolution of z for each system as reported in Figure A2, which shows major conformational fluctuations for four systems (oncogenic KRas-4B-Far and wild-type KRas-4B-Far, for 0 and 30% chol). For the cholesterol-free membrane systems, the protein majorly fluctuates between two distinct states in ranges of $2.4 \leq z \leq 4.3$ and $z > 4.3$ nm, rarely visiting lower z values.

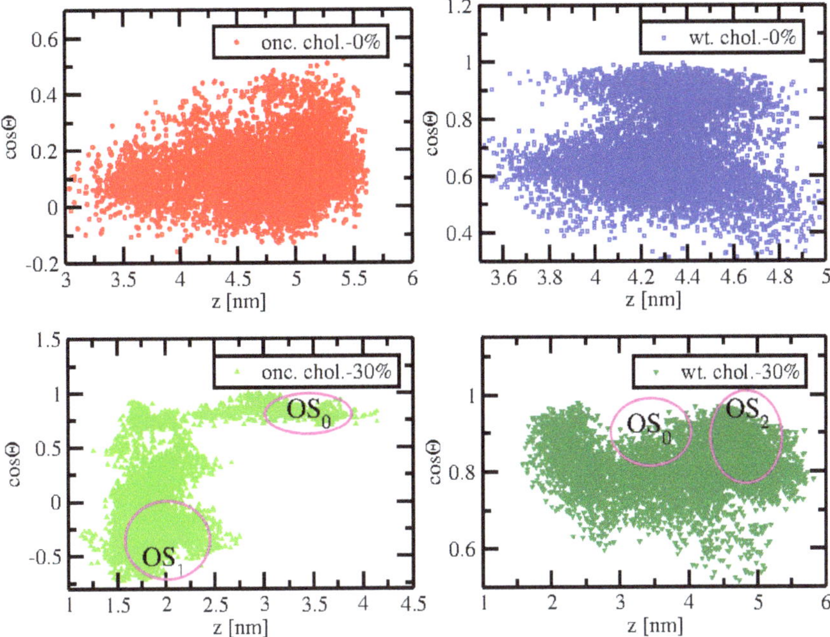

Figure 4. Density distributions of configurations of the oncogenic KRas-4B-Far and wild-type KRas-4B-Far systems with given values of coordinates z and cosΘ for 0 and 30% cholesterol. Observed OS_1, OS_2, and OS_0 have been encircled in their corresponding locations.

For the two cholesterol-free systems, are the orientation states of wild-type and oncogenic KRas-4B-Far always in its intermediate state OS_0 according to Prakash et al.'s work [20]? This is a question that we want to answer.

2.5. Reorientation of Mutant Kras-4b-Far on the Anionic Membranes

By adopting the two angles (Θ and Φ) defined above, we analyzed the corresponding density profiles. We present the reorientation of the mutant KRas-4B-Far when bound to the anionic membrane with 30% of the cholesterol in Figure 5. Results for the remaining three systems studied here are reported in Figures A3–A5.

From Figure 5, the reorientation of mutant KRas-4B-Far on the chol.-30% bilayer has been observed during the 500 ns simulations time, giving a hint on the low free energy barriers between two orientation states (OS_1-OS_0, and OS_0-OS_2). Mutant KRas-4B spends most of the time in the active OS_1 state, centered at (80°, 105°) on chol.-30% membrane, and fluctuates around (99°, 83°) when binding to the chol.-free bilayer, also in its OS_1 state. Wild-type KRas-4B protein, regardless of the cholesterol's content, prefers staying in its inactivate state, centered at (59°, 52.5°) and (53°, 37°) for chol.-0% and chol.-30%, respectively. This suggests that the orientation with the effector-binding loop occluded by the membrane (OS_2) is disfavored in the wild-type KRas-4B-Far protein.

Here, we could conclude that high cholesterol in the PM helps KRas-4B mutant stay in its constitutively active state, which suggests that high cholesterol intake can increase mortality and may promote cancer progression for cancer patients. Our findings agree with the experimental and clinical results [55,56,58,59].

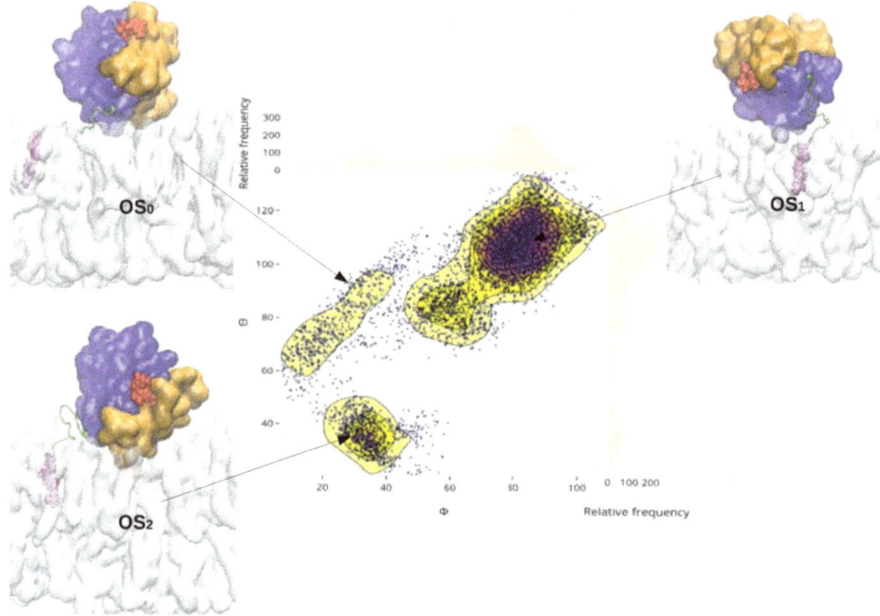

Figure 5. Reorientation of the oncogenic KRas-4B on the chol.-30% membrane. Density distribution of conformations projected onto a plane defined by the reaction coordinates Φ and Θ in degrees (°). The relative frequency of each coordinate is shown on the right and upsides. The membrane bilayer is shown as a white surface. Lobe 1 is highlighted in orange, lobe 2 in blue, HVR backbone in green, GTP in red, and FAR in violet. Water and ions are not shown here for the sake of clarity.

3. Methods

We performed four independent MD simulations of wild-type and mutated GTP-bound KRas-4B-Far attached to DOPC/DOPS (4:1) bilayers.

Eventually, some of the lipids were replaced by cholesterol molecules in such a way that two cholesterol percentages were considered: 0% and 30%. Each system contains a total of 304 lipid molecules fully solvated by 60,000 TIP3P water molecules and 48 potassium chloride at the human body concentration (0.15 M), yielding a system size of 222,000 atoms. All MD inputs were generated using a CHARMM-GUI web-based tool [84]. The force field was CHARMM36m for proteins [85] and CHARMM36 [86] for other molecules in each system. The crystal structure of KRas-4B with the partially disordered hypervariable region (pdb 5TB5) and GTP (pdb 5VQ2) were used to generate full-length GTP-bound KRas-4B-Far proteins. Two sequences of the wild-type and oncogenic KRas-4B-Far are presented in Figure A6.

After model building, each system was energy minimized for 5000 steps followed by three 250 ps simulations, and then four additional 500 ps equilibrium runs while gradually reducing the harmonic constraints on the systems. We used the NPT ensemble with the constant pressure of 1 atm maintained by the Parrinello–Rahman piston method with a damping coefficient of 5 ps^{-1} and temperature of 310.15 K controlled by the Nosé–Hoover thermostat method with a damping coefficient of 1 ps^{-1}. Meaningful production runs were performed with an NPT ensemble for 1 μs from the last configuration of equilibrium run for each system, for a total of 4 μs. Time steps of 2 fs were used in all production simulations and the particle mesh Ewald method with a Coulomb radius of 1.2 nm was employed to compute long-ranged electrostatic interactions. The cutoff for Lennard–Jones interactions was set to 1.2 nm. In all MD simulations, the GROMACS/2018.3 package was employed [87] and periodic boundary conditions in three directions of space have been taken.

4. Conclusions

In this work, we performed MD simulations on four systems of wild-type/oncogenic KRas-4B-Far protein binding to membranes with different cholesterol contents (0% and 30%) to study the influence of cholesterol on the orientation of KRas. KRas-4B-Far shows the condensing effect on the area per lipid of the anionic model membrane through strong interactions between its CD and HVR moieties with the head groups of the lipids. More flexibility in its CD structure of KRas-4B-Far has been observed when binding to the PM with high cholesterol concentration, for both wild-type and mutant KRas-4B-Far proteins. The reorientation of mutant KRas-4B-Far on the anionic chol.-30% model membrane has been observed during the 500 ns simulations time, giving a hint on the low free energy barriers between a pair of orientation states (e.g., OS_1-OS_0, and OS_0-OS_2).

It has been shown for the first time that cholesterol makes it much easier for the mutant KRas-4B-Far shifting between different orientation states. The high cholesterol content in the PM favors OS_1, exposing the effector-binding loop for signal transduction in cells from the atomic level. We propose that during the treatment of KRas-4B-related cancers, reducing the cholesterol level in the PM and sustaining cancer progression by controlling the plasma cholesterol intake should be taken into account in anti-cancer therapies. The present study of the role of cholesterol in Kras-4B orientation can provide one more direction and method for the treatment and prevention of cancer. By conducting four μs MD simulations, we confirm that high cholesterol in the PM helps KRas-4B mutant stays in its constitutively active state, which suggests that high cholesterol intake can increase mortality and may promote cancer progression for cancer patients.

Author Contributions: Data curation, H.L., J.M.; formal analysis, H.L., J.M.; funding acquisition, J.M.; investigation, H.L., J.M.; project administration, J.M.; software, H.L., J.M.; supervision, J.M.; writing—original draft preparation, H.L., J.M.; writing—review and editing, H.L., J.M. All authors have read and agreed to the published version of the manuscript.

Funding: We are thankful for the financial support provided by the Spanish Ministry of Science, Innovation and Universities (project number PGC2018-099277-B-C21, funds MCIU/AEI/FEDER, UE). Huixia Lu is a Ph.D. fellow from the China Scholarship Council (201607040059).

Acknowledgments: We acknowledge the use of computer resources from the "Barcelona Supercomputing Center-Red Espanola de Supercomputacion" through projects FI-2019-3-0008 and FI-2019-2-0004.

Conflicts of Interest: The authors declare no conflict of interest.

Abbreviations

The following abbreviations are used in this manuscript:

MD	molecular dynamics
PM	plasma membrane
DOPC	1,2-dioleoyl-sn-glycero-3-phosphocholine
DOPS	1,2-dioleoyl-sn-glycero-3-phospho-L-serine
GDP	guanosine diphosphate
GTP	guanosine triphosphate
GEF	guanine-nucleotide-exchange factors
GAP	GTPase-activating proteins
COSMIC	catalog of somatic mutations in cancer
KRas-4B-Far	farnesylated and demethylated KRas-4B
KRas-4B-FMe	farnesylated and methylated KRas-4B
HVR	hypervariable region
FAR	farnesyl group
CD	catalytic domain

PHOS	phosphorylated serine
PTMs	post-translational modifications
OS	orientation state
RMSD	root mean square deviation
PDEδ	phosphodiest-eraseδ
FTase	farnesyltrasferase
GGTase	geranylgeranyltransferase
RCE1	Ras-converting enzyme 1
ICMT	isoprenylcysteine carboxyl methyltransferase
PMPEases	prenylated/polyisoprenylated methylated protein methyl esterases
TSPO	translocator protein

Appendix A. Supporting Information

Appendix A.1. Area Per Lipid

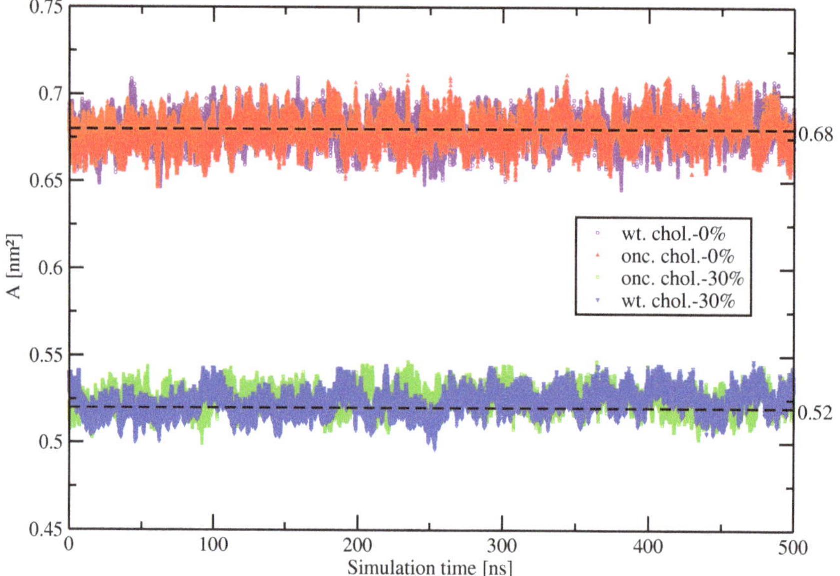

Figure A1. Area per lipid of four wild-type/mutant KRas-4B-Far systems with different content of cholesterol as a function of simulation time. The black dashed line indicates the average value for each system of the second half of the total 1000 ns production runs.

Appendix A.2. Orientation of Kras-4b on the Pm

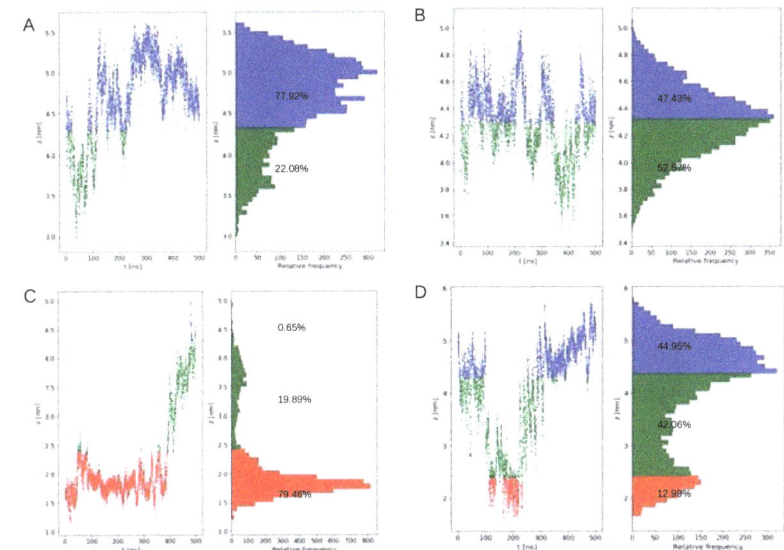

Figure A2. Time evolution of z during the last 500 ns MD simulations of the onc. and wt. KRas-4B-Far proteins for four cases studied in this work. The ratios of the different regions defined in Ref. [20] of the distance are shown on each panel. Three different pools of conformational states are depicted in different colors: $z \in [0, 2.4)$ in red, $z \in [2.4, 4.3]$ in green, and $z \in (4.3, 6]$ in blue. Panel (**A**) refers to onc. chol.-0%, panel (**B**) refers to the wt. chol.-0% system, panel (**C**) stands for the onc. chol.-30% system, and panel (**D**) represents the wt. chol.-30% system, respectively.

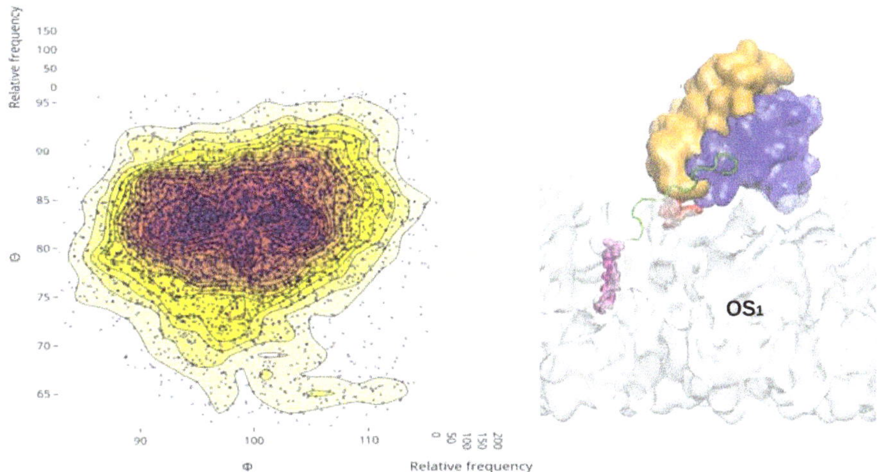

Figure A3. Orientation of the mutant KRas-4B-Far on the chol.-0% membrane. Colors defined as Figure 5.

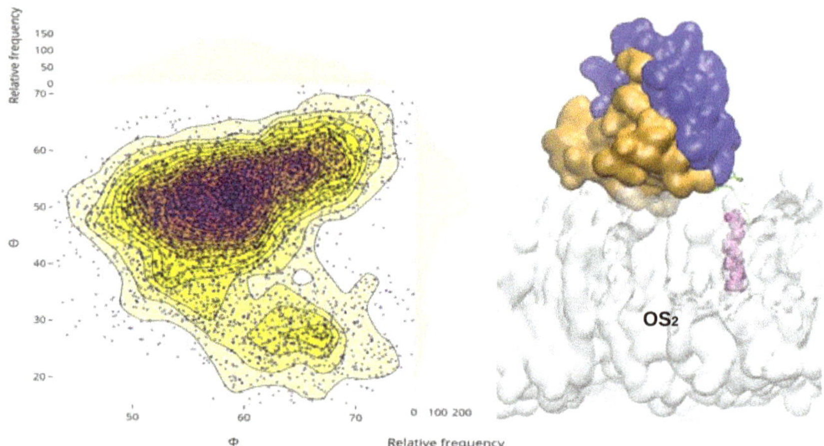

Figure A4. Orientation of the wild-type KRas-4B-Far on the chol.-0% membrane. Colors defined as Figure 5.

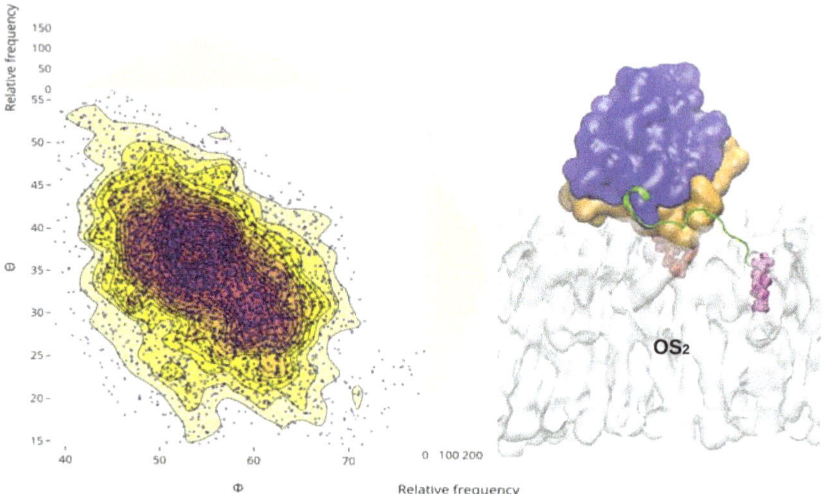

Figure A5. Orientation of the wild-type KRas-4B-Far on the anionic chol.-30% membrane. Colors defined as Figure 5.

Appendix A.3. Sequences of Wild-Type and Mutant Kras-4b-Far Proteins

Here in Figure A6, sequences for wild-type and mutated KRas-4B-Far are presented.

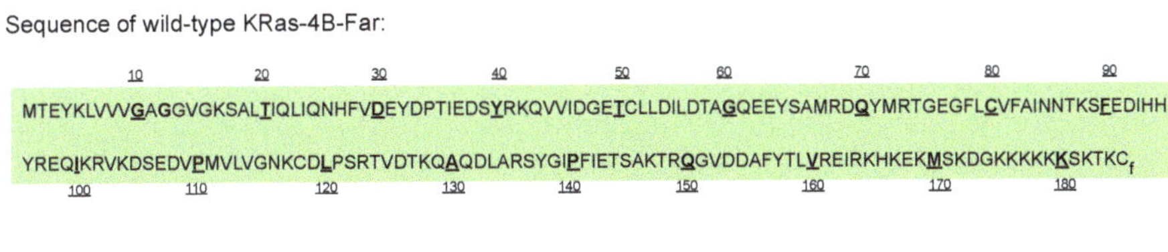

Sequence of oncogenic KRas-4B-Far:

MTEYKLVVVGADGVGKSALTIQLIQNHFVDEYDPTIEDSYRKQVVIDGETCLLDILDTAGQEEYSAMRDQYMRTGEGFLCVFAINNTKSFEDIHH

YREQIKRVKDSEDVPMVLVGNKCDLPSRTVDTKQAQDLARSYGIPFIETSAKTRQGVDDAFYTLVREIRKHKEKMSKDGKKKKKKS$_p$KTKC$_f$

Figure A6. Two sequences of oncogenic and wild-type KRas-4B-Far. Mutated sites are in red color. Here C_f denotes the farnesylated Cys-185 and S_p represents the phosphorylation adopted in the present work.

References

1. Mouritsen, O.G. *Life-as a Matter of Fat*; Springer: Berlin/Heidelberg, Germany, 2005.
2. Nagle, J.F.; Tristram-Nagle, S. Structure of lipid bilayers. *Biochim. Et Biophys. Acta (BBA)-Rev. Biomembr.* **2000**, *1469*, 159–195. [CrossRef]
3. Zhou, Y.; Liang, H.; Rodkey, T.; Ariotti, N.; Parton, R.G.; Hancock, J.F. Signal integration by lipid-mediated spatial cross talk between Ras nanoclusters. *Mol. Cell. Biol.* **2014**, *34*, 862–876. [CrossRef] [PubMed]
4. McLaughlin, S.; Murray, D. Plasma membrane phosphoinositide organization by protein electrostatics. *Nature* **2005**, *438*, 605–611. [CrossRef] [PubMed]
5. Ingólfsson, H.I.; Melo, M.N.; Van Eerden, F.J.; Arnarez, C.; Lopez, C.A.; Wassenaar, T.A.; Periole, X.; De Vries, A.H.; Tieleman, D.P.; Marrink, S.J. Lipid organization of the plasma membrane. *J. Am. Chem. Soc.* **2014**, *136*, 14554–14559. [CrossRef]
6. Zhang, Y.; Chen, X.; Gueydan, C.; Han, J. Plasma membrane changes during programmed cell deaths. *Cell Res.* **2018**, *28*, 9–21. [CrossRef]
7. Krapf, D. Compartmentalization of the plasma membrane. *Curr. Opin. Cell Biol.* **2018**, *53*, 15–21. [CrossRef]
8. Zhang, J.; Jin, R.; Jiang, D.; Chen, H.Y. Electrochemiluminescence-based capacitance microscopy for label-free imaging of antigens on the cellular plasma membrane. *J. Am. Chem. Soc.* **2019**, *141*, 10294–10299. [CrossRef]
9. Nussinov, R.; Tsai, C.J.; Jang, H. Oncogenic Ras isoforms signaling specificity at the membrane. *Cancer Res.* **2018**, *78*, 593–602. [CrossRef]
10. Bernards, A.; Settleman, J. GAP control: Regulating the regulators of small GTPases. *Trends Cell Biol.* **2004**, *14*, 377–385. [CrossRef]
11. Wennerberg, K.; Rossman, K.L.; Der, C.J. The Ras superfamily at a glance. *J. Cell Sci.* **2005**, *118*, 843–846. [CrossRef]
12. Schmidt, A.; Hall, A. Guanine nucleotide exchange factors for Rho GTPases: Turning on the switch. *Genes Dev.* **2002**, *16*, 1587–1609. [CrossRef] [PubMed]
13. Stephen, A.G.; Esposito, D.; Bagni, R.K.; McCormick, F. Dragging ras back in the ring. *Cancer Cell* **2014**, *25*, 272–281. [CrossRef] [PubMed]
14. Forbes, S.A.; Bindal, N.; Bamford, S.; Cole, C.; Kok, C.Y.; Beare, D.; Jia, M.; Shepherd, R.; Leung, K.; Menzies, A.; et al. COSMIC: Mining complete cancer genomes in the Catalogue of Somatic Mutations in Cancer. *Nucleic Acids Res.* **2010**, *39*, D945–D950. [CrossRef] [PubMed]
15. Hobbs, G.A.; Der, C.J.; Rossman, K.L. RAS isoforms and mutations in cancer at a glance. *J. Cell Sci.* **2016**, *129*, 1287–1292. [CrossRef] [PubMed]
16. Gelabert-Baldrich, M.; Soriano-Castell, D.; Calvo, M.; Lu, A.; Viña-Vilaseca, A.; Rentero, C.; Pol, A.; Grinstein, S.; Enrich, C.; Tebar, F. Dynamics of KRas on endosomes: Involvement of acidic phospholipids in its association. *FASEB J.* **2014**, *28*, 3023–3037. [CrossRef]

17. Cho, K.J.; van der Hoeven, D.; Zhou, Y.; Maekawa, M.; Ma, X.; Chen, W.; Fairn, G.D.; Hancock, J.F. Inhibition of acid sphingomyelinase depletes cellular phosphatidylserine and mislocalizes K-Ras from the plasma membrane. *Mol. Cell. Biol.* **2016**, *36*, 363–374. [CrossRef]
18. Gregory, M.C.; McLean, M.A.; Sligar, S.G. Interaction of KRas4b with anionic membranes: A special role for PIP2. *Biochem. Biophys. Res. Commun.* **2017**, *487*, 351–355. [CrossRef]
19. Zhou, Y.; Wong, C.O.; Cho, K.J.; Van Der Hoeven, D.; Liang, H.; Thakur, D.P.; Luo, J.; Babic, M.; Zinsmaier, K.E.; Zhu, M.X.; et al. Membrane potential modulates plasma membrane phospholipid dynamics and K-Ras signaling. *Science* **2015**, *349*, 873–876. [CrossRef]
20. Prakash, P.; Litwin, D.; Liang, H.; Sarkar-Banerjee, S.; Dolino, D.; Zhou, Y.; Hancock, J.F.; Jayaraman, V.; Gorfe, A.A. Dynamics of membrane-bound G12V-KRAS from simulations and single-molecule FRET in native nanodiscs. *Biophys. J.* **2019**, *116*, 179–183. [CrossRef]
21. Prakash, P.; Zhou, Y.; Liang, H.; Hancock, J.F.; Gorfe, A.A. Oncogenic K-Ras binds to an anionic membrane in two distinct orientations: A molecular dynamics analysis. *Biophys. J.* **2016**, *110*, 1125–1138. [CrossRef]
22. Prakash, P.; Gorfe, A.A. Probing the conformational and energy landscapes of KRAS membrane orientation. *J. Phys. Chem. B* **2019**, *123*, 8644–8652. [CrossRef] [PubMed]
23. Tsai, C.J.; Ma, B.; Sham, Y.Y.; Kumar, S.; Nussinov, R. Structured disorder and conformational selection. *Proteins Struct. Funct. Bioinform.* **2001**, *44*, 418–427. [CrossRef] [PubMed]
24. Lu, H.; Martí, J. Long-lasting Salt Bridges Provide the Anchoring Mechanism of Oncogenic Kirsten Rat Sarcoma Proteins at Cell Membranes. *J. Phys. Chem. Lett.* **2020**, *11*, 9938–9945. [CrossRef] [PubMed]
25. Cao, S.; Chung, S.; Kim, S.; Li, Z.; Manor, D.; Buck, M. K-Ras G-domain binding with signaling lipid phosphatidylinositol (4, 5)-phosphate (PIP2): Membrane association, protein orientation, and function. *J. Biol. Chem.* **2019**, *294*, 7068–7084. [CrossRef] [PubMed]
26. Yang, M.H.; Laurent, G.; Bause, A.S.; Spang, R.; German, N.; Haigis, M.C.; Haigis, K.M. HDAC6 and SIRT2 regulate the acetylation state and oncogenic activity of mutant K-RAS. *Mol. Cancer Res.* **2013**, *11*, 1072–1077. [CrossRef]
27. Lu, S.; Jang, H.; Gu, S.; Zhang, J.; Nussinov, R. Drugging Ras GTPase: A comprehensive mechanistic and signaling structural view. *Chem. Soc. Rev.* **2016**, *45*, 4929–4952. [CrossRef]
28. Ahearn, I.M.; Haigis, K.; Bar-Sagi, D.; Philips, M.R. Regulating the regulator: Post-translational modification of RAS. *Nat. Rev. Mol. Cell Biol.* **2012**, *13*, 39–51. [CrossRef]
29. Zhou, B.; Cox, A.D. Posttranslational Modifications of Small G Proteins. In *Ras Superfamily Small G Proteins: Biology and Mechanisms 1*; Springer: Vienna, Austria, 2014; pp. 99–131.
30. Abdelkarim, H.; Banerjee, A.; Grudzien, P.; Leschinsky, N.; Abushaer, M.; Gaponenko, V. The Hypervariable Region of K-Ras4B Governs Molecular Recognition and Function. *Int. J. Mol. Sci.* **2019**, *20*, 5718. [CrossRef]
31. Ahearn, I.; Zhou, M.; Philips, M.R. Posttranslational modifications of RAS proteins. *Cold Spring Harb. Perspect. Med.* **2018**, a031484. [CrossRef]
32. Konstantinopoulos, P.A.; Karamouzis, M.V.; Papavassiliou, A.G. Post-translational modifications and regulation of the RAS superfamily of GTPases as anticancer targets. *Nat. Rev. Drug Discov.* **2007**, *6*, 541–555. [CrossRef]
33. Zhang, S.Y.; Sperlich, B.; Li, F.Y.; Al-Ayoubi, S.; Chen, H.X.; Zhao, Y.F.; Li, Y.M.; Weise, K.; Winter, R.; Chen, Y.X. Phosphorylation weakens but does not inhibit membrane binding and clustering of K-Ras4B. *ACS Chem. Biol.* **2017**, *12*, 1703–1710. [CrossRef] [PubMed]
34. Alvarez-Moya, B.; Lopez-Alcala, C.; Drosten, M.; Bachs, O.; Agell, N. K-Ras4B phosphorylation at Ser181 is inhibited by calmodulin and modulates K-Ras activity and function. *Oncogene* **2010**, *29*, 5911–5922. [CrossRef] [PubMed]
35. Bivona, T.G.; Quatela, S.E.; Bodemann, B.O.; Ahearn, I.M.; Soskis, M.J.; Mor, A.; Miura, J.; Wiener, H.H.; Wright, L.; Saba, S.G.; et al. PKC regulates a farnesyl-electrostatic switch on K-Ras that promotes its association with Bcl-XL on mitochondria and induces apoptosis. *Mol. Cell* **2006**, *21*, 481–493. [CrossRef] [PubMed]
36. Kollár, P.; Rajchard, J.; Balounová, Z.; Pazourek, J. Marine natural products: Bryostatins in preclinical and clinical studies. *Pharm. Biol.* **2014**, *52*, 237–242. [CrossRef]
37. Barceló, C.; Paco, N.; Morell, M.; Alvarez-Moya, B.; Bota-Rabassedas, N.; Jaumot, M.; Vilardell, F.; Capella, G.; Agell, N. Phosphorylation at Ser-181 of oncogenic KRAS is required for tumor growth. *Cancer Res.* **2014**, *74*, 1190–1199. [CrossRef]

38. Jang, H.; Abraham, S.J.; Chavan, T.S.; Hitchinson, B.; Khavrutskii, L.; Tarasova, N.I.; Nussinov, R.; Gaponenko, V. Mechanisms of membrane binding of small GTPase K-Ras4B farnesylated hypervariable region. *J. Biol. Chem.* **2015**, *290*, 9465–9477. [CrossRef]
39. Alvarez-Moya, B.; Barceló, C.; Tebar, F.; Jaumot, M.; Agell, N. CaM interaction and Ser181 phosphorylation as new K-Ras signaling modulators. *Small GTPases* **2011**, *2*, 5911–5922. [CrossRef]
40. Dharmaiah, S.; Bindu, L.; Tran, T.H.; Gillette, W.K.; Frank, P.H.; Ghirlando, R.; Nissley, D.V.; Esposito, D.; McCormick, F.; Stephen, A.G.; et al. Structural basis of recognition of farnesylated and methylated KRAS4b by PDEδ. *Proc. Natl. Acad. Sci. USA* **2016**, *113*, E6766–E6775. [CrossRef]
41. Schmick, M.; Vartak, N.; Papke, B.; Kovacevic, M.; Truxius, D.C.; Rossmannek, L.; Bastiaens, P.I. KRas localizes to the plasma membrane by spatial cycles of solubilization, trapping and vesicular transport. *Cell* **2014**, *157*, 459–471. [CrossRef]
42. Zimmermann, G.; Papke, B.; Ismail, S.; Vartak, N.; Chandra, A.; Hoffmann, M.; Hahn, S.A.; Triola, G.; Wittinghofer, A.; Bastiaens, P.I.; et al. Small molecule inhibition of the KRAS–PDEδ interaction impairs oncogenic KRAS signalling. *Nature* **2013**, *497*, 638–642. [CrossRef]
43. Muratcioglu, S.; Jang, H.; Gursoy, A.; Keskin, O.; Nussinov, R. PDEδ binding to Ras isoforms provides a route to proper membrane localization. *J. Phys. Chem. B* **2017**, *121*, 5917–5927. [CrossRef] [PubMed]
44. Murarka, S.; Martín-Gago, P.; Schultz-Fademrecht, C.; Al Saabi, A.; Baumann, M.; Fansa, E.K.; Ismail, S.; Nussbaumer, P.; Wittinghofer, A.; Waldmann, H. Development of pyridazinone chemotypes targeting the PDEδ prenyl binding site. *Chem. Eur. J.* **2017**, *23*, 6083–6093. [CrossRef] [PubMed]
45. Ntai, I.; Fornelli, L.; DeHart, C.J.; Hutton, J.E.; Doubleday, P.F.; LeDuc, R.D.; van Nispen, A.J.; Fellers, R.T.; Whiteley, G.; Boja, E.S.; et al. Precise characterization of KRAS4b proteoforms in human colorectal cells and tumors reveals mutation/modification cross-talk. *Proc. Natl. Acad. Sci. USA* **2018**, *115*, 4140–4145. [CrossRef] [PubMed]
46. McMullen, T.P.; Lewis, R.N.; McElhaney, R.N. Cholesterol–phospholipid interactions, the liquid-ordered phase and lipid rafts in model and biological membranes. *Curr. Opin. Colloid Interface Sci.* **2004**, *8*, 459–468. [CrossRef]
47. Levitan, I.; Fang, Y.; Rosenhouse-Dantsker, A.; Romanenko, V. Cholesterol and ion channels. In *Cholesterol Binding and Cholesterol Transport Proteins*; Springer: Dordrecht, The Netherlands, 2010; pp. 509–549.
48. Ding, X.; Zhang, W.; Li, S.; Yang, H. The role of cholesterol metabolism in cancer. *Am. J. Cancer Res.* **2019**, *9*, 219.
49. Phillips, R. Membranes by the Numbers. In *Physics of Biological Membranes*; Springer: Cham, Switzerland, 2018; pp. 73–105.
50. Liu, S.L.; Sheng, R.; Jung, J.H.; Wang, L.; Stec, E.; O'Connor, M.J.; Song, S.; Bikkavilli, R.K.; Winn, R.A.; Lee, D.; et al. Orthogonal lipid sensors identify transbilayer asymmetry of plasma membrane cholesterol. *Nat. Chem. Biol.* **2017**, *13*, 268–274. [CrossRef]
51. Lange, Y.; Swaisgood, M.; Ramos, B.; Steck, T. Plasma membranes contain half the phospholipid and 90% of the cholesterol and sphingomyelin in cultured human fibroblasts. *J. Biol. Chem.* **1989**, *264*, 3786–3793.
52. Das, A.; Brown, M.S.; Anderson, D.D.; Goldstein, J.L.; Radhakrishnan, A. Three pools of plasma membrane cholesterol and their relation to cholesterol homeostasis. *Elife* **2014**, *3*, e02882. [CrossRef]
53. Lange, Y.; Slayton, J.M. Interaction of cholesterol and lysophosphatidylcholine in determining red cell shape. *J. Lipid Res.* **1982**, *23*, 1121–1127.
54. Lu, H.; Martí, J. Effects of cholesterol on the binding of the precursor neurotransmitter tryptophan to zwitterionic membranes. *J. Chem. Phys.* **2018**, *149*, 164906. [CrossRef]
55. Boughter, C.T.; Monje-Galvan, V.; Im, W.; Klauda, J.B. Influence of cholesterol on phospholipid bilayer structure and dynamics. *J. Phys. Chem. B* **2016**, *120*, 11761–11772. [CrossRef] [PubMed]
56. Kim, S.W.; Wiers, C.E.; Tyler, R.; Shokri-Kojori, E.; Jang, Y.J.; Zehra, A.; Freeman, C.; Ramirez, V.; Lindgren, E.; Miller, G.; et al. Influence of alcoholism and cholesterol on TSPO binding in brain: PET [11 C] PBR28 studies in humans and rodents. *Neuropsychopharmacology* **2018**, *43*, 1832–1839. [CrossRef] [PubMed]
57. Miller, K.D.; Siegel, R.L.; Lin, C.C.; Mariotto, A.B.; Kramer, J.L.; Rowland, J.H.; Stein, K.D.; Alteri, R.; Jemal, A. Cancer treatment and survivorship statistics, 2016. *CA Cancer J. Clin.* **2016**, *66*, 271–289. [CrossRef] [PubMed]
58. Chen, H.; Qin, S.; Wang, M.; Zhang, T.; Zhang, S. Association between cholesterol intake and pancreatic cancer risk: evidence from a meta-analysis. *Sci. Rep.* **2015**, *5*, 8243. [CrossRef] [PubMed]

59. Huang, B.Z.; Chang, J.I.; Li, E.; Xiang, A.H.; Wu, B.U. Influence of statins and cholesterol on mortality among patients with pancreatic cancer. *JNCI: J. Natl. Cancer Inst.* **2017**, *109*. [CrossRef]
60. Karplus, M.; Petsko, G. Molecular dynamics simulations in biology. *Nature* **1990**, *347*, 631–639. [CrossRef]
61. Karplus, M.; McCammon, J. Molecular dynamics simulations of biomolecules. *Nat. Struct. Biol.* **2002**, *9*, 646–652. [CrossRef]
62. Nagy, G.; Gordillo, M.; Guàrdia, E.; Martí, J. Liquid water confined in carbon nanochannels at high temperatures. *J. Phys. Chem. B* **2007**, *111*, 12524–12530. [CrossRef]
63. Poger, D.; Mark, A.E. On the validation of molecular dynamics simulations of saturated and cis-monounsaturated phosphatidylcholine lipid bilayers: A comparison with experiment. *J. Chem. Theory Comput.* **2010**, *6*, 325–336. [CrossRef]
64. Lütgebaucks, C.; Macias-Romero, C.; Roke, S. Characterization of the interface of binary mixed DOPC: DOPS liposomes in water: The impact of charge condensation. *J. Chem. Phys.* **2017**, *146*, 044701. [CrossRef]
65. Novakova, E.; Giewekemeyer, K.; Salditt, T. Structure of two-component lipid membranes on solid support: An x-ray reflectivity study. *Phys. Rev. E* **2006**, *74*, 051911. [CrossRef] [PubMed]
66. Chaban, V. Computationally efficient prediction of area per lipid. *Chem. Phys. Lett.* **2014**, *616*, 25–29. [CrossRef]
67. Petrache, H.I.; Tristram-Nagle, S.; Gawrisch, K.; Harries, D.; Parsegian, V.A.; Nagle, J.F. Structure and fluctuations of charged phosphatidylserine bilayers in the absence of salt. *Biophys. J.* **2004**, *86*, 1574–1586. [CrossRef]
68. Lu, H.; Martí, J. Binding and dynamics of melatonin at the interface of phosphatidylcholine-cholesterol membranes. *PLoS ONE* **2019**, *14*, e0224624. [CrossRef]
69. Litz, J.P.; Thakkar, N.; Portet, T.; Keller, S.L. Depletion with cyclodextrin reveals two populations of cholesterol in model lipid membranes. *Biophys. J.* **2016**, *110*, 635–645. [CrossRef]
70. Kufareva, I.; Abagyan, R. Methods of protein structure comparison. In *Homology Modeling*; Springer: Berlin/Heidelberg, Germany; Humana Press: Totowa, NJ, USA, **2011**; pp. 231–257.
71. Abankwa, D.; Gorfe, A.A.; Inder, K.; Hancock, J.F. Ras membrane orientation and nanodomain localization generate isoform diversity. *Proc. Natl. Acad. Sci. USA* **2010**, *107*, 1130–1135. [CrossRef]
72. Kapoor, S.; Triola, G.; Vetter, I.R.; Erlkamp, M.; Waldmann, H.; Winter, R. Revealing conformational substates of lipidated N-Ras protein by pressure modulation. *Proc. Natl. Acad. Sci. USA* **2012**, *109*, 460–465. [CrossRef]
73. Mazhab-Jafari, M.T.; Marshall, C.B.; Smith, M.J.; Gasmi-Seabrook, G.M.; Stathopulos, P.B.; Inagaki, F.; Kay, L.E.; Neel, B.G.; Ikura, M. Oncogenic and RASopathy-associated K-RAS mutations relieve membrane-dependent occlusion of the effector-binding site. *Proc. Natl. Acad. Sci. USA* **2015**, *112*, 6625–6630. [CrossRef]
74. Yan, S.; Qu, X.; Xu, L.; Che, X.; Ma, Y.; Zhang, L.; Teng, Y.; Zou, H.; Liu, Y. Bufalin enhances TRAIL-induced apoptosis by redistributing death receptors in lipid rafts in breast cancer cells. *Anti-Cancer Drugs* **2014**, *25*, 683–689. [CrossRef]
75. Kuzu, O.F.; Noory, M.A.; Robertson, G.P. The role of cholesterol in cancer. *Cancer Res.* **2016**, *76*, 2063–2070. [CrossRef]
76. Ravnskov, U.; Rosch, P.J.; McCully, K.S. Statins do not protect against cancer: Quite the opposite. *J. Clin. Oncol.* **2015**, *33*, 810–811. [CrossRef] [PubMed]
77. Ravnskov, U.; McCully, K.; Rosch, P. The statin-low cholesterol-cancer conundrum. *QJM Int. J. Med.* **2012**, *105*, 383–388. [CrossRef] [PubMed]
78. Nielsen, S.F.; Nordestgaard, B.G.; Bojesen, S.E. Statin use and reduced cancer-related mortality. *N. Engl. J. Med.* **2012**, *367*, 1792–1802. [CrossRef] [PubMed]
79. Radišauskas, R.; Kuzmickienė, I.; Milinavičienė, E.; Everatt, R. Hypertension, serum lipids and cancer risk: A review of epidemiological evidence. *Medicina* **2016**, *52*, 89–98. [CrossRef]
80. Murai, T. Cholesterol lowering: Role in cancer prevention and treatment. *Biol. Chem.* **2015**, *396*, 1–11. [CrossRef]
81. Khan, I.; Spencer-Smith, R.; O'Bryan, J.P. Targeting the $\alpha4$–$\alpha5$ dimerization interface of K-RAS inhibits tumor formation in vivo. *Oncogene* **2019**, *38*, 2984–2993. [CrossRef]
82. Spencer-Smith, R.; Koide, A.; Zhou, Y.; Eguchi, R.R.; Sha, F.; Gajwani, P.; Santana, D.; Gupta, A.; Jacobs, M.; Herrero-Garcia, E.; et al. Inhibition of RAS function through targeting an allosteric regulatory site. *Nat. Chem. Biol.* **2017**, *13*, 62–68. [CrossRef]

83. Ambrogio, C.; Köhler, J.; Zhou, Z.W.; Wang, H.; Paranal, R.; Li, J.; Capelletti, M.; Caffarra, C.; Li, S.; Lv, Q.; et al. KRAS dimerization impacts MEK inhibitor sensitivity and oncogenic activity of mutant KRAS. *Cell* **2018**, *172*, 857–868. [CrossRef]
84. Jo, S.; Kim, T.; Iyer, V.G.; Im, W. CHARMM-GUI: A web-based graphical user interface for CHARMM. *J. Comput. Chem.* **2008**, *29*, 1859–1865. [CrossRef]
85. Huang, J.; Rauscher, S.; Nawrocki, G.; Ran, T.; Feig, M.; de Groot, B.L.; Grubmüller, H.; MacKerell, A.D. CHARMM36m: An improved force field for folded and intrinsically disordered proteins. *Nat. Methods* **2017**, *14*, 71–73. [CrossRef]
86. Huang, J.; MacKerell, A.D., Jr. CHARMM36 all-atom additive protein force field: Validation based on comparison to NMR data. *J. Comput. Chem.* **2013**, *34*, 2135–2145. [CrossRef] [PubMed]
87. Lemkul, J. From proteins to perturbed Hamiltonians: A suite of tutorials for the GROMACS-2018 molecular simulation package [article v1. 0]. *Living J. Comput. Mol. Sci.* **2018**, *1*, 5068. [CrossRef]

Publisher's Note: MDPI stays neutral with regard to jurisdictional claims in published maps and institutional affiliations.

membranes

Article

A Study of the Interaction of a New Benzimidazole Schiff Base with Synthetic and Simulated Membrane Models of Bacterial and Mammalian Membranes

Alberto Aragón-Muriel [1], Yamil Liscano [2], David Morales-Morales [3], Dorian Polo-Cerón [1,*] and Jose Oñate-Garzón [2,*]

1 Laboratorio de Investigación en Catálisis y Procesos (LICAP), Departamento de Química, Facultad de Ciencias Naturales y Exactas, Universidad del Valle, Cali 760031, Colombia; alberto.aragon@correounivalle.edu.co
2 Grupo de Investigación en Química y Biotecnología (QUIBIO), Facultad de Ciencias Básicas, Universidad Santiago de Cali, Cali 760035, Colombia; yamil.liscano00@usc.edu.co
3 Instituto de Química, Universidad Nacional Autónoma de México, Cd. Universitaria, Circuito Exterior, Coyoacán, Mexico D.F. 04510, Mexico; damor@unam.mx
* Correspondence: dorian.polo@correounivalle.edu.co (D.P.-C.); jose.onate00@usc.edu.co (J.O.-G.)

Abstract: Biological membranes are complex dynamic systems composed of a great variety of carbohydrates, lipids, and proteins, which together play a pivotal role in the protection of organisms and through which the interchange of different substances is regulated in the cell. Given the complexity of membranes, models mimicking them provide a convenient way to study and better understand their mechanisms of action and their interactions with biologically active compounds. Thus, in the present study, a new Schiff base (*Bz-Im*) derivative from 2-(*m*-aminophenyl)benzimidazole and 2,4-dihydroxybenzaldehyde was synthesized and characterized by spectroscopic and spectrometric techniques. Interaction studies of (*Bz-Im*) with two synthetic membrane models prepared with 1,2-dimyristoyl-sn-glycero-3-phosphocholine (DMPC) and DMPC/1,2-dimyristoyl-sn-glycero-3-phosphoglycerol (DMPG) 3:1 mixture, imitating eukaryotic and prokaryotic membranes, respectively, were performed by applying differential scanning calorimetry (DSC). Molecular dynamics simulations were also developed to better understand their interactions. In vitro and in silico assays provided approaches to understand the effect of *Bz-Im* on these lipid systems. The DSC results showed that, at low compound concentrations, the effects were similar in both membrane models. By increasing the concentration of *Bz-Im*, the DMPC/DMPG membrane exhibited greater fluidity as a result of the interaction with *Bz-Im*. On the other hand, molecular dynamics studies carried out on the erythrocyte membrane model using the phospholipids POPE (1-palmitoyl-2-oleoyl-sn-glycero-3-phosphoethanolamine), SM (N-(15Z-tetracosenoyl)-sphing-4-enine-1-phosphocholine), and POPC (1-palmitoyl-2-oleoyl-sn-glycero-3-phosphocholine) revealed that after 30 ns of interaction, both hydrophobic interactions and hydrogen bonds were responsible for the affinity of *Bz-Im* for PE and SM. The interactions of the imine with POPG (1-Palmitoyl-2-Oleoyl-sn-Glycero-3-Phosphoglycerol) in the *E. coli* membrane model were mainly based on hydrophobic interactions.

Keywords: model membranes; molecular dynamics; calorimetry; Schiff base; imine; benzimidazole; 2,4-dihydroxybenzaldehyde

Citation: Aragón-Muriel, A.; Liscano, Y.; Morales-Morales, D.; Polo-Cerón, D.; Oñate-Garzón, J. A Study of the Interaction of a New Benzimidazole Schiff Base with Synthetic and Simulated Membrane Models of Bacterial and Mammalian Membranes. *Membranes* **2021**, *11*, 449. https://doi.org/10.3390/membranes11060449

Academic Editors: Jordi Marti and Carles Calero

Received: 2 May 2021
Accepted: 10 June 2021
Published: 16 June 2021

1. Introduction

Biological membranes are essential for life since they regulate the entry and exit of nutrients, neurotransmitters, and drugs [1]. Biological membranes contain three main types of lipids: phospholipids, glycolipids, and cholesterol. Phospholipids are, in turn, divided into different groups according to the structural properties of the polar head: phosphatidylcholine (PC), sphingomyeline (SM), and phosphatidylethanolamine (PE) are common lipids present in eukaryotic cell membranes [2].

Since drugs operate through different mechanisms when their main targets are intracellular and, therefore, they must penetrate the cell membrane to exert their pharmacological action, it is essential to understand drug–membrane interactions [3]. However, the complexity of the structure and functionality of cell membranes, as well as the highly dynamic nature of lipid–lipid and lipid–protein interactions, make drug–membrane system studies difficult [4].

Thus, artificial model membrane systems were developed to facilitate the understanding of the effects of membrane lipids on drug transport and absorption in cells, drug activity, and even drug toxicity [5,6]. Within the different types of model membranes, liposomes are highly suitable for permeability research and drug delivery systems. Additionally, they allow for the use of various thermoanalytical and spectroscopic techniques—such as isothermal titration calorimetry (ITC), differential scanning calorimetry (DSC), Fourier transform infrared spectroscopy (FT-IR), fluorescence spectroscopy, and nuclear magnetic resonance (NMR) methods—to study biophysical interactions of the drug–membrane complex [7–11].

In studies that make use of model membranes, saturated or unsaturated PC species are used to mimic eukaryotic cells, while the PC/PG model is used to mimic bacterial membranes [12]. Studies have revealed that the effect of azole compounds on model membranes [13,14], including membranes based on the PC/PG species, is controlled by drug–membrane interactions which depend on the length, unsaturation, and head group of the phospholipids, as well as the surface charge of the target cell [15]. In particular, compounds derived from benzimidazole interact with model membranes of human erythrocytes using the passive diffusion method [16]. In silico studies demonstrated that hydroxyl groups present in derivatives of benzimidazole decrease the hydrophobic character of DPPC (dipalmitoylphosphatidylcholine) model membranes and interact with the phosphate group of the polar heads present in the membrane [17].

Studies including compounds with the benzimidazole motif on their structures are interesting for the scientific community, not only because of their known antibacterial and cytotoxic properties [18–20], but because of the high conjugation that they exhibit when forming Schiff bases, improving their electronic characteristics and often conferring fluorescent properties that facilitate the monitoring of morphology in microorganisms subjected to these types of drugs [21].

Hence, based on the antibacterial and cytotoxic properties that Schiff bases obtained from (1*H*-benzimidazol-2-yl)anilines have demonstrated [22–24], this article describes the synthesis and characterization of 4-(((3-(1*H*-benzo[*d*]imidazol-2-yl)phenyl)imino)methyl) benzene-1,3-diol and the study of its possible mechanism of interaction with bacterial and mammalian membrane models by analyzing the thermodynamic profiles of the phase transition by DSC. In addition, in order to explore the *Bz-Im* effect on the thermotropic behavior of bacterial and mammalian systems, proper membrane models consistent with experimental membrane models were developed. Thus, the interaction of the imine towards model membranes of human erythrocytes and *E. coli* were described from results by molecular dynamics (MD) simulations.

2. Materials and Methods

2.1. Synthesis of Benzimidazole Schiff Base

2.1.1. Materials

All chemical reagents used for the synthesis of benzimidazole and subsequent imine were used as received and without further purification. Elemental analyses were performed using Flash EA 1112 Series CHN Analyzer. A Shimadzu Affinity 1 FT-IR spectrometer was used to obtain the infrared spectra. IR data are reported using the following abbreviations: vs = very strong; s = strong; m = medium; w = weak; sh = shoulder; br = broad. ^{1}H and $^{13}C\{^{1}H\}$ NMR spectra were obtained on a Bruker Avance II 400 spectrometer using DMSO-d_6 as a solvent at 25 °C. The following abbreviations were used: s = singlet; d = doublet; t = triplet; m = multiplet. The mass spectrum of the benzimidazole was recorded on a

Shimadzu-GCMS-QP2010 at 70 eV by electronic impact (EI) ionization, while the mass spectrum of the derived imine was obtained by direct analysis in real time (DART) ionization system on a JEOL AccuTOF JMS-T100LC.

2.1.2. Synthesis of 2-(m-aminophenyl)benzimidazole (*Bz*)

The condensation reaction of *o*-phenylenediamine with *m*-aminobenzoic acid was carried out following a similar methodology to that previously reported [25]. *o*-phenylenediamine (0.54 g, 5 mmol), *m*-aminobenzoic acid (0.69 g, 5 mmol), and polyphosphoric acid were mixed and stirred for 2.5 h at 180 °C. After this time, the resulting reaction mixture was allowed to cool and then neutralized with sodium carbonate (20%) before the resulting solution was filtered. The violet precipitate was then washed with distilled water, purified with activated carbon, and recrystallized in ethanol to give the final product as a beige powder. Yield: 0.87 g, 83%. $C_{13}H_{11}N_3$ (209.25 g·mol^{-1}): Calc. C, 74.62; H, 5.30; N, 20.08. Found: C, 74.58; H, 5.36; N, 20.09%. IR (ATR cm^{-1}): 3826 w, 3739 w, 3425 w, 3321 w, 3205 w, 2328 w, 1683 w, 1616 vs. 1566 s, 1510 s, 1463 s, 1394 m, 1346 m, 1234 m, 1062 m, 945 w, 893 m, 835 w, 750 vs. ^{1}H NMR (DMSO-d_6) δ (ppm) 12.70 (s, 1H), 7.62 (d, *J* = 7.0 Hz, 1H), 7.49 (d, *J* = 7.0 Hz, 1H), 7.43 (s, 1H), 7.28 (d, *J* = 7.5 Hz, 1H), 7.23–7.11 (m, 3H), 6.68 (d, *J* = 7.8 Hz, 1H), 5.30 (s, 2H). MS (EI, *m*/*z*): 209.

2.1.3. Synthesis of 4-(((3-(1*H*-benzo[*d*]imidazol-2-yl)phenyl)imino)methyl)benzene-1,3-diol (*Bz-Im*)

As was the case for the intermediate *Bz*, the imine *Bz-Im* was synthesized based on previously reported procedures [26,27]. 2-(*m*-aminophenyl)benzimidazole (0.52 g, 2.5 mmol), 2,4-dihydroxybenzaldehyde (0.35 g, 2.5 mmol), and methanol were mixed and set to reflux under stirring for 2 h. After this time, the yellow precipitate obtained was washed with cold water and then dried under vacuum for 4 h. Yield: 0.72 g, 87%. $C_{20}H_{15}N_3O_2$ (329.36 g·mol^{-1}): Calc. C, 72.94; H, 4.59; N, 12.76. Found: C, 72.83; H, 4.56; N, 12.91%. IR (ATR cm^{-1}): 3381 w, 3046 w, 1891 w, 1600 s, 1567 sh, 1512 w, 1494 vs. 1451 m, 1384 s, 1268 sh, 1259 s, 1222 m, 1143 s, 1114 m, 963 w, 910 m, 856 s, 807 vs. 725 vs. ^{1}H NMR (DMSO-d_6) δ (ppm) 13.46 (s, 1H), 12.98 (s, 1H), 10.34 (s, 1H), 8.94 (s, 1H), 8.22–8.01 (m, 2H), 7.70–7.45 (m, 5H), 7.24 (s, 2H), 6.45 (dd, *J* = 8.4, 2.3 Hz, 1H), 6.35 (d, *J* = 2.3 Hz, 1H). ^{13}C{^{1}H} NMR (DMSO-d_6) δ (ppm) 163.7, 163.5, 163.2, 151.3, 149.3, 144.2, 135.5, 135.1, 131.83, 130.6, 124.6, 123.2, 122.8, 122.3, 119.6, 119.4, 112.6, 111.9, 108.5, 102.9. MS (DART+) *m*/*z*: 330.

2.2. Interaction with Models of Synthetic Membranes

2.2.1. Membrane Preparation

Model membranes mimicking mammalian and bacterial membranes were prepared following the methodology reported previously [28]. Thus, DMPC and DMPG lipids at a molar ratio of 3:1 were dissolved in chloroform/methanol (2:1 *v*/*v*) to imitate gram-negative bacterial membranes [2], while the DMPC lipid alone was dissolved in chloroform/methanol (2:1 *v*/*v*) to imitate zwitterionic human cell membranes. The lipid mixture was first dried under a stream of nitrogen and then under vacuum for a further three hours. The hydration process was performed by preparing different concentrations of *Bz-Im* using HEPES buffer (25 mM HEPES, pH 7.0; 100 mM NaCl and 0.2 mM EDTA), which were added to the existing dry lipid mixture and vigorously shaken with a vortex for 2 min before incubation for 10 min at 37 °C above the phase transition temperature (T_m). The multilamellar vesicles (MLVs) were obtained after repeating the shaking and incubation process three times [29].

2.2.2. Differential Scanning Calorimetry

For the acquisition of thermograms by DSC analysis, a TA instrument DSC Q25 was used. Multillamellar vesicles (MLVs) were prepared using 2 mg of lipids hydrated with *Bz-Im* diluted in HEPES buffer to give three *Bz-Im*–lipid ratios: 1:50, 1:25, and 1:10. HEPES buffer was used as a reference solution. The samples were placed and subsequently sealed within standard aluminum DSC pans, and were analyzed over a range of 10 to 35 °C at a heating rate of 1 °C/min. Trios software (TA Instruments) was used to obtain the phase transition temperature (T_m), the transition enthalpy (ΔH), and the full width at half-maximum from thermograms (FWHM, $\Delta T_m/2$).

2.3. Molecular Dynamics (MD) Studies

2.3.1. Construction of the 3D Structure of *Bz-Im*

The 3D structure of *Bz-Im* was drawn using Chemdraw software (https://chemdrawdirect.perkinelmer.cloud/js/sample/index.html, access date: 8 March 2021). The *Bz-Im* structure was optimized using the universal force field (UFF) [30] and the steepest descent algorithm with Avogadro version 1.2 software.

2.3.2. Erythrocyte Membrane Construction

The zwitterionic model membrane of erythrocyte cells, which was mimicked by zwitterionic phosphatidylcholine for DSC assays, was built with the CHARMM-GUI [31] platform using, as a basis, the phospholipid composition mentioned by Texeira et al. [32]. The phospholipid proportions used were 20 units of POPE, 40 units of SM, and 40 units of POPC, which were distributed both in the upper and lower layer of the membrane. For the placement of the ions, the Monte Carlo method was used with a concentration of 0.15 M NaCl. In addition, a water thickness of 22.5 Å, and a force field for the entire CHARMM36m system was used [33]. The files were prepared to minimize energy, balance, and dynamics with GROMACS [34] at 310 K.

2.3.3. Construction of Gram-Negative Bacterial Membrane Models

The membrane model systems for *E. coli* were constructed based on the data reported by Epand et al. [35], using the same distribution published by Liscano et al. [36] for a gram-negative membrane model system (POPE = 80 units and POPG = 20 units), with CHARMM-GUI software [31].

2.3.4. Implementing Molecular Dynamics

The minimization energy of the erythrocyte model membrane and *E. coli* membrane system with the ligand *Bz-Im* was adjusted with the steepest descent algorithm in 5000 steps using the Verlet cutoff scheme. Equilibration was performed for 2 ns using the Berendsen algorithm to equilibrate the temperature and pressure of the system. Molecular dynamics were run for 10 ns at 310 K using the Nose–Hoover and Parrinello–Rahman algorithms to adjust temperature and pressure.

Gromacs software version 2020.1 [34] was used for the molecular dynamics simulation of gram-negative bacterial and erythrocyte membrane models. The CHARMM36m force field [33] was used for the simulation. For the localization of the ions, the Monte Carlo method was used with 0.15 M NaCl in water 22.5 Å thick. For the energy minimization, the steepest descent algorithm was used, running for 1 ns. Using the Berendsen algorithm, the system was adjusted to a temperature of 310 K with an equilibration of 2 fs/step for 300 ps to 155,000 n-steps. Once the system was equilibrated the molecular dynamics were run for 30 ns for both erythrocyte and *E. coli* model membranes, using the Nose–Hoover and Parrinello–Rahman algorithms to adjust the temperature and pressure.

2.3.5. Interaction Analysis

Gromacs was used to obtain the hydrogen bonds between *Bz-Im* and the phospholipids of each membrane model system within 30 ns. PyMOL PDB files were obtained for each membrane system at six different times: 1, 5, 10, 20, 25, and 30 ns. These files were used to visualize and analyze the interactions between the different components of each model system and the *Bz-Im* using Discovery Studio Visualizer software.

3. Results and Discussion

3.1. Schiff Base (Imine) Characterization

The synthesis of Schiff base 4-(((3-(1*H*-benzo[*d*]imidazol-2-yl)phenyl)imino)methyl) benzene-1,3-diol **(*Bz-Im*)** was carried out from the reaction between 2,4-dihydroxybenzaldehyde and 2-(*m*-aminophenyl)benzimidazole **(*Bz*)** (Scheme 1).

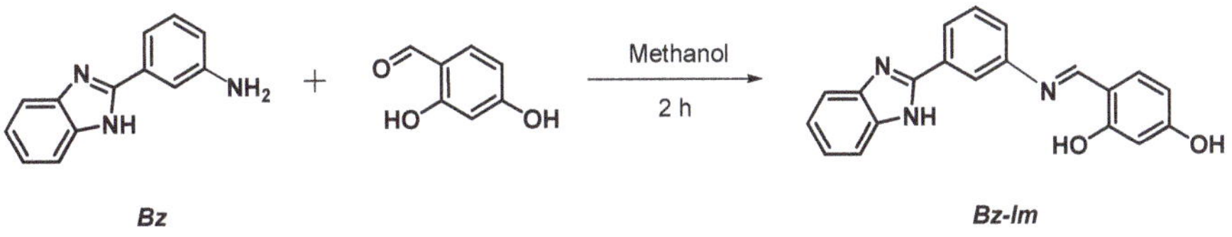

Scheme 1. Synthesis of *Bz-Im*.

Compound *Bz-Im* was obtained in high yield (87%) as a yellow powder, and its structure was unequivocally determined by mass spectrometry, elemental analyses (C, H, and N), infrared spectroscopy, and nuclear magnetic resonance spectroscopy. The FT-IR spectra of *Bz-Im* showed bands corresponding to υ (3425 and 3321 cm^{-1}) stretching and deformation δ (1616 cm^{-1}) vibrations of the amino group in *Bz* disappeared after the formation of the imine, while a new band was observed at 1600 cm^{-1}, characteristic for this type of compound, confirming the formation of the N=CH bond [37]. In the infrared spectrum, the characteristic band of the stretching vibration $\upsilon_{C\text{-}O}$ at 1259 cm^{-1} was also observed (phenolic fragment). In addition, the ^1H NMR spectrum of *Bz-Im* showed the characteristic signal of the imine group at 8.94 ppm, while signals due to the aromatic protons were observed around 8.25–6.25 ppm [22,27]. Additionally, the $^{13}C\{^1H\}$ NMR spectra exhibited a typical signal due to the imine carbon at ~163 ppm. This one-dimensional NMR analysis was further supported with two-dimensional studies that can be consulted in Figures S4, S6, and S7 of the supplementary material. Finally, analysis by mass spectrometry (DART+) afforded a spectrum exhibiting the peak due to the molecular ion [M+1] at 330 m/z (Figure S9). Elemental analysis results were also in agreement with the proposed structure.

3.2. Model Membrane Studies

3.2.1. Thermotropic Behavior of Synthetic Model Membranes

The membrane models included mammalian-like membranes consisting of the phospholipid DMPC and bacterial-like membranes consisting of a 3:1 ratio of DMPC:DMPG. By gradually heating the vesicles without compound, the acquisition of thermotropic profiles was achieved (Figure 1), where a pre-transition endothermic peak was observed at 12.94 °C for the DMPC systems and at 12.71 °C for the DMPC: DMPG 3:1 mixture. PCs have a fairly bulky headgroup, creating a size mismatch with their acyl chains, especially below the main phase transition [38]. As the temperature increases, the main transition peak emerges at 23.02 °C for the two systems mentioned above—these results being consistent with those reported previously [28].

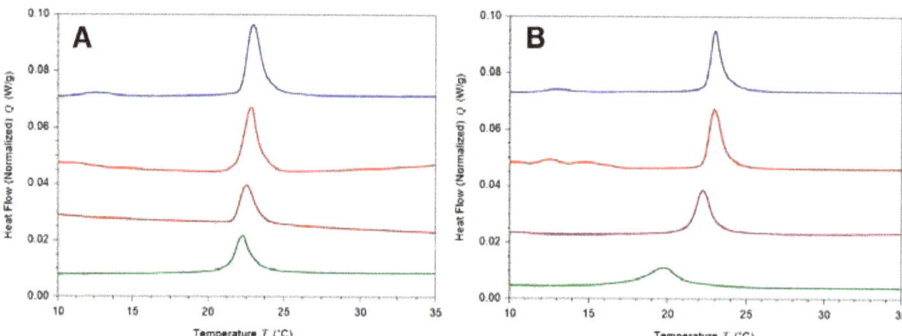

Figure 1. Thermotropic profile of MLVs made up of (**A**) DMPC and (**B**) DMPC-DMPG (3:1), under different amounts of imine. Compound–lipid molar ratios: 0:1 (—); 1:50 (—); 1:25 (—); and 1:10 (—).

By addition of *Bz-Im* in a 1:50 compound–lipid molar ratio, changes were observed in the pre-transition of both systems (Table 1), suggesting that this compound affects the transition from a flat membrane phase (Lβ) to a ripple phase (Pβ) as a result of the changes in size mismatch with the phosphatydilcholine headgroup hydration [38]. Interestingly, in the pre-transition from the DMPC: DMPG mixture, two subtle peaks emerged (Figure 1B). Riske et al. [38] described that up to 20% fluid lipid population were detected between the pre- and the main transitions. In addition, gauche conformers are introduced into the acyl lipid chains in the pre-transition [39]. By increasing the concentration of *Bz-Im* (1:25 compound–lipid molar ratio), the pre-transition is abolished, the T_m is changed moderately, and the size of the peak of the main transition decreases as the width of the peak increases in both lipid systems (Figure 1). The surface of the bilayer must be considered as a set of several phospholipid "clusters", where all the molecules of each cluster exhibit a simultaneous behavior in the transition. This cooperative property of the melting process defines more sharp and symmetrical curves at the transition peak. In this way, *Bz-Im* is able to penetrate into the bilayer since the "clusters" noticeably increase in number [40]. $\Delta T_m/2$ is a relative measure of molecular cooperativity and it linearly increases with the concentration of *Bz-Im*, ranging from 0.98 to 1.14 °C and from 0.73 to 1.08 °C for DMPC and DMPC:DMPG systems, respectively, suggesting the insertion of "free volumes" into the bilayer structure [40]. Therefore, the full width at half-maximum of the main transition peak is a variable that indicates how cooperative the phospholipids are when they undergo a transition [41]. Similarly, the enthalpy of transition is considerably and moderately reduced in DMPC and DMPC/DMPG, respectively (Table 1), suggesting that the addition of anionic lipids to the zwitterionic lipids avoids greater alterations to the interactions between lipid acyl chains, i.e., the disruption of *trans-gauche* isomerization and the inter- and intramolecular van der Waals interactions [7]. This can be explained by a strong adhesion of the *Bz-Im* on the anionic surface by phosphatidylglycerol at this concentration, where the small size and the reduced flexibility of *Bz-Im* prevents their hydrophobic moieties from being inserted inside the bilayer.

Table 1. T_m, $\Delta T_m/2$, and enthalpy transition (ΔH) values of MLVs constituted by DMPC and DMPC/DMPG (3:1) before and after the addition of *Bz-Im* at different *Bz-Im*–lipid molar ratios.

MLV	Compound–Lipid Molar Ratio	Pretransition Temperature (°C)	T_m (°C)	$\Delta T_m/2$ (°C)	ΔH ($J \cdot g^{-1}$)
DMPC	0:1	12.94	23.02	0.98	1.48
DMPC-Compound	1:50	11.58	22.94	1.09	1.05
	1:25	-	22.60	1.14	0.78
	1:10	-	22.38	1.23	0.76
DMPC/DMPG (3:1)	0:1	12.71	23.02	0.73	1.71
DMPC/DMPG (3:1)-Compound	1:50	12.49	22.98	0.90	1.46
	1:25	-	22.24	1.08	1.38
	1:10	-	19.72	2.34	1.13

At the maximum concentration of *Bz-Im*, the size of the peaks representing the main transition were decreased. Interestingly, a pronounced change in T_m and in the width of the peak was observed only in the DMPC/DMPG mixture (Figure 1), suggesting that *Bz-Im* at this concentration increases the fluidity and the lateral phase separation in membranes that mimic those of bacteria. In fact, fluidity is known to increase in response to an increase in the lateral diffusion rates of lipid molecules [42], and has been related to alterations within the hydrophobic nucleus of the bilayer [43]. Thus, it is likely that more moles of *Bz-Im* bind to a smaller unit area in DMPC/DMPG than to an entire surface area of DMPC, reaching a lower threshold concentration due to the presence of DMPG, which is related to the degree of insertion of compounds inside the bilayer [44].

3.2.2. Analysis of Molecular Dynamics

In order to understand the results of the thermotropic profile of the membrane models, the interaction between *Bz-Im* and the molecular models of erythrocyte zwitterionic membranes was analyzed by molecular dynamics. Figure 2 shows the root mean square deviation (RMSD) of *Bz-Im* in the erythrocyte membrane over 30 ns. From 0 to 10 ns a continuous variation of the RMSD is observed (Figure 2A), suggesting the whole structure fluctuates, or it might reflect only large displacements of a small structural subset within an overall rigid structure [45] as a result of the loss of bonds between phospholipids and the formation of *Bz-Im*-membrane interactions during the insertion and adjustment of the compound inside the polar head of the phospholipids in these first nanoseconds (Figure 2B). A stabilization of the structural configuration of *Bz-Im* is observed from 10 ns that coincides with the penetration of *Bz-Im* inside the polar region of the membrane, indicating that it does not bring any considerable changes to the overall conformation of the system over 30 ns MD trajectories.

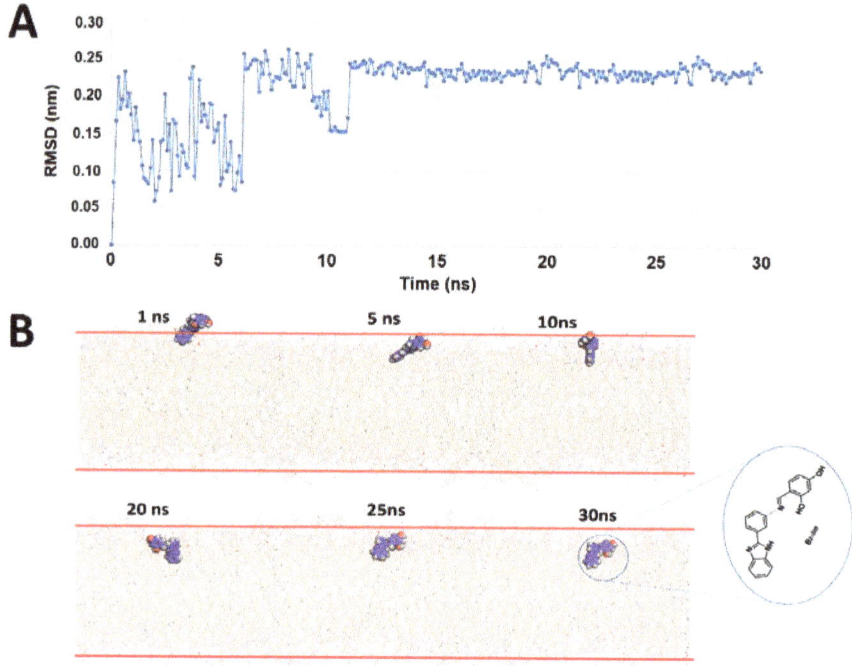

Figure 2. Behavior of *Bz-Im* in the erythrocyte membrane model system for 30 ns. (**A**) Root mean square deviation (RMSD) of *Bz-Im* in the erythrocyte membrane system at 30 ns. (**B**) *Bz-Im* movement from the surface of the erythrocyte membrane during the 30 ns.

Figure 3 shows the interactions of *Bz-Im* with the components of the erythrocyte membrane model system, observing a greater number of hydrogen bond-type interactions with water molecules in the first nanoseconds after starting the simulation due to the interaction with the membrane-water interfacial region. These interactions with water decrease as time progresses from 10 to 30 ns. Conversely, interactions with SM and POPE increase as time progresses since they are the phospholipids that most interact with *Bz-Im* in comparison with POPC (Figure 3) due to the presence of the amine group in POPE which forms the additional bonds [46]. In addition, PC and SM have the same polar head but differ in their interfacial structures due to a decrease in headgroup size of the SM causing closer molecular packing. The increased interactions at the membrane interface could influence increased affinity of *Bz-Im* for SM as compared with POPC [47]. Both hydrophobic interactions and hydrogen bonds are responsible for the affinity of *Bz-Im* for PE and SM (Figure 3A), suggesting that the C-N bonds are oriented towards the core of the bilayer and the O-H groups are oriented towards the water phase (Figure 2B).

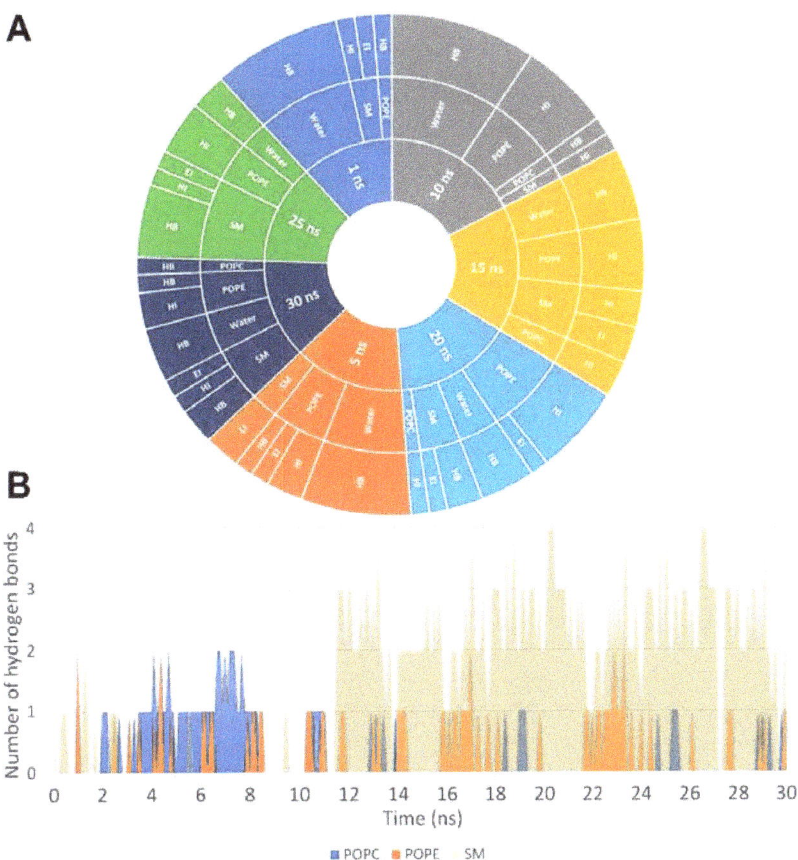

Figure 3. Intermolecular interactions between *Bz-Im* and the erythrocyte membrane system (**A**) at 1 ns (blue color), 5 ns (orange color), 10 ns (gray color), 15 ns (yellow color), 20 ns (light blue color), 25 ns (green color), and 30 ns (dark blue color). HB, hydrogen bond; HI, hydrophobic interaction; E, electrostatic interactions (**B**) Hydrogen bonds between *Bz-Im* and the phospholipids of the system. SM, sphingomyelin; POPC, phosphatidylcholine; POPE, phosphatidylethanolamine.

Figure 3B shows the number of hydrogen bonds between *Bz-Im* and phospholipids through time. Interestingly, interactions with SM emerge from 11 nanoseconds, remaining up to 30 ns, suggesting an initial selectivity for PE, which would directly interact with the NH_3^+ moiety and not with the quaternary amine group of choline. When replacing PE by PC, this could arise from the sole removal of the hydrogen-bonding capability of the headgroup [48]. However, there are also hydrogen bonds between *Bz-Im* and POPC from 2 to 8 ns while it was entering the membrane. This could be due to the larger size of the PC polar head, compared with those of PE and SM, occupying a greater volume and exhibiting a better probability of initial contact with *Bz-Im*. Once it is internalized, a strong interaction with sphingomyelin is maintained. It was revealed that the OH-group or NH-function of SM play an important role in hydrogen bonding interactions with foreign compounds [49].

Based on the above, *Bz-Im* can be buried up to the interfacial region of the outer monolayer of a zwitterionic membrane, decreasing the van der Waals interactions between the phospholipids, while interactions that require less heat to undergo the transition are formed (Figure 1A). Hence, it is likely that *Bz-Im* exhibits cytotoxic activity against mammalian cells, a property that could be explored and exploited for tumor cells (Supplementary materials, Section 2).

Figure 4 shows the behavior of *Bz-Im* on the *E. coli* model membrane for 30 ns. The RMSD reveals that during the first 6 ns there is a great structural variation of *Bz-Im* related to its location between the surface of the membrane and the water phase, which causes intermittent contact with both water molecules and POPE or POPG. Between 10 and 20 ns there is a slight structural variation of *Bz-Im* and during this time the molecule remains submerged in the membrane. Again, *Bz-Im* re-emerges on the membrane surface between the aqueous and lipid phase, which is reflected in a greater structural deviation between 20 and 21 ns as a result of different solvation changes. Finally, *Bz-Im* is internalized again inside the head group of phospholipids between 28 and 30 ns. Unlike the erythrocyte membrane, the position of the *Bz-Im* in the *E. coli* membrane model fluctuated highly during the 30 ns. To explain this, it must be considered that the erythrocyte membrane is made up of only 20% POPE while that of *E. coli* has 80% POPE. The NH_3^+ group of PE binds with oxygen from unesterified phosphate by very close contacts [50]. Subsequently, the bonds between adjacent phosphates form a very compact network of PE polar heads in the surface of the membrane, hindering the access of *Bz-Im* and reorienting it within the lipid phase over the first 20 ns. On the other hand, the glycerol moiety of PG mimics the solvation water of the phosphate group [51]. This internal hydrogen bonding makes the hydrogen bonding between the foreign compounds and the phosphate less favorable than when the phosphate is linked to cholines. Thus, phosphate must be shielded by the glycerol moiety in PG, avoiding the formation of hydrogen bonds with compounds at the expense of dehydration [52].

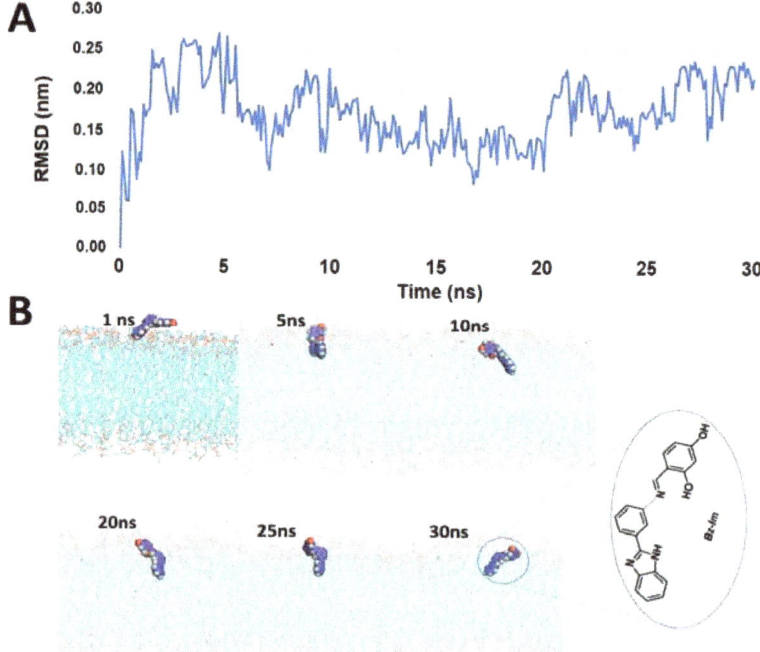

Figure 4. Behavior of *Bz-Im* in the *E. coli* membrane model system for 30 ns. (**A**) Root mean square deviation (RMSD) of *Bz-Im* in the *E. coli* membrane model system for 30 ns. (**B**) *Bz-Im* movement from the surface of the *E. coli* membrane model system during the 30 ns.

Figure 5A shows a greater interaction with water molecules through hydrogen bonds in the first 15 nanoseconds. From 10 ns there is a tendency to form hydrophobic bonds with both phospholipids; this type of interaction is maintained until 30 ns, suggesting that *Bz-Im* must penetrate at least partially into the hydrophobic core of the phospholipid bilayer. As the system is highly dynamic, it is probable that the hydrophobic moiety of the aromatic heterocyclic ring interacts with the acyl chains in a region close to the interface (Figure 4B). The partial insertion into the hydrophobic core would be responsible for the increase in the fluidity described in bacterial model membranes at a 1:10 *Bz-Im*:lipid molar ratio (Figure 1B), since interaction of compounds with the phospholipid acyl chains has been related to a net fluidizing effect of the apolar part of the bilayer [43]. Finally, Figure 5B shows the number of hydrogen bonds formed between *Bz-Im* and the phospholipids of the *E. coli* model membrane; a similar interaction with POPE and POPG was observed, except for at 10 and 22 ns when interaction peaks with POPE of up to six hydrogen bonds were found. The amine group of PE which forms additional bonds [46] would be favorably oriented towards the polar groups of *Bz-Im* at both times.

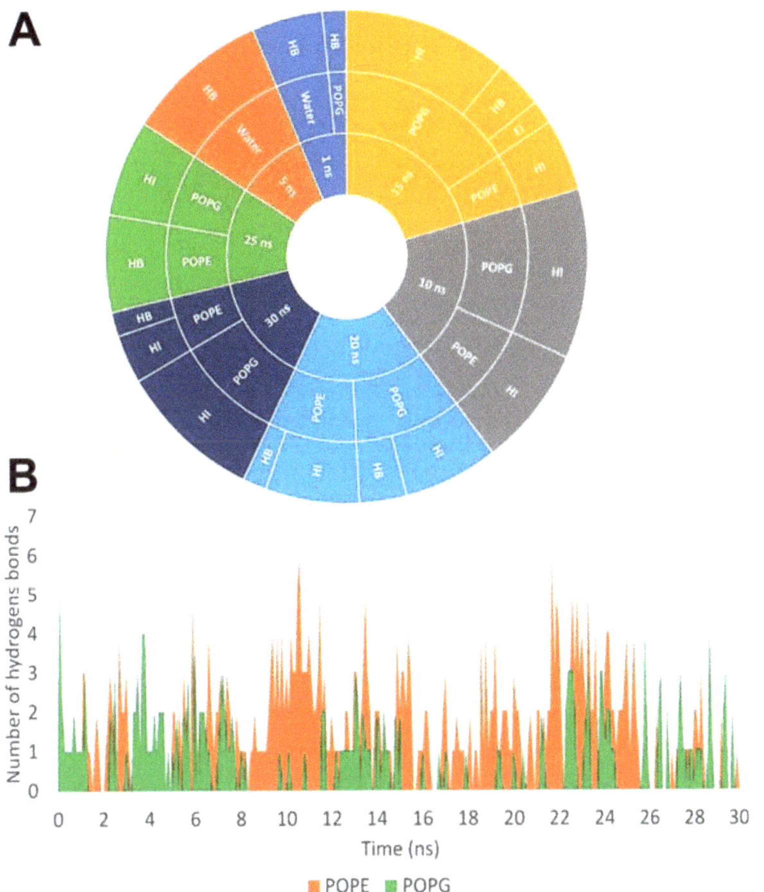

Figure 5. Intermolecular interactions between *Bz-Im* and the *E. coli* membrane model (**A**) at 1 ns (blue color), 5 ns (orange color), 10 ns (gray color), 15 ns (yellow color), 20 ns (light blue color), 25 ns (green color), and 30 ns (dark blue color). HB, hydrogen bond; HI, hydrophobic interaction; EI, electrostatic interactions. (**B**) Hydrogen bonds between *Bz-Im* and the phospholipids of the system. POPE, phosphatidylethanolamine; POPG, phosphatidylglycerol.

4. Conclusions

A benzimidazole-derived imine was successfully synthesized and characterized by spectroscopic and spectrometric techniques. The thermotropic profiles indicate that this compound can bind both to DMPC and DMPC:DMPG (3:1), which mimic the mammalian and bacterial membranes, respectively. Our results suggest that *Bz-Im* can increase the fluidity in membranes that mimic those of bacteria, which might be correlated with their potential antibacterial activity, representing a valuable contribution towards the further design of antimicrobial compounds based on benzimidazole-derived imine analogues. Preliminary evidence shows that compound *Bz-Im* has affinity for PC phospholipids, suggesting that this molecule may have effects against normal human cells, that is, the *Bz-Im* compound could be cytotoxic toward these cells. At 30 ns of simulation, hydrogen bonding interactions between *Bz-Im* and SM prevail in erythrocyte membrane models, while in *E. coli* membrane models the hydrophobic interactions between *Bz-Im* and PG/PE play an important role on the fluidizing effect exhibited in bacterial membrane models. Although 30 ns is a relatively short simulation time, it is sufficient to understand the behavior of the system since a trend is clearly defined from 10 ns. Finally, this study serves as a prototype for better understanding of the interactions between these kinds of molecules and biological membranes, as well as opening prospects for future work in this area.

Supplementary Materials: The following are available online at https://www.mdpi.com/article/10.3390/membranes11060449/s1. Section 1: Spectral data. Figure S1: Comparative FT-IR spectra of Bz and *Bz-Im*. Figure S2: NMR-1H spectra of Bz. Figure S3: NMR-1H spectra of *Bz-Im*. Figure S4: NMR-COSY spectra of *Bz-Im*. Figure S5: NMR-13C{1H} spectra of *Bz-Im*. Figure S6: NMR-HSQC spectra of *Bz-Im*. Figure S7: NMR-HMBC spectra of *Bz-Im*. Figure S8: Mass spectra (EI) of Bz. Figure S9: Mass spectra (DART+) of *Bz-Im* [M+1]; Section 2: Figure S10: Root-mean-square deviation (RMSD) of *Bz-Im* in the system. Figure S11: Hydrogen bonds between *Bz-Im* and the phospholipids of the system.

Author Contributions: Conceptualization, A.A.-M. and Y.L.; methodology, A.A.-M., Y.L., and J.O.-G.; software, Y.L.; validation, J.O.-G. and D.P.-C.; formal analysis, A.A.-M., D.M.-M., Y.L., and J.O.-G.; investigation, A.A.-M. and Y.L.; resources, J.O.-G. and D.P.-C.; data curation, A.A.-M. and Y.L.; writing—original draft preparation, A.A.-M.; writing—review and editing, D.M.-M., J.O.-G., and D.P.-C.; visualization, A.A.-M., Y.L., and J.O.-G.; supervision, D.P.-C.; project administration, D.P.-C.; funding acquisition, D.M.-M., J.O.-G., and D.P.-C. All authors have read and agreed to the published version of the manuscript.

Funding: This research was funded by the Universidad del Valle, by Dirección General de Investigaciones of Universidad Santiago de Cali under call No. 01-2021 and grant number DGI-COCEIN No 512-621120-1529, and the "Doctorados Nacionales Colciencias Convocatoria 727–2015" program. D.M.-M. would like to thank PAPIIT-DGAPA-UNAM (PAPIIT IN210520) and CONACYT A1-S-33933 for their generous financial support.

Institutional Review Board Statement: Not applicable.

Conflicts of Interest: The authors declare no conflict of interest.

References

1. Xu, Y.; Tillman, T.S.; Tang, P. *Membranes and Drug Action*, 1st ed.; Elsevier Inc.: Amsterdam, The Netherlands, 2009; ISBN 9780123695215.
2. Van Meer, G.; Voelker, D.R.; Feigenson, G.W. Membrane lipids: Where they are and how they behave. *Nat. Rev. Mol. Cell Biol.* **2008**, *9*, 112–124. [CrossRef] [PubMed]
3. Schubert, T.; Römer, W. How synthetic membrane systems contribute to the understanding of lipid-driven endocytosis. *Biochim. Biophys. Acta Mol. Cell Res.* **2015**, *1853*, 2992–3005. [CrossRef] [PubMed]
4. Peetla, C.; Stine, A.; Labhasetwar, V. Biophysical interactions with model lipid membranes: Applications in drug discovery and drug delivery. *Mol. Pharm.* **2009**, *6*, 1264–1276. [CrossRef]
5. Essaid, D.; Rosilio, V.; Daghildjian, K.; Solgadi, A.; Vergnaud, J.; Kasselouri, A.; Chaminade, P. Artificial plasma membrane models based on lipidomic profiling. *Biochim. Biophys. Acta Biomembr.* **2016**, *1858*, 2725–2736. [CrossRef] [PubMed]

6. Knobloch, J.; Suhendro, D.K.; Zieleniecki, J.L.; Shapter, J.G.; Köper, I. Membrane-drug interactions studied using model membrane systems. *Saudi J. Biol. Sci.* **2015**, *22*, 714–718. [CrossRef] [PubMed]
7. Oñate-Garzón, J.; Ausili, A.; Manrique-Moreno, M.; Torrecillas, A.; Aranda, F.J.; Patiño, E.; Gomez-Fernández, J.C. The increase in positively charged residues in cecropin D-like *Galleria mellonella* favors its interaction with membrane models that imitate bacterial membranes. *Arch. Biochem. Biophys.* **2017**, *629*, 54–62. [CrossRef]
8. Rivera-Sánchez, S.P.; Agudelo-Góngora, H.A.; Oñate-Garzón, J.; Flórez-Elvira, L.J.; Correa, A.; Londoño, P.A.; Londoño-Mosquera, J.D.; Aragón-Muriel, A.; Polo-Cerón, D.; Ocampo-Ibáñez, I.D. Antibacterial Activity of a Cationic Antimicrobial Peptide against Multidrug-Resistant Gram-Negative Clinical Isolates and Their Potential Molecular Targets. *Molecules* **2020**, *25*, 5035. [CrossRef]
9. Correa, W.; Manrique-Moreno, M.; Patiño, E.; Peláez-Jaramillo, C.; Kaconis, Y.; Gutsmann, T.; Garidel, P.; Heinbockel, L.; Brandenburg, K. *Galleria mellonella* native and analogue peptides Gm1 and ΔgmI. I) Biophysical characterization of the interaction mechanisms with bacterial model membranes. *Biochim. Biophys. Acta Biomembr.* **2014**, *1838*, 2728–2738. [CrossRef]
10. Haralampiev, I.; Alonso de Armiño, D.J.; Luck, M.; Fischer, M.; Abel, T.; Huster, D.; Di Lella, S.; Scheidt, H.A.; Müller, P. Interaction of the small-molecule kinase inhibitors tofacitinib and lapatinib with membranes. *Biochim. Biophys. Acta Biomembr.* **2020**, *1862*, 183414. [CrossRef]
11. Kaur, N.; Fischer, M.; Kumar, S.; Gahlay, G.K.; Scheidt, H.A.; Mithu, V.S. Role of cationic head-group in cytotoxicity of ionic liquids: Probing changes in bilayer architecture using solid-state NMR spectroscopy. *J. Colloid Interface Sci.* **2021**, *581*, 954–963. [CrossRef] [PubMed]
12. Lind, T.K.; Skoda, M.W.A.; Cárdenas, M. Formation and Characterization of Supported Lipid Bilayers Composed of Phosphatidylethanolamine and Phosphatidylglycerol by Vesicle Fusion, a Simple but Relevant Model for Bacterial Membranes. *ACS Omega* **2019**, *4*, 10687–10694. [CrossRef]
13. Britt, H.M.; Prakash, A.S.; Appleby, S.; Mosely, J.A.; Sanderson, J.M. Lysis of membrane lipids promoted by small organic molecules: Reactivity depends on structure but not lipophilicity. *Sci. Adv.* **2020**, *6*, eaaz8598. [CrossRef]
14. Dadhich, R.; Singh, A.; Menon, A.P.; Mishra, M.; Athul, C.D.; Kapoor, S. Biophysical characterization of mycobacterial model membranes and their interaction with rifabutin: Towards lipid-guided drug screening in tuberculosis. *Biochim. Biophys. Acta Biomembr.* **2019**, *1861*, 1213–1227. [CrossRef]
15. Alves, A.C.; Ribeiro, D.; Nunes, C.; Reis, S. Biophysics in cancer: The relevance of drug-membrane interaction studies. *Biochim. Biophys. Acta Biomembr.* **2016**, *1858*, 2231–2244. [CrossRef]
16. Castillo, I.; Suwalsky, M.; Gallardo, M.J.; Troncoso, V.; Sánchez-Eguía, B.N.; Santiago-Osorio, E.; Aguiñiga, I.; González-Ugarte, A.K. Structural and functional effects of benzimidazole/thioether-copper complexes with antitumor activity on cell membranes and molecular models. *J. Inorg. Biochem.* **2016**, *156*, 98–104. [CrossRef]
17. Lopes-de-Campos, D.; Nunes, C.; Sarmento, B.; Jakobtorweihen, S.; Reis, S. Metronidazole within phosphatidylcholine lipid membranes: New insights to improve the design of imidazole derivatives. *Eur. J. Pharm. Biopharm.* **2018**, *129*, 204–214. [CrossRef] [PubMed]
18. Chen, M.; Su, S.; Zhou, Q.; Tang, X.; Liu, T.; Peng, F.; He, M.; Luo, H.; Xue, W. Antibacterial and antiviral activities and action mechanism of flavonoid derivatives with a benzimidazole moiety. *J. Saudi Chem. Soc.* **2021**, *25*, 101194. [CrossRef]
19. Aragón-Muriel, A.; Liscano-Martínez, Y.; Rufino-Felipe, E.; Morales-Morales, D.; Oñate-Garzón, J.; Polo-Cerón, D. Synthesis, biological evaluation and model membrane studies on metal complexes containing aromatic N,O-chelate ligands. *Heliyon* **2020**, *6*, e04126. [CrossRef] [PubMed]
20. Alasmary, F.A.S.; Snelling, A.M.; Zain, M.E.; Alafeefy, A.M.; Awaad, A.S.; Karodia, N. Synthesis and evaluation of selected benzimidazole derivatives as potential antimicrobial agents. *Molecules* **2015**, *20*, 15206–15223. [CrossRef] [PubMed]
21. Saluja, P.; Sharma, H.; Kaur, N.; Singh, N.; Jang, D.O. Benzimidazole-based imine-linked chemosensor: Chromogenic sensor for Mg2+ and fluorescent sensor for Cr3+. *Tetrahedron* **2012**, *68*, 2289–2293. [CrossRef]
22. Dutta Gupta, S.; Revathi, B.; Mazaira, G.I.; Galigniana, M.D.; Subrahmanyam, C.V.S.; Gowrishankar, N.L.; Raghavendra, N.M. 2,4-dihydroxy benzaldehyde derived Schiff bases as small molecule Hsp90 inhibitors: Rational identification of a new anticancer lead. *Bioorg. Chem.* **2015**, *59*, 97–105. [CrossRef] [PubMed]
23. Ganga Raju, M.; Saritha, L.; Dutta Gupta, S.; Divya, N. Antimicrobial, Anti-Inflammatory and Anti-Parkinson's Screening of Imine Analogues through HSP90 Inhibition. *J. Chem. Pharm. Res.* **2017**, *9*, 258–266.
24. Mahmood, K.; Hashmi, W.; Ismail, H.; Mirza, B.; Twamley, B.; Akhter, Z.; Rozas, I.; Baker, R.J. Synthesis, DNA binding and antibacterial activity of metal(II) complexes of a benzimidazole Schiff base. *Polyhedron* **2019**, *157*, 326–334. [CrossRef]
25. Braña, M.F.; Castellano, J.M.; Yunta, M.J.R. Synthesis of benzimidazo-substituted 3-quinolinecarboxylic acids as antibacterial agents. *J. Heterocycl. Chem.* **1990**, *27*, 1177–1180. [CrossRef]
26. Roopashree, B.; Gayathri, V.; Gopi, A.; Devaraju, K.S. Syntheses, characterizations, and antimicrobial activities of binuclear ruthenium(III) complexes containing 2-substituted benzimidazole derivatives. *J. Coord. Chem.* **2012**, *65*, 4023–4040. [CrossRef]
27. Suman, G.R.; Bubbly, S.G.; Gudennavar, S.B.; Muthu, S.; Roopashree, B.; Gayatri, V.; Nanje Gowda, N.M. Structural investigation, spectroscopic and energy level studies of Schiff base: 2-[(3′-N-salicylidenephenyl)benzimidazole] using experimental and DFT methods. *J. Mol. Struct.* **2017**, *1139*, 247–254. [CrossRef]
28. Aragón-Muriel, A.; Ausili, A.; Sánchez, K.; Rojas, A.O.E.; Londoño Mosquera, J.; Polo-Cerón, D.; Oñate-Garzón, J. Studies on the Interaction of Alyteserin 1c Peptide and Its Cationic Analogue with Model Membranes Imitating Mammalian and Bacterial Membranes. *Biomolecules* **2019**, *9*, 527. [CrossRef]

29. Oñate-Garzón, J.; Manrique-Moreno, M.; Trier, S.; Leidy, C.; Torres, R.; Patiño, E. Antimicrobial activity and interactions of cationic peptides derived from *Galleria mellonella* cecropin D-like peptide with model membranes. *J. Antibiot.* **2017**, *70*, 238–245. [CrossRef]
30. Rappe, A.K.; Casewit, C.J.; Colwell, K.S.; Goddard, W.A.; Skiff, W.M. UFF, a full periodic table force field for molecular mechanics and molecular dynamics simulations. *J. Am. Chem. Soc.* **1992**, *114*, 10024–10035. [CrossRef]
31. Jo, S.; Kim, T.; Iyer, V.G.; Im, W. CHARMM-GUI: A web-based graphical user interface for CHARMM. *J. Comput. Chem.* **2008**, *29*, 1859–1865. [CrossRef]
32. Teixeira, V.; Feio, M.J.; Bastos, M. Role of lipids in the interaction of antimicrobial peptides with membranes. *Prog. Lipid Res.* **2012**, *51*, 149–177. [CrossRef]
33. Huang, J.; Mackerell, A.D. CHARMM36 all-atom additive protein force field: Validation based on comparison to NMR data. *J. Comput. Chem.* **2013**, *34*, 2135–2145. [CrossRef]
34. Abraham, M.J.; Murtola, T.; Schulz, R.; Páll, S.; Smith, J.C.; Hess, B.; Lindah, E. Gromacs: High performance molecular simulations through multi-level parallelism from laptops to supercomputers. *SoftwareX* **2015**, *1–2*, 19–25. [CrossRef]
35. Epand, R.F.; Savage, P.B.; Epand, R.M. Bacterial lipid composition and the antimicrobial efficacy of cationic steroid compounds (Ceragenins). *Biochim. Biophys. Acta Biomembr.* **2007**, *1768*, 2500–2509. [CrossRef]
36. Liscano, Y.; Salamanca, C.H.; Vargas, L.; Cantor, S.; Laverde-Rojas, V.; Oñate-Garzón, J. Increases in hydrophilicity and charge on the polar face of alyteserin 1c helix change its selectivity towards gram-positive bacteria. *Antibiotics* **2019**, *8*, 238. [CrossRef] [PubMed]
37. Chandrakala, M.; Sheshadri, B.S.; Nanje Gowda, N.M.; Murthy, K.G.S.; Nagasundara, K.R. Synthesis and spectral studies of 2-salicylidine-4-aminophenyl benzimidazole and its reaction with divalent Zn, Cd and Hg: Crystal structure of the cadmium bromide complex. *J. Chem. Res.* **2010**, 576–580. [CrossRef]
38. Riske, K.A.; Barroso, R.P.; Vequi-Suplicy, C.C.; Germano, R.; Henriques, V.B.; Lamy, M.T. Lipid bilayer pre-transition as the beginning of the melting process. *Biochim. Biophys. Acta Biomembr.* **2009**, *1788*, 954–963. [CrossRef] [PubMed]
39. Meyer, H.W. Pretransition-ripples in bilayers of dipalmitoylphosphatidylcholine: Undulation or periodic segments? A freeze-fracture study. *Biochim. Biophys. Acta Lipids Lipid Metab.* **1996**, *1302*, 138–144. [CrossRef]
40. Di Foggia, M.; Bonora, S.; Tinti, A.; Tugnoli, V. DSC and Raman study of DMPC liposomes in presence of Ibuprofen at different pH. *J. Therm. Anal. Calorim.* **2017**, *127*, 1407–1417. [CrossRef]
41. Ohline, S.M.; Campbell, M.L.; Turnbull, M.T.; Kohler, S.J. Differential scanning calorimetric study of bilayer membrane phase transitions: A biophysical chemistry experiment. *J. Chem. Educ.* **2001**, *78*, 1251–1256. [CrossRef]
42. Bilge, D.; Kazanci, N.; Severcan, F. Acyl chain length and charge effect on Tamoxifen-lipid model membrane interactions. *J. Mol. Struct.* **2013**, *1040*, 75–82. [CrossRef]
43. Carrillo, C.; Teruel, J.A.; Aranda, F.J.; Ortiz, A. Molecular mechanism of membrane permeabilization by the peptide antibiotic surfactin. *Biochim. Biophys. Acta Biomembr.* **2003**, *1611*, 91–97. [CrossRef]
44. Huang, H.W. Action of antimicrobial peptides: Two-state model. *Biochemistry* **2000**, *39*, 8347–8352. [CrossRef]
45. Martínez, L. Automatic identification of mobile and rigid substructures in molecular dynamics simulations and fractional structural fluctuation analysis. *PLoS ONE* **2015**, *10*, e0119264. [CrossRef]
46. Bera, I.; Klauda, J.B. Molecular Simulations of Mixed Lipid Bilayers with Sphingomyelin, Glycerophospholipids, and Cholesterol. *J. Phys. Chem. B* **2017**, *121*, 5197–5208. [CrossRef] [PubMed]
47. Björkbom, A.; Róg, T.; Kaszuba, K.; Kurita, M.; Yamaguchi, S.; Lönnfors, M.; Nyholm, T.K.M.; Vattulainen, I.; Katsumura, S.; Slotte, J.P. Effect of sphingomyelin headgroup size on molecular properties and interactions with cholesterol. *Biophys. J.* **2010**, *99*, 3300–3308. [CrossRef] [PubMed]
48. Hakizimana, P.; Masureel, M.; Gbaguidi, B.; Ruysschaert, J.M.; Govaerts, C. Interactions between phosphatidylethanolamine headgroup and LmrP, a multidrug transporter: A conserved mechanism for proton gradient sensing? *J. Biol. Chem.* **2008**, *283*, 9369–9376. [CrossRef] [PubMed]
49. Blom, T.S.; Koivusalo, M.; Kuismanen, E.; Kostiainen, R.; Somerharju, P.; Ikonen, E. Mass spectrometric analysis reveals an increase in plasma membrane polyunsaturated phospholipid species upon cellular cholesterol loading. *Biochemistry* **2001**, *40*, 14635–14644. [CrossRef]
50. Hauser, H.; Pascher, I.; Pearson, R.H.; Sundell, S. Preferred conformation and molecular packing of phosphatidylethanolamine and phosphatidylcholine. *BBA Rev. Biomembr.* **1981**, *650*, 21–51. [CrossRef]
51. Zhang, Y.P.; Lewis, R.N.A.H.; McElhaney, R.N. Calorimetric and spectroscopic studies of the thermotropic phase behavior of the n-saturated 1,2-diacylphosphatidylglycerols. *Biophys. J.* **1997**, *72*, 779–793. [CrossRef]
52. Lairion, F.; Disalvo, E.A. Effect of phloretin on the dipole potential of phosphatidylcholine, phosphatidylethanolamine, and phosphatidylglycerol monolayers. *Langmuir* **2004**, *20*, 9151–9155. [CrossRef] [PubMed]

MDPI

Article

Phosphatidylserine Exposed Lipid Bilayer Models for Understanding Cancer Cell Selectivity of Natural Compounds: A Molecular Dynamics Simulation Study

Navaneethan Radhakrishnan [1], Sunil C. Kaul [2], Renu Wadhwa [2,*] and Durai Sundar [1,3,*]

[1] DAILAB, Department of Biochemical Engineering and Biotechnology, Indian Institute of Technology (IIT) Delhi, New Delhi 110016, India; navaneethan@dbeb.iitd.ac.in
[2] AIST-INDIA DAILAB, DBT-AIST International Center for Translational and Environmental Research (DAICENTER), National Institute of Advanced Industrial Science and Technology (AIST), Tsukuba 305-8565, Japan; s-kaul@aist.go.jp
[3] School of Artificial Intelligence, Indian Institute of Technology (IIT) Delhi, New Delhi 110016, India
* Correspondence: renu-wadhwa@aist.go.jp (R.W); sundar@dbeb.iitd.ac.in (D.S.)

Abstract: Development of drugs that are selectively toxic to cancer cells and safe to normal cells is crucial in cancer treatment. Evaluation of membrane permeability is a key metric for successful drug development. In this study, we have used in silico molecular models of lipid bilayers to explore the effect of phosphatidylserine (PS) exposure in cancer cells on membrane permeation of natural compounds Withaferin A (Wi-A), Withanone (Wi-N), Caffeic Acid Phenethyl Ester (CAPE) and Artepillin C (ARC). Molecular dynamics simulations were performed to compute permeability coefficients. The results indicated that the exposure of PS in cancer cell membranes facilitated the permeation of Wi-A, Wi-N and CAPE through a cancer cell membrane when compared to a normal cell membrane. In the case of ARC, PS exposure did not have a notable influence on its permeability coefficient. The presented data demonstrated the potential of PS exposure-based models for studying cancer cell selectivity of drugs.

Keywords: phosphatidylserine; cancer cells; MD simulation; membrane permeability; withaferin A; withanone; CAPE; artepillin C

Citation: Radhakrishnan, N.; Kaul, S.C.; Wadhwa, R.; Sundar, D. Phosphatidylserine Exposed Lipid Bilayer Models for Understanding Cancer Cell Selectivity of Natural Compounds: A Molecular Dynamics Simulation Study. *Membranes* **2022**, *12*, 64. https://doi.org/10.3390/membranes12010064

Academic Editors: Jordi Marti and Carles Calero

Received: 16 November 2021
Accepted: 27 December 2021
Published: 1 January 2022

1. Introduction

Cancer cells are highly complex and extremely difficult to treat due to the involvement of multifactorial signaling pathways involved in the process of carcinogenesis. These involve genetic and somatic aberrations that are highly heterogeneous, often leading to intra-tumor heterogeneity. Molecular targeted therapies use small molecules to target one or more proteins involved in the regulation of cell cycle progression, cancer signaling pathways, angiogenesis, growth arrest and/or apoptosis in cancer cells. By activating or inhibiting the target proteins, such small molecules/drugs block cancer cell proliferation and tumor growth. Many of these small molecules can also affect cell migration and invasion capability, and therefore are used for blocking cancer metastasis. However, most cancer drugs have very low therapeutic indices and are used near their maximum-tolerated doses to attain clinically meaningful results [1]. Adverse side effects of chemotherapeutic drugs pose a major concern in cancer chemotherapy [2]. These can be attributed to their effect on normal cells, due to low or lack of selectivity to cancer cells [1]. Hence, there is a need to develop innovative strategies to predict and measure the cancer cell selective effects of chemotherapeutic drugs.

Interactions of drugs with the cell membrane are critical, as the drugs must cross the lipid bilayer of the cells to reach their intra-cellular targets. Malignant transformation has been shown to involve alterations in the lipid profile of cell membranes [3,4]. Changes in the levels of different types of lipid molecules in cell membranes have been reported in various types of cancers [5–10]. A common hallmark observed across several types of cancers appears to be the loss of asymmetry in the distribution of different types of lipid molecules between the two leaflets of the cell membrane [11–13]. The basic structure of the cell membrane consists of a lipid bilayer that is mainly composed of phospholipids. Phosphatidylcholines (PC), phosphatidylethanolamines (PE), phosphatidylserines (PS), phosphatidylinositols (PI) and sphingomyelins (SM) form the majority of the phospholipids in the cell membrane [14]. A non-uniform distribution of these lipids across the two leaflets of the bilayer is a characteristic feature of a normal eukaryotic cell membrane [14,15]. PS and PE are usually present in the inner leaflet, while the outer leaflet is mostly composed of PC and SM [14,15]. This asymmetric distribution is actively maintained by the ATP-dependent enzymes, flippases and floppases [16]. PS and PE are transported from outer leaflet to inner leaflet by flippases, while PC and SM are transported in the opposite direction by floppases [16]. An absence of such asymmetric distribution of phospholipids has been reported in cancers. Exposure of phosphatidylserine and phosphatidylethanolamine molecules on the outer leaflet has been reported in cancer cells [11–13]. The altered distribution of lipids in cancer cell membranes makes their structural and biophysical properties different to that of normal cells. Such changes could modulate drug penetration, thereby influencing drug activity [4].

Molecular dynamics simulations using atomistic models of lipid bilayers made up of PC, SM, PS, PE and/or cholesterol have been used by different studies to investigate the effects of asymmetric lipid distribution on membrane properties [17–20]. Studies on the permeability of lipid bilayers to small molecules using molecular dynamics simulations have demonstrated the potential of in silico membrane models in analyzing the membrane permeation of drugs [21–24].

In the present study, molecular dynamics simulations involving in silico atomistic models of lipid bilayers have been used to explicitly explore the effect of PS exposure in cancer cells on membrane permeation of natural compounds reported to have anti-cancer properties. Atomistic lipid bilayer models of cancer and normal cell membranes used in this study are based on the relative distribution of two kinds of phospholipids: the most common phospholipid, phosphatidylcholine (PC), and the anionic phospholipid which is exposed exclusively in cancer cells, phosphatidylserine (PS). Both the bilayer models were built using PC and PS in the ratio 2:1 that roughly equates to their reported proportion [14]. The normal cell membrane model was built to have all the 1-palmitoyl-2-oleoyl-sn-glycero-3-phosphoserine (POPS) molecules in the inner leaflet, whereas the cancer cell membrane model was built with POPS molecules in both leaflets of the membrane. For simplicity, other types of phospholipids and sterols were not included in the model.

The natural anti-cancer compounds chosen for this study included Withaferin A (Wi-A), Withanone (Wi-N), Caffeic Acid Phenethyl Ester (CAPE) and Artepillin C (ARC) (Figure 1). These are bioactive molecules from natural sources known for their therapeutic potential with lesser side effects compared to synthetic drugs. Wi-A and Wi-N are secondary metabolites from Ashwagandha that have been extensively studied for their anti-cancer activities [25–28]. Molecular modeling approaches accompanied with in vitro assays have revealed the multi-modal anti-cancer activities of Wi-A and Wi-N [25–27,29]. Although Wi-A and Wi-N are closely related structural analogs, several previous studies have reported their different levels of cytotoxicity in cancer and normal cells [30,31]. Whereas Wi-A showed stronger cytotoxicity to both cancer and normal cells, Wi-N exhibited milder toxicity to cancer cells and was safer for normal cells [31]. CAPE, a bioactive compound isolated from New Zealand honeybee propolis, was earlier studied for its anti-cancer activities and reported to cause the death of cancer cells selectively [26]. Another propolis-derived bioactive compound, ARC—particularly enriched in Brazilian honeybee propolis—has also been reported for its anti-cancer activity [32–34].

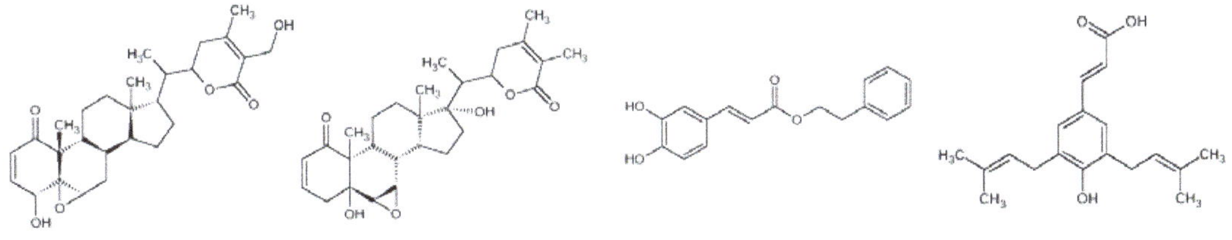

Withaferin A (Wi-A) **Withanone (Wi-N)** **Caffeic acid Phenethyl Ester (CAPE)** **Artepillin C (ARC)**

Figure 1. The 2D pictorial representation of natural compounds: Withaferin A (Wi-A), Withanone (Wi-N), Caffeic Acid Phenethyl Ester (CAPE) and Artepillin C (ARC).

In this study, we examined the permeation of Wi-A, Wi-N, CAPE and ARC using the models of cancer and normal cell membranes, with particular reference to the effect of PS exposure in cancer cells. Computational free energy profiles of drugs across the membrane provide a good understanding of their permeation mechanisms [35,36]. Potential of mean force (PMF) values derived through molecular dynamics simulations demonstrate the free energy landscape of permeation of drugs through the membrane. Classical molecular dynamics simulations are not suitable for efficiently sampling the configuration space for PMF calculations, as the systems might remain trapped in local free energy minima leaving out the events involving large energy barriers. Umbrella sampling methods coupled with molecular dynamics simulations have been proven to be well suited for generating PMF profiles [37]. Steered molecular dynamics (SMD) simulations, in which an external force is applied to an atom of the drug molecule to pull the molecule through the membrane, can be used for generating initial configurations for umbrella sampling. SMD simulations and umbrella sampling methods have been successfully used by different studies for deriving the PMF profiles of lipid membrane traversal of the small molecules [22,23,38–40]. Here, SMD and umbrella sampling simulations were employed to derive the effect of PS exposure in the cancer membrane on the permeation of chosen molecules.

2. Materials and Methods

2.1. Generation of Lipid Bilayer Systems

Two atomistic lipid bilayer systems were generated using CHARMMGUI [41]. Each of the generated systems contained 48 molecules of POPC (1-palmitoyl-2-oleoyl-sn-glycero-3-phosphocholine), 24 molecules of POPS (1-palmitoyl-2-oleoyl-sn-glycero-3-phosphoserine) and ~5000 water molecules. Table 1 shows the distribution of lipids in the leaflets of the membrane. NaCl was used to neutralize the system and an additional 0.15 M NaCl was added to maintain the salt concentration.

Table 1. The distribution of POPC and POPS molecules in cancer and normal membrane models.

Membrane	No. of POPC Molecules		No. of POPS Molecules	
	Outer Leaflet	Inner Leaflet	Outer Leaflet	Inner Leaflet
Normal	36	12	0	24
Cancer	24	24	12	12

2.2. Equilibration of Lipid Bilayer Systems

Classical molecular dynamics simulations were used for equilibrating the generated cancer and normal membrane systems. CHARMM36 force field parameters were used for all molecules in the systems [42]. TIP3P water model was used. All molecular dynamics simulations were performed in GROMACS 2020 using leap-frog integrator and Verlet cutoff scheme [43]. PME method with a cut-off distance of 1.2 nm was used for calculating coulomb interactions [44]. The systems were periodic in all directions.

Energy minimization was performed for 5000 steps using steepest descent algorithm. During energy minimization, position restraints were applied to phosphate atoms with a force constant of 1000 kJ/mol/nm^2. After energy minimization, simulation was conducted in canonical (NVT) ensemble for 100 ps with a timestep of 1 fs, during which a temperature of 310 K was reached using a velocity rescale thermostat with a time constant of 1 ps. During NVT equilibration, position restraints were applied to phosphate atoms with a force constant of 1000 kJ/mol/nm^2.

After NVT equilibration, the systems were equilibrated in isothermal–isobaric (NPT) ensemble in six stages: from NPT-1 to NPT-6 for 502.5 ns. Position restraints applied to the phosphate atoms were gradually relieved during NPT-1 to NPT-4, with no restraints applied in NPT-5 and NPT-6. The force constants used for position restraints on phosphate atoms were 600 kJ/mol/nm^2, 400 kJ/mol/nm^2, 200 kJ/mol/nm^2 and 50 kJ/mol/nm^2 from NPT-1 to NPT-4, respectively. Timestep was set to 1 fs in NPT-1 and NPT-2, while from NPT-3 onwards it was set to 2 fs. From NPT-1 to NPT-5, temperature was maintained at 310 K using a velocity rescale thermostat with a time constant of 1 ps, while pressure was maintained at 1 atm by semi-isotropic coupling using Berendsen barostat with a time constant of 5 ps. During NPT-6, Nose–Hoover thermostat with a time constant of 1 ps and Parrinello–Rahman barostat with a time constant of 5 ps were used to maintain the temperature and the pressure at 310 K and 1 atm, respectively. NPT-1 to NPT-6 were run for a duration of 250 ps, 250 ps, 1 ns, 1 ns, 300 ns and 200 ns, respectively. The production simulation during which membrane properties were assessed was run in NPT ensemble for 200 ns. No position restraints were used during production and the time step was 2 fs. Thermostat and barostat used were the same as for the NPT-6 equilibration. Snapshots were saved every 10 ps for analysis.

2.3. *Calculation of Membrane Properties*

Area per lipid and order parameters of lipid tails were calculated using MEMBPLUGIN through VMD [45,46]. Area per lipid was computed by selecting a triad of atoms for phospholipids, projecting their x and y coordinates into a plane, dividing them into polygons using a Voronoi diagram [47] and then calculating the area of the polygons. The formula used for order parameter (S_{CD}) calculation is as follows [48]:

$$S_{CD} = -\frac{1}{2} 3\ \cos^2\theta\ -\ 1, \tag{1}$$

where θ is the instantaneous angle between the C–H bond and the bilayer normal.

2.4. *Steered Molecular Dynamics Simulations and Umbrella Sampling*

To compute PMF and diffusivities of the molecules, configurations and energies of drug molecules had to be sampled along the bilayer normal. Steered molecular dynamics simulations were performed to generate the initial configurations for sampling. The force field parameters of small molecules Wi-A, Wi-N, CAPE and ARC were generated using CGENFF server [42].

The small molecule was inserted into the water at around 2 nm from the lipid head groups. After insertion of the small molecule into the system, energy minimization was performed using steepest descent algorithm for 5000 steps, followed by NVT and NPT equilibration for 50 ps and 100 ps, respectively. Position restraints were applied to the small molecule in Z-direction during NVT and NPT equilibration. Steered MD simulations were performed in NVT ensemble using the pull-code of GROMACS. Nose–Hoover thermostat with a time constant of 1 picosecond and Parrinello–Rahman barostat with a time constant of 5 ps were used to maintain the temperature and pressure at 310 K and 1 atm, respectively, during the pulling simulation. The small molecule was pulled along Z-direction from water, towards the hydrophobic core of the membrane, using an umbrella potential with a force constant of 100 kJ/mol/nm^2 and a pulling rate of 0.00025 nm/ps. After pulling, configurations were sampled every 0.2 nm along the reaction coordinate: the Z-component of distance (z). A total of 22 windows spaced 0.2 nm apart were taken for umbrella sampling (US).

For each window, US simulations were performed in NPT ensemble, with a harmonic force constant of 100 kJ/mol/nm^2. Nosé–Hoover thermostat was used for maintaining temperature at 310 K, with a semi-isotropic Parrinello–Rahman barostat used to maintain pressure at 1 atm. Each window was run for 40 ns, and data from the last 25 ns were used for analysis. Positions and forces along the reaction coordinate were saved every 2 fs. For the cancer cell membrane, US simulations were performed along only one leaflet, as the membrane has a symmetric distribution of lipids between the two leaflets. For the normal cell membrane, US sampling simulations were performed along both leaflets independently.

2.5. Calculation of PMF and Permeability Coefficients

PMF profiles were estimated using Weighted Histogram Analysis Method (WHAM) implementation in GROMACS [27]. For the normal membrane model, PMF values were calculated for both leaflets separately, starting from bulk water to membrane core, and the PMF curves were smoothened by using moving average spanning 0.5 Å. In the case of the cancer membrane, PMF values of one leaflet were duplicated for the other, with moving average smoothing conducted the same way as for the normal membrane. The convergence of PMF profiles was confirmed by checking PMF values at different intervals of umbrella sampling simulations (Supplementary Materials: Figure S1). The permeability coefficient (P) was calculated using the Inhomogeneous Solubility Diffusion model [35,36]. Accordingly, P was derived from effective resistivity (R_{eff}) using the relation,

$$P = \frac{1}{R_{eff}}, \tag{2}$$

and R_{eff} was calculated using,

$$R_{eff} = \int_{z_1}^{z_2} R(z)\, dz, \tag{3}$$

where z is a collective variable describing the relative position of the solute along the reaction coordinate and $R(z)$ is the resistivity at z. $R(z)$ was calculated from free energies ($\Delta G(z)$) and diffusion coefficients ($D(z)$) along z using the equation,

$$R(z) = \frac{e^{\beta \Delta G(z)}}{D(z)} \tag{4}$$

where β is the inverse of the Boltzmann constant times the temperature.

Diffusion coefficients were calculated using the method proposed in 2005 by Hummer [49], from the autocorrelation function ($C_{zz}(t)$) of z and variance ($var(z)$) of z as follows.

$$D(z) = \frac{var(z)^2}{\int_0^{\infty} C_{zz}(t)\, dt} \tag{5}$$

$C_{zz}(t)$ was calculated using the "analyze" module of GROMACS.

3. Results and Discussion

The phospholipid PS is usually present in the inner leaflet of the cell membranes of normal cells [14,15]. Exposure of PS on the cell membrane of cancer cells has been reported in a wide range of cancers [11,12,50,51]. Here, atomistic models of cancer and normal membranes based on a relative PS distribution between the outer and the inner leaflets of the membrane were built to study the permeation of some natural compounds (Wi-A, Wi-N, CAPE and ARC) shown to possess anti-cancer activity. The built membrane models consisted of only two types of lipid molecules, POPC and POPS, in contrast to the diversity of lipid molecules in the cell membrane. Usage of such simplistic models reduced the simulation time needed for proper equilibration and production, compared to complex models. The non-inclusion of other types of lipid molecules of the biological cell membrane in our models may have some limitations. However, the models fitted the purpose of simulation, which was to study the influence of PS exposure on the membrane outer surface on drug permeation. Simplistic models composed of one/two kinds of lipid molecules have been conservatively used in the past for drug–membrane interaction studies and have been shown to produce results comparable with experimental studies [52]. The lipid bilayer systems were equilibrated for roughly 500 ns and production simulations were run for 200 ns prior to introducing drug molecules into the systems. Figure 2 shows the cancer and normal lipid bilayer systems after equilibration. Different properties of the lipid membranes were computed during the production runs, to check if the membranes were adequately equilibrated.

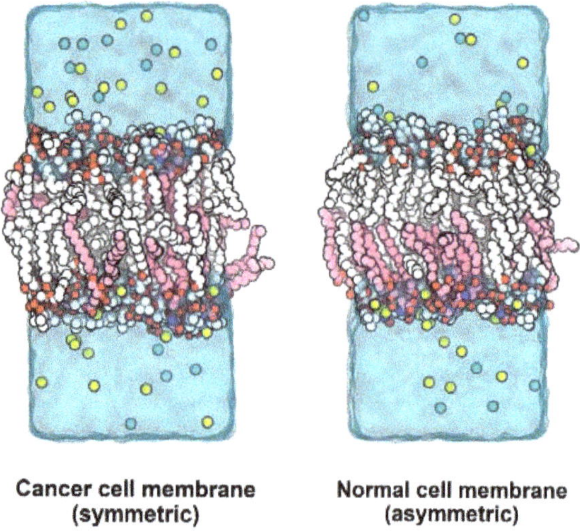

Figure 2. Equilibrated bilayer systems: Carbon atoms of POPC are shown in white, carbon atoms of POPS in magenta, sodium atoms in yellow and chloride atoms in cyan, with hydrogen atoms not shown. The outer (extracellular) leaflet is at the top and the inner (cytoplasmic) leaflet at the bottom. The normal cell membrane model contains POPS only in the inner leaflet, whereas the cancer cell membrane model contains POPS in both leaflets.

3.1. Structural Properties of Lipid Bilayer Systems

Densities of different system components across the bilayer normalwere measured to verify the localization of different molecules in the systems. The densities of different components of the systems along the direction of normal to the membranes (Z-direction) are shown in Figure 3. On the horizontal axis, "0" indicates the center of the hydrophobic core of the membrane, while negative values point towards the outer leaflet and positive values point towards the inner leaflet. The peaks at "−2" and "+2" indicate the polar heads of lipid molecules. Figure 3d shows the asymmetric distribution of POPC and POPS molecules in the normal membrane in contrast to the symmetric distribution in the cancer membrane (Figure 3b). The area per lipid denotes the average cross-sectional area of lipid molecules in the membrane leaflets. Time evolution of the average area per lipid is a good criterion to check if the system has reached a steady state [53]. The time-dependent area per lipid of the leaflets of cancer and normal membranes is shown in Figure 4. As there are no noticeable drifts in area per lipid over the course of simulation, it was assumed that the bilayers were stable. Lipid-tail-order parameters are a measure of alignment of lipid tails with respect to the bilayer normal. The use of order parameter profiles is a widely used method for characterizing the structure of hydrocarbon region in lipid bilayers. Order parameters were calculated for sn1 and sn2 tails of POPC and POPS molecules in the membrane using Equation (1) described in the Section 2. Figure 5 shows the order parameters of lipid molecules in cancer and normal membranes. Order parameter plots of sn2 tails are characteristically different from that of sn1 tails because of the low order parameters around carbons 9 and 10 of sn2 tails, attributed to the presence of a double bond in the oleoyl group. The values were close to the previously reported values in the literature [20,54]. In view of these data, the membrane models were considered equilibrated enough and used for drug permeation studies.

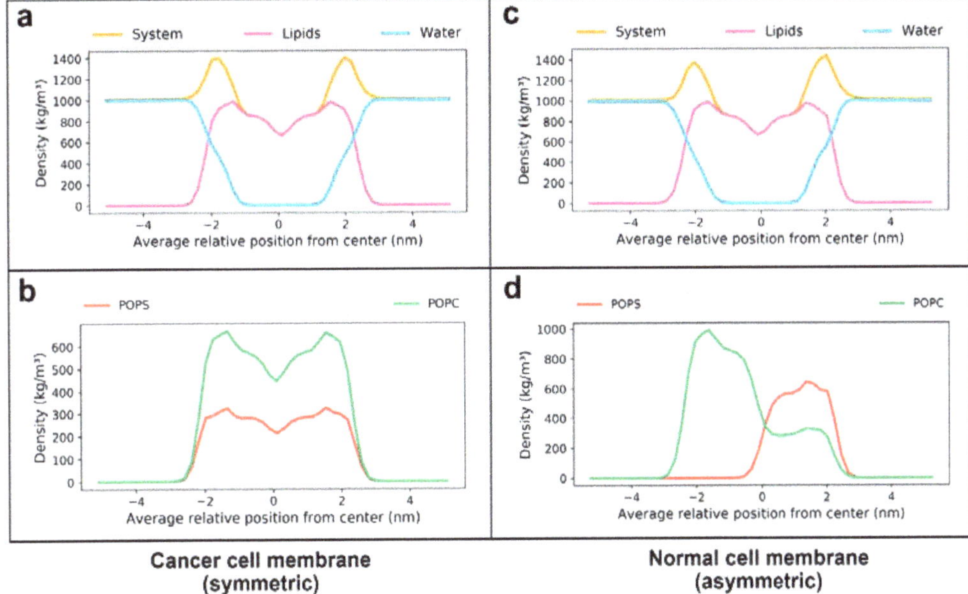

Figure 3. Densities of components of the equilibrated membrane systems: (**a**,**b**) cancer cell membrane and (**c**,**d**) normal cell membrane. The densities of the whole system (yellow), lipids (magenta) water (cyan), POPC (green) and POPS (orange) along the reaction coordinate "z" are shown. On the horizontal axis, the value 0 indicates the center hydrophobic core of the membrane, negative values indicate the outer (extracellular) leaflet and positive values indicate the inner (cytoplasmic) leaflet.

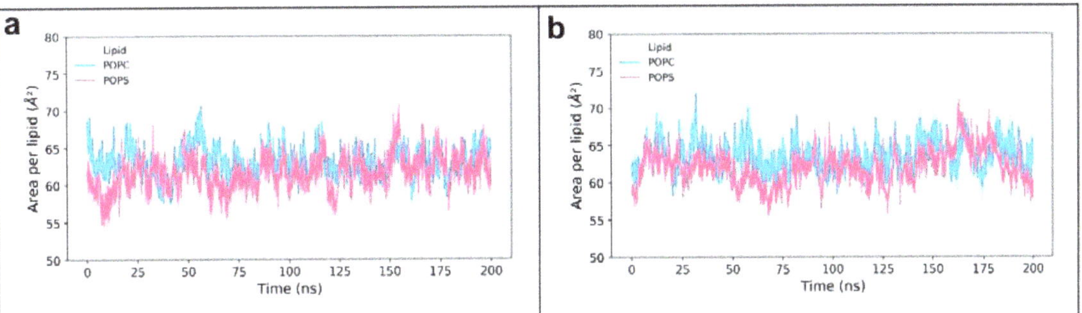

Figure 4. Time-dependent area per lipid values of the membrane models: (**a**) normal membrane and (**b**) cancer membrane. POPC (cyan) and POPS (magenta).

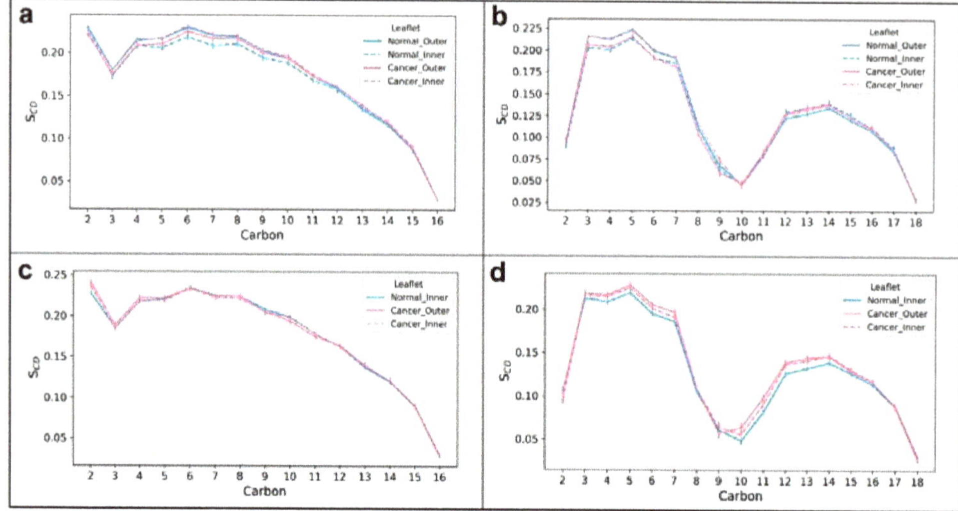

Figure 5. Order parameters of hydrophobic tails of lipid molecules in equilibrated systems: (**a**) POPC sn1, (**b**) POPC sn2, (**c**) POPS sn1 and (**d**) POPS sn2. The outer and inner leaflets of a normal membrane are shown in solid cyan and dashed cyan lines, respectively. The outer and inner leaflets of a cancer membrane are shown in solid magenta and dashed magenta lines, respectively.

3.2. PMF Profiles of the Natural Compounds

SMD and umbrella sampling simulations were used to generate the PMF profiles of the selected small molecules Wi-A, Wi-N, CAPE and ARC. The PMF profiles show how the free energy changes as a function of position of the molecule through the membrane along the direction of the bilayer normal. Here, the Z-axis is the reaction coordinate along which PMF is defined. Figure 6 shows the computed PMF curves of the small molecules. The raw PMF values, with their standard deviation before moving average smoothing, are shown in Supplementary Materials: Figure S2. The magenta solid curves in Figure 6 indicate PMF in the cancer membrane and the cyan-dashed curves indicate PMF in the normal membrane. The highest point in PMF (ΔG_{max}) indicates the energy barrier that has to be overcome for the passage of small molecules through membranes. Wi-A, Wi-N and CAPE had ΔG_{max} in the center of the hydrophobic core of the membranes (Figure 6). Hence, the hydrophobic core of the membranes formed the principal hindrance for the passage of these three molecules. The ΔG_{max} values of Wi-A, Wi-N and CAPE were in the order CAPE < Wi-A < Wi-N (Figure 6), predicting that CAPE is able to traverse through the membrane easily compared to Wi-A and Wi-N. Furthermore, Wi-A showed easier traversal compared to Wi-N. Previous studies that have used atomistic models of "POPC + cholesterol" bilayers to study the permeability of Wi-A and Wi-N have reported easier permeation of Wi-A through the cell membrane compared to Wi-N [23]. Our results based on POPC + POPS models were in line with the earlier report [23]. The ΔG_{max} of Wi-A, Wi-N and CAPE was higher in the normal membrane compared to that in the cancer membrane (Figure 6). Hence, PS exposure in the outer leaflet may have facilitated easier permeation of Wi-A, Wi-N and CAPE through the membrane by reducing the energy barrier in the hydrophobic core. In the case of ARC, ΔG_{max} occurred at the polar head group regions of lipid molecules (Figure 6). Unlike Wi-A, Wi-N and CAPE, the lipid head groups caused the principal hindrance for ARC permeation. The magnitude of ΔG_{max} of ARC was much lower than that of the other three drug molecules in comparison (Figure 6), predicting that ARC may traverse through the membrane easier than Wi-A, Wi-N and CAPE.

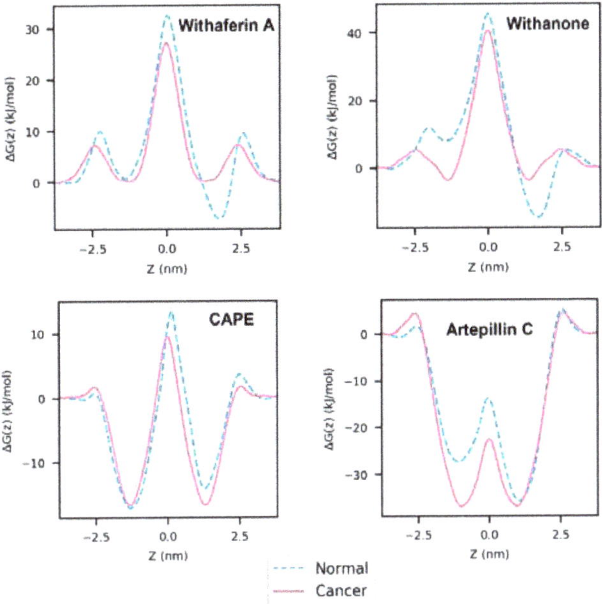

Figure 6. PMF profiles of Withaferin A (Wi-A), Withanone (Wi-N), Caffeic Acid Phenethyl Ester (CAPE) and Artepillin C (ARC). The solid magenta line indicates the PMF profile in the cancer cell membrane model and the dashed cyan line indicates the PMF profile in the normal cell membrane model. On the horizontal axis, the value 0 indicates the center of the hydrophobic core of the membrane, negative values indicate the outer (extracellular) leaflet and positive values indicate the inner (cytoplasmic) leaflet.

As Wi-A and Wi-N are structural analogs, they have comparable PMF landscapes in cancer and normal membrane models. They show interesting differences in PMF between $z \approx -2.5$ and $z \approx -1.5$ (Figure 6). It is the region along the reaction coordinate where the centers of masses of drug molecules pass from bulk water to membrane. At this cell–water interface, the difference between PMF values of Wi-N in cancer and normal membranes was prominent, while Wi-A had proximate values in cancer and normal membranes (Figure 6). Differences in PMF values in cancer and normal membranes for Wi-N indicated differential binding of Wi-N to the outer leaflet of cancer and normal membranes. Lower values of PMF in the cancer cell membrane implied that more Wi-N molecules could bind to the cancer membrane compared to the normal cell membrane. Hence, it can be inferred that the exposure of PS could potentially facilitate selective binding of Wi-N to cancer cells. The propolis compounds CAPE and ARC had dissimilar structures, as did their PMF landscapes. The hydrophobic cores of both cancer and normal membranes had a higher affinity for ARC compared to CAPE, indicated by the lowest values in its PMF profile (Figure 6). This is due to the presence of the hydrophobic diprenyl groups in its structure.

The simulation trajectories were visualized, with reference to the PMF profiles, to gain insights into the structural aspects of permeation. Snapshots of the last frame from different umbrella sampling windows shown in Figures S3–S6 in Supplementary Materials indicate the converged orientations of Wi-A, Wi-N, CAPE and ARC, respectively, along the reaction coordinate at different regions of the membrane during permeation. Figures S3–S6: subfigures B and E in Supplementary Materials show the preferred orientation of molecules inside the outer and inner leaflets of the normal membrane, respectively, associated with the troughs in the PMF curves. Figures S3–S6: subfigure H in Supplementary Materials shows the preferred orientation of molecules inside the cancer membrane. In the preferred orientation, the polar groups of the natural compounds are oriented towards the head groups of lipids, with hydrophobic groups of natural compounds oriented towards the lipid hydrocarbon region. The passage of natural compounds through the membrane core involves a flip in the orientation of natural compounds from their polar groups oriented towards the head groups of outer leaflet lipids to that of lower leaflet lipids. In the case of Wi-A, Wi-N and CAPE, the primary barrier to permeation—related to ΔG_{max}—is associated with the traversal of their polar groups through the hydrophobic core of the membrane. In light of this fact, the hydrophobic cores of cancer and normal membranes were inspected. It was found that the lipid head polar groups of the two leaflets of cancer membrane were slightly closer to the hydrophobic core in the cancer membrane compared to that in the normal membrane (Supplementary Materials: Figure S7). This could be the basis for the lower ΔG_{max} of Wi-A, Wi-N and CAPE in the cancer membrane. In the case of ARC, the primary barrier to permeation is associated with the passage of its hydrophobic diprenyl groups through the hydrophilic head region of lipids.

3.3. Resistivity Profiles of the Natural Compounds

Resistivity profiles of the drug molecules were calculated from PMF (Figure 6) and diffusivity profiles (Supplementary Materials: Figure S8) using Equation 4. The calculated resistivity profiles are shown in Figure 7. The curves of Wi-A, Wi-N and CAPE were similar in shape, showing the highest resistance for permeation in the hydrophobic core of the membrane (Figure 7). The resistance offered by the hydrophobic core of the cancer membrane was less compared to that of the normal membrane for Wi-A, Wi-N and CAPE, indicating higher permeation rates in the cancer membrane. ARC had resistivity peaks at polar regions of the membrane (Figure 7). ARC had higher resistance in the inner leaflet of the normal membrane, where POPS molecules were high in number compared to that in the cancer membrane. It implied that the presence of POPS head groups increased the resistance for the passage of ARC through polar regions of the membrane.

3.4. Permeability Coefficients of the Natural Compounds

Permeability coefficients (P) of the drug molecules were calculated using the Inhomogeneous Solubility Diffusion model (ISD) [35,36]. Table 2 shows the permeability coefficients of Wi-A, Wi-N, CAPE and ARC in cancer and normal membranes. The permeability coefficients were compared with the computed octanol–water partition coefficients (XLogP3-AA) of the drug molecules obtained from the Pubchem database [55,56]. The octanol–water partition coefficient is a measure of lipophilicity of the molecule. The calculated permeability coefficients were directly correlated with the octanol–water partition coefficients obeying Meyer–Overton's rule [57], confirming the validity of the methods used in our study (Table 2). The calculated permeability coefficients were in the order Wi-N < Wi-A < CAPE < ARC (Table 2). The lower permeability coefficients of Wi-A and Wi-N were anticipated due to their bigger size and the presence of polar groups all over their structures. The lower permeability of Wi-N compared to that of its analog Wi-A is in line with the higher IC50 values of Wi-N compared to that of Wi-A, as reported in previous studies (IC50 for Wi-A and Wi-N for most cancer cells is 0.3–0.5 μM and >10–20 μM, respectively) [26,29,30]. The higher permeability coefficients of ARC and CAPE is due to the presence of hydrophobic diprenyl group and phenethyl group in their structures, respectively. The higher lipophilicity (XLogP3-AA in Table 2) of ARC and CAPE also implies that they undergo higher membrane retention compared to Wi-A and Wi-N. The relatively extreme negative PMF values of ARC within the membrane core surrounded by PMF peaks at lipid polar regions (Figure 6) implies that more ARC molecules will become trapped inside the membrane core. This might be an underlying cause of the low efficacy of ARC (IC50 = 275 μM) [58].

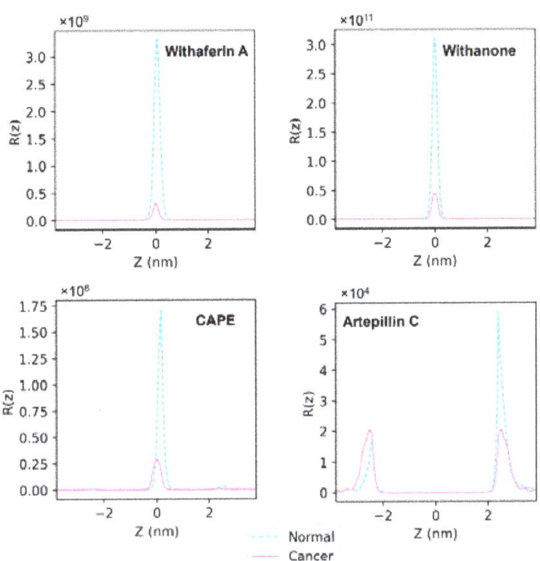

Figure 7. Resistivity profiles of Withaferin A (Wi-A), Withanone (Wi-N), Caffeic Acid Phenethyl Ester (CAPE) and Artepillin C (ARC). The solid magenta line indicates the resistivity profile in the cancer cell membrane model and the dashed cyan line indicates the resistivity profile in the normal cell membrane model. On the horizontal axis, the value 0 indicates the center of the hydrophobic core of the membrane, negative values indicate the outer (extracellular) leaflet and positive values indicate the inner (cytoplasmic) leaflet.

Table 2. Calculated permeability coefficients of the drug molecules. Note: XLOGP3-AA indicates the computed octanol–water partition coefficients retrieved from the Pubchem database.

	Cancer Cell Membrane		Normal Cell Membrane		XLOGP3-AA
	P (cm/s)	*log P*	*P* (cm/s)	*log P*	
Withanone (Wi-N)	7.64×10^{-6}	−5.12	1.33×10^{-6}	−5.88	3.1
Withaferin A (Wi-A)	1.16×10^{-3}	−2.94	1.06×10^{-4}	−3.98	3.8
Caffeic Acid Phenethyl Ester (CAPE)	8.37×10^{-1}	−0.08	2.31×10^{-1}	−0.64	4.2
Artepillin C (ARC)	4.67	0.67	4.14	0.62	5.4

Wi-N, Wi-A and CAPE had notable differences in their permeability coefficients for the cancer and the normal cell membrane models, whereas ARC had closer values (Table 2). This implied that the differential distribution of PS in the membrane models does not have a notable influence on the permeation rate of ARC compared to that of Wi-A, Wi-N and CAPE. The permeability coefficients of Wi-N, Wi-A and CAPE were higher in cancer membranes compared to normal membranes, implying higher permeations rates due to the presence of PS in both leaflets (Table 2). The structural analogs Wi-N and Wi-A had similar fold differences in permeation rates between cancer and normal membranes (Table 2). As inferred from the PMF curves, the presence of PS in the outer leaflet lowered the free energy in the cell–water interface of the cancer membrane, aiding more molecules to bind to it. Hence, PS exposure may render higher tumor selectivity for Wi-N compared to Wi-A.

3.5. Implications of the Study

With the use of atomistic models and umbrella sampling methods to perform molecular dynamics simulations, the possible role of PS exposure in cancer cells in modulating the permeation of selected anti-cancer compounds have been interpreted. Several carrier systems using nanoparticles, synthetic polymers, liposomes, peptides and antibodies have been reported in the literature for targeted delivery of drugs to cancer cells [59–64]. However, successful clinical translations of such systems are limited. The present study relied on the fact that studying the molecules, which are inherently selective to cancer cells in action, will aid in understanding the molecular mechanisms that confer them selectivity. Though there are numerous factors that govern selectivity of a drug towards cancer cells, we have focused only on the effect of PS exposure on modulating membrane permeation. The exposure of PS caused alterations in the free energy landscapes underlying the traversal of the drug molecules through the membrane, thereby influencing the permeation. The molecular models and methods employed in this study have explained the selectivity of Wi-A, Wi-N and CAPE by differences in PMF landscapes and permeability coefficients between cancer and normal membrane models. In the case of ARC, PS exposure did not seem to have notable effects on the permeation rates. The interesting case is the structural analogs Wi-A and Wi-N, whose differential selectivity was explained by differences in free energy landscapes underlying membrane permeation. The used models demonstrate the possible contribution of PS exposure to drug selectivity. Hence, studying the influence of PS exposure on drug permeation offers, though not complete, a reasonable understanding of tumor selectivity of the drugs.

This study used simplistic molecular models to evaluate the possible contribution of PS exposure to drug selectivity and, hence, the limitations associated with the used materials and methods must be understood. The effect of PS exposure has been explored in the absence of other diverse types of lipid molecules, which might have led to a reduction in accuracy. The combination effect of different types of lipids and their distribution on the permeability of small molecules will differ from the observed sole effect of PS distribution in our binary lipid mixture. The models are not completely reflective of the cellular membrane, however, they aided in exploring one specific aspect that contributes to the differences in permeability between cancer and normal cells. The effect of PS exposure using complex models containing other major lipids will be explored in the future in light of the findings from this study. The proportion of charged species of small molecules in the tumor microenvironment and their interaction with the cell membrane also influence the permeation, and thereby their activity. Here, only the neutral species of the small molecules were studied.

4. Conclusions

Permeability of drugs through the cell membrane is crucial for its bioactivity. In this study, we have built cancer and normal membrane models based on PS distribution between membrane leaflets and assessed the permeation of natural compounds (Wi-A, Wi-N, CAPE and ARC) that have been shown to possess anti-cancer potential. It has been shown that PS exposure may influence drug permeation and thereby the drug activity. The cancer cell selectivity of these compounds was clearly evident. This study highlighted the effect of PS exposure in the cancer cell membrane on selective action of the chosen molecules. Though the simplistic models based on PS exposure might not be sufficient to explain the tumor selectivity of all drugs, this study explored a potential niche area, which may aid in the development or optimization of drugs that are inherently selective to cancer cells in action. With a high number of drugs failing in clinical trial due to lack of selectivity, the molecular differences between cancer and normal cell membranes can be exploited to improve the selectivity of potent drugs. Similar specialized in silico models of cancer cell membranes can be easily developed in the future to assess the tumor selectivity of existing drugs and to screen compound libraries to find cancer selective drugs.

Supplementary Materials: The following supporting information can be downloaded at: https://www.mdpi.com/article/10.3390/membranes12010064/s1, Figure S1: PMF curves of Withaferin A (Wi-A), Withanone (Wi-N), Caffeic Acid Phenethyl Ester (CAPE) and Artepillin C (ARC) calculated after 10, 20, 30 and 40 ns: (A) PMF values in outer (extracellular) leaflet of normal membrane, (B) PMF values in inner (cytoplasmic) leaflet of normal membrane and (C) PMF values in outer leaflet of cancer membrane. Figure S2: PMF values of Withaferin A (Wi-A), Withanone (Wi-N), Caffeic Acid Phenethyl Ester (CAPE) and Artepillin C (ARC) before moving average smoothing: Magenta markers indicates PMF values in the cancer cell membrane model and cyan markers indicate PMF values in the normal cell membrane model. '0' in horizontal axis indicates the center hydrophobic core of the membrane, negative values indicate the outer (extracellular) leaflet, positive values indicate the inner (cytoplasmic) leaflet. Error bars show standard deviation. Figure S3: Snapshots of last frame from umbrella sampling windows showing the converged orientations of Withaferin A (Wi-A) associated with permeation through (A–F) normal and (G–I) cancer membranes: Outer (extracellular) leaflet is shown at the top and inner (cytoplasmic) leaflet is shown at the bottom. (B,E) Orientations corresponding to the lowest points in PMF in the outer and inner leaflets of the normal membrane model. (H) Orientation corresponding to the lowest points in PMF in the leaflets of the cancer membrane model. Orientations are shown for only one leaflet in cancer membrane, as the cancer membrane model is symmetric. Colors are as per CPK rules. Figure S4: Snapshots of last frame from umbrella sampling windows showing the converged orientations of Withanone (Wi-N) associated with permeation through (A–F) normal and (G–I) cancer membranes: Outer (extracellular) leaflet is shown at the top and inner (cytoplasmic) leaflet is shown at the bottom. (B,E) Orientations corresponding to the lowest points in PMF in the outer and inner leaflets of the normal membrane model. (H) Orientation corresponding to the lowest points in PMF in the leaflets of the cancer membrane model. Orientations are shown for only one leaflet in cancer membrane, as the cancer membrane model is symmetric. Colors are as per CPK rules. Figure S5: Snapshots of last frame from umbrella sampling windows showing the converged orientations of Caffeic Acid Phenethyl Ester (CAPE) associated with permeation through (A–F) normal and (G–I) cancer membranes: Outer (extracellular) leaflet is shown at the top and inner (cytoplasmic) leaflet is shown at the bottom. (B,E) Orientations corresponding to the lowest points in PMF in the outer and inner leaflets of the normal membrane model. (H) Orientation corresponding to the lowest points in PMF in the leaflets of the cancer membrane model. Orientations are shown for only one leaflet in cancer membrane, as the cancer membrane model is symmetric. Colors are as per CPK rules. Figure S6: Snapshots of last frame from umbrella sampling windows showing the converged orientations of Artepillin C (ARC) associated with permeation through (A–F) normal and (G–I) cancer membranes: Outer (extracellular) leaflet is shown at the top and inner (cytoplasmic) leaflet is shown at the bottom. (B,E) Orientations corresponding to the lowest points in PMF in the outer and inner leaflets of the normal membrane model. (H) Orientation corresponding to the lowest points in PMF in the leaflets of the cancer membrane model. Orientations are shown for only one leaflet in cancer membrane, as the cancer membrane model is symmetric. Colors are as per CPK rules. Figure S7: Density of polar groups of cancer (magenta) and normal (normal) membranes along the reaction coordinate 'z'. '0' in the horizontal axis indicates the center hydrophobic core of the membrane, negative values indicate the outer (extracellular) leaflet, positive values indicate the inner (cytoplasmic) leaflet. Figure S8: Diffusivities of Withaferin A (Wi-A), Withanone (Wi-N), Caffeic Acid Phenethyl Ester (CAPE) and Artepillin C (ARC): Magenta markers indicates diffusivity values in the cancer cell membrane model and cyan markers indicate diffusivity values in the normal cell membrane model. '0' in horizontal axis indicates the center hydrophobic core of the membrane, negative values indicate the outer (extracellular) leaflet, positive values indicate the inner (cytoplasmic) leaflet. Error bars show standard deviation.

Author Contributions: Conceptualization, N.R., S.C.K., R.W. and D.S.; Methodology, N.R. and D.S.; Writing—original draft, N.R.; Writing—review and editing, N.R., S.C.K., R.W. and R.W. All authors have read and agreed to the published version of the manuscript.

Funding: This study was supported by the funds granted by DBT (India) and AIST (Japan).

Data Availability Statement: The data generated during this study are included in this article.

Acknowledgments: We acknowledge the High-Performance Computing (HPC) Facility of IIT Delhi for computational resources. NR was awarded the BINC Junior Research Fellowship (JRF), Department of Biotechnology (DBT), Govt. of India.

Conflicts of Interest: The authors declare no conflict of interest.

References

1. Chari, R.V. Targeted cancer therapy: Conferring specificity to cytotoxic drugs. *Acc. Chem. Res.* **2008**, *41*, 98–107. [CrossRef] [PubMed]
2. Coates, A.; Abraham, S.; Kaye, S.B.; Sowerbutts, T.; Frewin, C.; Fox, R.; Tattersall, M. On the receiving end—Patient perception of the side-effects of cancer chemotherapy. *Eur. J. Cancer Clin. Oncol.* **1983**, *19*, 203–208. [CrossRef]
3. Escribá, P.V. Membrane-lipid therapy: A new approach in molecular medicine. *Trends Mol. Med.* **2006**, *12*, 34–43. [CrossRef]
4. Alves, A.C.; Ribeiro, D.; Nunes, C.; Reis, S. Biophysics in cancer: The relevance of drug-membrane interaction studies. *Biochim. Biophys. Acta (BBA)-Biomembr.* **2016**, *1858*, 2231–2244. [CrossRef] [PubMed]
5. Azordegan, N.; Fraser, V.; Le, K.; Hillyer, L.M.; Ma, D.W.; Fischer, G.; Moghadasian, M.H. Carcinogenesis alters fatty acid profile in breast tissue. *Mol. Cell. Biochem.* **2013**, *374*, 223–232. [CrossRef]
6. Merchant, T.; Kasimos, J.; De Graaf, P.; Minsky, B.; Gierke, L.; Glonek, T. Phospholipid profiles of human colon cancer using 31 P magnetic resonance spectroscopy. *Int. J. Colorectal Dis.* **1991**, *6*, 121–126. [CrossRef] [PubMed]
7. Selkirk, J.K.; Elwood, J.; Morris, H. Study on the proposed role of phospholipid in tumor cell membrane. *Cancer Res.* **1971**, *31*, 27–31. [PubMed]
8. Hildebrand, J.; Marique, D.; Vanhouche, J. Lipid composition of plasma membranes from human leukemic lymphocytes. *J. Lipid Res.* **1975**, *16*, 195–199. [CrossRef]
9. Liebes, L.F.; Pelle, E.; Zucker-Franklin, D.; Silber, R. Comparison of lipid composition and 1, 6-diphenyl-1, 3, 5-hexatriene fluorescence polarization measurements of hairy cells with monocytes and lymphocytes from normal subjects and patients with chronic lymphocytic leukemia. *Cancer Res.* **1981**, *41*, 4050–4056. [PubMed]
10. Baro, L.; Hermoso, J.; Nunez, M.; Jimenez-Rios, J.; Gil, A. Abnormalities in plasma and red blood cell fatty acid profiles of patients with colorectal cancer. *Br. J. Cancer* **1998**, *77*, 1978–1983. [CrossRef]
11. Ran, S.; Downes, A.; Thorpe, P.E. Increased exposure of anionic phospholipids on the surface of tumor blood vessels. *Cancer Res.* **2002**, *62*, 6132–6140. [PubMed]
12. Ran, S.; Thorpe, P.E. Phosphatidylserine is a marker of tumor vasculature and a potential target for cancer imaging and therapy. *Int. J. Radiat. Oncol. Biol. Phys.* **2002**, *54*, 1479–1484. [CrossRef]
13. Stafford, J.H.; Thorpe, P.E. Increased exposure of phosphatidylethanolamine on the surface of tumor vascular endothelium. *Neoplasia* **2011**, *13*, 299–308. [CrossRef] [PubMed]
14. Verkleij, A.; Zwaal, R.; Roelofsen, B.; Comfurius, P.; Kastelijn, D.; Van Deenen, L. The asymmetric distribution of phospholipids in the human red cell membrane. A combined study using phospholipases and freeze-etch electron microscopy. *Biochim. Biophys. Acta (BBA)-Biomembr.* **1973**, *323*, 178–193. [CrossRef]
15. Yamaji-Hasegawa, A.; Tsujimoto, M. Asymmetric distribution of phospholipids in biomembranes. *Biol. Pharm. Bull.* **2006**, *29*, 1547–1553. [CrossRef] [PubMed]
16. Clark, M.R. Flippin'lipids. *Nat. Immunol.* **2011**, *12*, 373–375. [CrossRef] [PubMed]
17. Gurtovenko, A.A.; Vattulainen, I. Lipid transmembrane asymmetry and intrinsic membrane potential: Two sides of the same coin. *J. Am. Chem. Soc.* **2007**, *129*, 5358–5359. [CrossRef]
18. Gurtovenko, A.A.; Vattulainen, I. Membrane potential and electrostatics of phospholipid bilayers with asymmetric transmembrane distribution of anionic lipids. *J. Phys. Chem. B* **2008**, *112*, 4629–4634. [CrossRef]
19. Falkovich, S.G.; Martinez-Seara, H.; Nesterenko, A.M.; Vattulainen, I.; Gurtovenko, A.A. What Can We Learn about Cholesterol's Transmembrane Distribution Based on Cholesterol-Induced Changes in Membrane Dipole Potential? *J. Phys. Chem. Lett.* **2016**, *7*, 4585–4590. [CrossRef] [PubMed]
20. López Cascales, J.; Otero, T.; Smith, B.D.; Gonzalez, C.; Marquez, M. Model of an asymmetric DPPC/DPPS membrane: Effect of asymmetry on the lipid properties. A molecular dynamics simulation study. *J. Phys. Chem. B* **2006**, *110*, 2358–2363. [CrossRef]
21. Orsi, M.; Sanderson, W.E.; Essex, J.W. Permeability of small molecules through a lipid bilayer: A multiscale simulation study. *J. Phys. Chem. B* **2009**, *113*, 12019–12029. [CrossRef] [PubMed]
22. Thai, N.Q.; Theodorakis, P.E.; Li, M.S. Fast Estimation of the Blood–Brain Barrier Permeability by Pulling a Ligand through a Lipid Membrane. *J. Chem. Inf. Model.* **2020**, *60*, 3057–3067. [CrossRef] [PubMed]
23. Wadhwa, R.; Yadav, N.S.; Katiyar, S.P.; Yaguchi, T.; Lee, C.; Ahn, H.; Yun, C.-O.; Kaul, S.C.; Sundar, D. Molecular dynamics simulations and experimental studies reveal differential permeability of withaferin-A and withanone across the model cell membrane. *Sci. Rep.* **2021**, *11*, 1–15. [CrossRef] [PubMed]
24. Dickson, C.J.; Hornak, V.; Bednarczyk, D.; Duca, J.S. Using Membrane Partitioning Simulations to Predict Permeability of Forty-Nine Drug-Like Molecules. *J. Chem. Inf. Model.* **2019**, *59*, 236–244. [CrossRef]

25. Bhargava, P.; Malik, V.; Liu, Y.; Ryu, J.; Kaul, S.C.; Sundar, D.; Wadhwa, R. Molecular insights into withaferin-A-induced senescence: Bioinformatics and experimental evidence to the role of NFκB and CARF. *J. Gerontol. Ser. A* **2019**, *74*, 183–191. [CrossRef]
26. Sari, A.N.; Bhargava, P.; Dhanjal, J.K.; Putri, J.F.; Radhakrishnan, N.; Shefrin, S.; Ishida, Y.; Terao, K.; Sundar, D.; Kaul, S.C. Combination of withaferin-A and CAPE provides superior anticancer potency: Bioinformatics and experimental evidence to their molecular targets and mechanism of action. *Cancers* **2020**, *12*, 1160. [CrossRef]
27. Garg, S.; Huifu, H.; Kumari, A.; Sundar, D.; Kaul, S.C.; Wadhwa, R. Induction of senescence in cancer cells by a novel combination of Cucurbitacin B and withanone: Molecular mechanism and therapeutic potential. *J. Gerontol. Ser. A* **2020**, *75*, 1031–1041. [CrossRef] [PubMed]
28. Malik, V.; Kumar, V.; Kaul, S.C.; Wadhwa, R.; Sundar, D. Computational Insights into the Potential of Withaferin-A, Withanone and Caffeic Acid Phenethyl Ester for Treatment of Aberrant-EGFR Driven Lung Cancers. *Biomolecules* **2021**, *11*, 160. [CrossRef] [PubMed]
29. Yu, Y.; Katiyar, S.P.; Sundar, D.; Kaul, Z.; Miyako, E.; Zhang, Z.; Kaul, S.C.; Reddel, R.R.; Wadhwa, R. Withaferin-A kills cancer cells with and without telomerase: Chemical, computational and experimental evidences. *Cell Death Dis.* **2017**, *8*, e2755. [CrossRef] [PubMed]
30. Gao, R.; Shah, N.; Lee, J.-S.; Katiyar, S.P.; Li, L.; Oh, E.; Sundar, D.; Yun, C.-O.; Wadhwa, R.; Kaul, S.C. Withanone-rich combination of Ashwagandha withanolides restricts metastasis and angiogenesis through hnRNP-K. *Mol. Cancer Ther.* **2014**, *13*, 2930–2940. [CrossRef] [PubMed]
31. Vaishnavi, K.; Saxena, N.; Shah, N.; Singh, R.; Manjunath, K.; Uthayakumar, M.; Kanaujia, S.P.; Kaul, S.C.; Sekar, K.; Wadhwa, R. Differential Activities of the Two Closely Related Withanolides, Withaferin A and Withanone: Bioinformatics and Experimental Evidences. *PLoS ONE* **2012**, *7*, e44419. [CrossRef]
32. Kimoto, T.; Arai, S.; Kohguchi, M.; Aga, M.; Nomura, Y.; Micallef, M.J.; Kurimoto, M.; Mito, K. Apoptosis and suppression of tumor growth by artepillin C extracted from Brazilian propolis. *Cancer Detect. Prev.* **1998**, *22*, 506–515. [CrossRef]
33. Szliszka, E.; Zydowicz, G.; Mizgala, E.; Krol, W. Artepillin C (3,5-diprenyl-4-hydroxycinnamic acid) sensitizes LNCaP prostate cancer cells to TRAIL-induced apoptosis. *Int. J. Oncol.* **2012**, *41*, 818–828. [CrossRef] [PubMed]
34. Souza, R.P.; Bonfim-Mendonça, P.S.; Damke, G.M.; de-Assis Carvalho, A.R.; Ratti, B.A.; Dembogurski, D.S.; da-Silva, V.R.; Silva, S.O.; Da-Silva, D.B.; Bruschi, M.L. Artepillin C induces selective oxidative stress and inhibits migration and invasion in a comprehensive panel of human cervical cancer cell lines. *Anti-Cancer Agents Med. Chem.* **2018**, *18*, 1750–1760. [CrossRef] [PubMed]
35. Diamond, J.M.; Katz, Y. Interpretation of nonelectrolyte partition coefficients between dimyristoyl lecithin and water. *J. Membr. Biol.* **1974**, *17*, 121–154. [CrossRef]
36. Marrink, S.-J.; Berendsen, H.J. Simulation of water transport through a lipid membrane. *J. Phys. Chem.* **1994**, *98*, 4155–4168. [CrossRef]
37. Torrie, G.M.; Valleau, J.P. Nonphysical sampling distributions in Monte Carlo free-energy estimation: Umbrella sampling. *J. Comput. Phys.* **1977**, *23*, 187–199. [CrossRef]
38. Meng, F.; Xu, W. Drug permeability prediction using PMF method. *J. Mol. Model.* **2013**, *19*, 991–997. [CrossRef] [PubMed]
39. Lee, C.T.; Comer, J.; Herndon, C.; Leung, N.; Pavlova, A.; Swift, R.V.; Tung, C.; Rowley, C.N.; Amaro, R.E.; Chipot, C. Simulation-based approaches for determining membrane permeability of small compounds. *J. Chem. Inf. Model.* **2016**, *56*, 721–733. [CrossRef]
40. DeMarco, K.R.; Bekker, S.; Clancy, C.E.; Noskov, S.Y.; Vorobyov, I. Digging into lipid membrane permeation for cardiac ion channel blocker d-sotalol with all-atom simulations. *Front. Pharmacol.* **2018**, *9*, 26. [CrossRef]
41. Jo, S.; Kim, T.; Iyer, V.G.; Im, W. CHARMM-GUI: A web-based graphical user interface for CHARMM. *J. Comput. Chem.* **2008**, *29*, 1859–1865. [CrossRef]
42. Vanommeslaeghe, K.; Hatcher, E.; Acharya, C.; Kundu, S.; Zhong, S.; Shim, J.; Darian, E.; Guvench, O.; Lopes, P.; Vorobyov, I. CHARMM general force field: A force field for drug-like molecules compatible with the CHARMM all-atom additive biological force fields. *J. Comput. Chem.* **2010**, *31*, 671–690. [CrossRef] [PubMed]
43. Berendsen, H.J.; van der Spoel, D.; van Drunen, R. GROMACS: A message passing parallel molecular dynamics implementation. *Comput. Phys. Commun.* **1995**, *91*, 43–56. [CrossRef]
44. Essmann, U.; Perera, L.; Berkowitz, M.L.; Darden, T.; Lee, H.; Pedersen, L.G. A smooth particle mesh Ewald method. *J. Chem. Phys.* **1995**, *103*, 8577–8593. [CrossRef]
45. Guixà-González, R.; Rodriguez-Espigares, I.; Ramírez-Anguita, J.M.; Carrió-Gaspar, P.; Martinez-Seara, H.; Giorgino, T.; Selent, J. MEMBPLUGIN: Studying membrane complexity in VMD. *Bioinformatics* **2014**, *30*, 1478–1480. [CrossRef]
46. Humphrey, W.; Dalke, A.; Schulten, K. VMD: Visual molecular dynamics. *J. Mol. Gr.* **1996**, *14*, 33–38. [CrossRef]
47. Barber, C.B.; Dobkin, D.P.; Huhdanpaa, H. The quickhull algorithm for convex hulls. *ACM Trans. Math. Softw. (TOMS)* **1996**, *22*, 469–483. [CrossRef]
48. Vermeer, L.S.; De Groot, B.L.; Réat, V.; Milon, A.; Czaplicki, J. Acyl chain order parameter profiles in phospholipid bilayers: Computation from molecular dynamics simulations and comparison with 2 H NMR experiments. *Eur. Biophys. J.* **2007**, *36*, 919–931. [CrossRef]
49. Hummer, G. Position-dependent diffusion coefficients and free energies from Bayesian analysis of equilibrium and replica molecular dynamics simulations. *New J. Phys.* **2005**, *7*. [CrossRef]

50. Utsugi, T.; Schroit, A.J.; Connor, J.; Bucana, C.D.; Fidler, I.J. Elevated expression of phosphatidylserine in the outer membrane leaflet of human tumor cells and recognition by activated human blood monocytes. *Cancer Res.* **1991**, *51*, 3062–3066.
51. Riedl, S.; Rinner, B.; Asslaber, M.; Schaider, H.; Walzer, S.; Novak, A.; Lohner, K.; Zweytick, D. In search of a novel target—Phosphatidylserine exposed by non-apoptotic tumor cells and metastases of malignancies with poor treatment efficacy. *Biochim. Biophys. Acta (BBA)-Biomembr.* **2011**, *1808*, 2638–2645. [CrossRef] [PubMed]
52. Lopes, D.; Jakobtorweihen, S.; Nunes, C.; Sarmento, B.; Reis, S. Shedding light on the puzzle of drug-membrane interactions: Experimental techniques and molecular dynamics simulations. *Prog. Lipid Res.* **2017**, *65*, 24–44. [CrossRef]
53. Moradi, S.; Nowroozi, A.; Shahlaei, M. Shedding light on the structural properties of lipid bilayers using molecular dynamics simulation: A review study. *RSC Adv.* **2019**, *9*, 4644–4658. [CrossRef]
54. Klauda, J.B.; Venable, R.M.; Freites, J.A.; O'Connor, J.W.; Tobias, D.J.; Mondragon-Ramirez, C.; Vorobyov, I.; MacKerell, A.D., Jr.; Pastor, R.W. Update of the CHARMM all-atom additive force field for lipids: Validation on six lipid types. *J. Phys. Chem. B* **2010**, *114*, 7830–7843. [CrossRef]
55. Kim, S.; Thiessen, P.A.; Bolton, E.E.; Chen, J.; Fu, G.; Gindulyte, A.; Han, L.; He, J.; He, S.; Shoemaker, B.A. PubChem substance and compound databases. *Nucleic Acids Res.* **2016**, *44*, D1202–D1213. [CrossRef] [PubMed]
56. Cheng, T.; Zhao, Y.; Li, X.; Lin, F.; Xu, Y.; Zhang, X.; Li, Y.; Wang, R.; Lai, L. Computation of octanol-water partition coefficients by guiding an additive model with knowledge. *J. Chem. Inf. Model.* **2007**, *47*, 2140–2148. [CrossRef] [PubMed]
57. Missner, A.; Pohl, P. 110 years of the Meyer–Overton rule: Predicting membrane permeability of gases and other small compounds. *ChemPhysChem* **2009**, *10*, 1405–1414. [CrossRef] [PubMed]
58. Bhargava, P.; Grover, A.; Nigam, N.; Kaul, A.; Ishida, Y.; Kakuta, H.; Kaul, S.C.; Terao, K.; Wadhwa, R. Anticancer activity of the supercritical extract of Brazilian green propolis and its active component, artepillin C: Bioinformatics and experimental analyses of its mechanisms of action. *Int. J. Oncol.* **2018**, *52*, 925–932. [CrossRef]
59. Li, J.; Shen, Z.; Ma, X.; Ren, W.; Xiang, L.; Gong, A.; Xia, T.; Guo, J.; Wu, A. Neuropeptide Y Y1 receptors meditate targeted delivery of anticancer drug with encapsulated nanoparticles to breast cancer cells with high selectivity and its potential for breast cancer therapy. *ACS Appl. Mater. Interfaces* **2015**, *7*, 5574–5582. [CrossRef] [PubMed]
60. Zhao, X.B.; Lee, R.J. Tumor-selective targeted delivery of genes and antisense oligodeoxyribonucleotides via the folate receptor. *Adv. Drug Deliv. Rev.* **2004**, *56*, 1193–1204. [CrossRef] [PubMed]
61. Mahmoodzadeh, F.; Jannat, B.; Ghorbani, M. Chitosan-based nanomicelle as a novel platform for targeted delivery of methotrexate. *Int. J. Biol. Macromol.* **2019**, *126*, 517–524. [CrossRef]
62. Chatzisideri, T.; Leonidis, G.; Sarli, V. Cancer-targeted delivery systems based on peptides. *Future Med. Chem.* **2018**, *10*, 2201–2226. [CrossRef] [PubMed]
63. Zhou, X.; Zhang, M.; Yung, B.; Li, H.; Zhou, C.; Lee, L.J.; Lee, R.J. Lactosylated liposomes for targeted delivery of doxorubicin to hepatocellular carcinoma. *Int. J. Nanomed.* **2012**, *7*, 5465. [CrossRef]
64. Ayatollahi, S.; Salmasi, Z.; Hashemi, M.; Askarian, S.; Oskuee, R.K.; Abnous, K.; Ramezani, M. Aptamer-targeted delivery of Bcl-xL shRNA using alkyl modified PAMAM dendrimers into lung cancer cells. *Int. J. Biochem. Cell Biol.* **2017**, *92*, 210–217. [CrossRef] [PubMed]

membranes

MDPI

Article

Microfluidics Approach to the Mechanical Properties of Red Blood Cell Membrane and Their Effect on Blood Rheology

Claudia Trejo-Soto [1,*], Guillermo R. Lázaro [2] and Ignacio Pagonabarraga [2,3,4] and Aurora Hernández-Machado [2,5,6]

1 Instituto de Física, Pontificia Universidad Católica de Valparaiso, Casilla 4059, Chile
2 Departament de Física de la Materia Condensada, Universitat de Barcelona, Av. Diagonal 645, 08028 Barcelona, Spain; grolazaro@gmail.com (G.R.L.); ipagonabarraga@ub.edu (I.P.); a.hernandezmachado@gmail.com (A.H.-M.)
3 CECAM, Centre Europeén de Calcul Atomique et Moleéculaire, École Polytechnique Feédeérale de Lausanne (EPFL), Batochime—Avenue Forel 2, 1015 Lausanne, Switzerland
4 Universitat de Barcelona Institute of Complex Systems (UBICS), Universitat de Barcelona, 08028 Barcelona, Spain
5 Centre de Recerca Matemàtica, Edifici C, Campus de Bellaterra, 08193 Barcelona, Spain
6 Institute of Nanoscience and Nanotechnology (IN2UB), University of Barcelona, 08028 Barcelona, Spain
* Correspondence: claudia.trejo@pucv.cl

Abstract: In this article, we describe the general features of red blood cell membranes and their effect on blood flow and blood rheology. We first present a basic description of membranes and move forward to red blood cell membranes' characteristics and modeling. We later review the specific properties of red blood cells, presenting recent numerical and experimental microfluidics studies that elucidate the effect of the elastic properties of the red blood cell membrane on blood flow and hemorheology. Finally, we describe specific hemorheological pathologies directly related to the mechanical properties of red blood cells and their effect on microcirculation, reviewing microfluidic applications for the diagnosis and treatment of these diseases.

Keywords: membrane elasticity; red blood cells; hemodynamics; hemorheology; microfluidics

Citation: Trejo-Soto, C.; Lázaro, G.R.; Pagonabarraga, I.; Hernández-Machado, A. Microfluidics Approach to the Mechanical Properties of Red Blood Cell Membrane and Their Effect on Blood Rheology. *Membranes* **2022**, *12*, 217. https://doi.org/10.3390/membranes12020217

Academic Editors: Jordi Marti and Carles Calero

Received: 15 December 2021
Accepted: 18 January 2022
Published: 13 February 2022

1. Introduction

The membrane is a fundamental structure in all living organisms, as it defines the cell as an entity. It separates the external environment from the cell's inner region, which contains all the organelles and molecular machinery. The elastic behavior of more complex membranes, such as those present in mammalian cells, is still subject to lively debate in the literature. In this context, most research has focused on the study of human red blood cells as a mechanical model system [1,2], due to its structural simplicity and the lack of a nucleus and any internal structure.

From a theoretical point of view, our knowledge about membranes' molecular compositions and functioning has continuously increased from the pioneer biological model of the fluid mosaic by Singer and Nicolson [3]. In the last 40 years, membranes have also been studied by physicists, providing a complementary picture about membrane behavior and the properties of vesicles and cells. The subject was first approached by Canham in 1970 [4] and then by Helfrich in 1973 [5,6], and based on their models an outstanding number of membrane phenomena have been understood and explained from a physical perspective. Additionally, in spite of membranes' intrinsic complexity, physical models have explained a high number of phenomena observed experimentally, inviting an extensive theoretical exploration of biological membranes.

Erythrocytes or red blood cells (RBCs hereafter) present a remarkable capability to deform and pass through very thin capillaries, and in microcirculation they acquire strange shapes, the benefits of which are still unknown; see the article of G. Tomaiuolo 2009 [7].

The dynamics of RBCs in shear flow have been studied and an unsteady tumbling solid-like motion has been observed when cells are suspended in plasma [8]. Additionally, at high shear stress they exhibit a drop-like tank-treading motion characterized by a steady orientation and membrane rotation about the internal fluid [9–11]. They also develop a number of different morphologies if their membranes are altered or damaged, as known from a number of anemias, malaria, or during blood storage [12]. The delicate membrane equilibrium at the molecular scale ultimately affects the mechanisms taking place at a much larger scale, such as cell shape and blood properties.

Initially, RBCs were studied from a numerical point of view, considering confined geometries to simulate the circulatory system conditions. In the present decade, the developments in microfluidics technologies have contributed significantly to the experimental study of RBCs' membrane properties [13] and the rheological properties of blood and its relation to the RBCs [14,15]. The combination of biological studies with microfluidics has been fundamental in the biomedical research of the biomechanical properties of the RBCs in health and disease [16] and the development of new Point of Care Diagnostics (PoCD) techniques using blood [17–19].

Considering that the blood is the most important fluid in our body and that blood circulation plays a fundamental role in maintaining an appropriate environment in the body's tissues, ensuring the optimal functioning of cells, understanding the flow properties of blood is crucial. These properties depend on the composition of blood and the particular properties of its constituents. The blood is known to be a complex mixture of blood plasma and blood cells. It presents a non-Newtonian behavior, even if blood plasma behaves as a Newtonian fluid by itself. The study of blood flow can be approached from two points of view: we can study the fluid dynamics of the blood flow as a continuous fluid with its constitutive equations or we can study the flow properties and rheology of blood and its components' contributions, from single cells to their collective behavior.

This review is dedicated to describing the general features of RBC membranes and their effect on blood flow, using numerical and experimental studies based on microfluidics technologies. Section 2 is dedicated to describing the composition of the cell membrane in order to understand its constitution. In Section 3, we discuss current numerical modeling techniques of RBC membranes. In Section 4, we review the composition of blood, and the mechanical properties of RBCs. Here, we define how these properties affect the behavior of blood and its consequences. In Section 5, we refer to the past and current studies of the characteristics of RBCs and blood flow at the microscale and their effect on blood rheology from a single cell to the collective behavior. Finally, Section 6 is dedicated to describing our interest in hemorheology, exposing its high importance in the diagnosis of diseases related to blood viscosity and the properties of RBC membranes, focusing on novel microfluidics applications for diagnosis and treatment.

2. Cell Membranes

Cell membranes represent an essential element in the development of living organisms. They constitute the cells' boundaries, separating the interior of the cell from the external environment. Membranes enclose the organelles and components that together form the basic units of life. However, membrane functionality is not limited to its simple structural role, but membranes are also responsible for the interactions of the cell with neighboring cells. These interactions are mediated by a certain type of transmembrane proteins that coordinate the cell signaling, enabling the cell's response to environmental pressures. Additionally, membranes maintain ion gradients which allow the synthesis of ATP, the basic energetic molecule [20]. The plasma membrane is the most important membrane of the cell, but other types of membranes are present in organelles such as the nucleus, the Golgi apparatus, the endoplasmic reticulum and the mitochondria. All the membranes of the cell constitute around 30% of the total protein activity [21].

All biological membranes share a common structure and composition in spite of being part of different entities, and regardless of their function. Membranes are composed of different lipid molecules that assemble to form bilayers. Lipid bilayers are selectively permeable to the exchange of polar molecules and host a high density of transmembrane proteins, which essentially define the specific membrane functionality. Lipids are bound by relatively weak, non-covalent interactions that allow a rapid lateral interchange of positions, leading to a significant surface diffusion over the membrane plane, $D \approx 10^{-12}$ m^2/s, refs. [12,22]. The lipids practically behave as a fluid in the bilayer plane, a property with important implications for the cell activity. The membrane is connected with the inner cytoskeleton, a three-dimensional mesh formed by actin filaments which provides compactness and structural ordering, and determines the cell shape, which in turn depends on the type of cell and its function. In some cells, an exterior cytoskeleton also exists, and it connects with neighboring cells in order to facilitate a coordinate response of the tissue. The membrane equilibrium is controlled by a number of active processes, including a flip-flop rearrangement of the different lipid species of the bilayer, remodeling of the cytoskeleton, or the balance of lipid densities during vesiculation processes (e.g., during endo- and exocytosis), which are able to occur due to the existence of lipid reservoirs in the interior of the cell.

Lipids represent up to 50% of the total mass of the membrane in mammalian cells [23,24]. They are amphiphilic molecules with a polar head (which prefers to contact and interact with other polar molecules, such as water) and a tail formed by two hydrocarbon chains which present a strong hydrophobicity, and therefore the tails avoid the interacting with water. If lipids are immersed in water, they tend to self-assemble to avoid the hydrophobic interactions with the surrounding water. Two basic structures can be formed by these aggregates. Sometimes they assemble to form micelles, a closed structure, with all the tails in the inner, free-water region, and the lipid heads oriented to the exterior, in contact with water. Another possibility is the formation of bilayers and vesicles, when two lipid monolayers fold in opposite directions, so that the heads form two parallel sheets whereas the tails are trapped in the intermediate region, without contact with the aqueous environment, see Figure 1. Lipids rearrange to avoid the presence of edges, forming closed surfaces in which the water is at both the inner and outer regions, but there is no direct interaction with the tails. The strong hydrophobicity causes these closed structures to be much energetically favorable, thus ensuring large stability under thermal fluctuations and other mechanical disruptions [23].

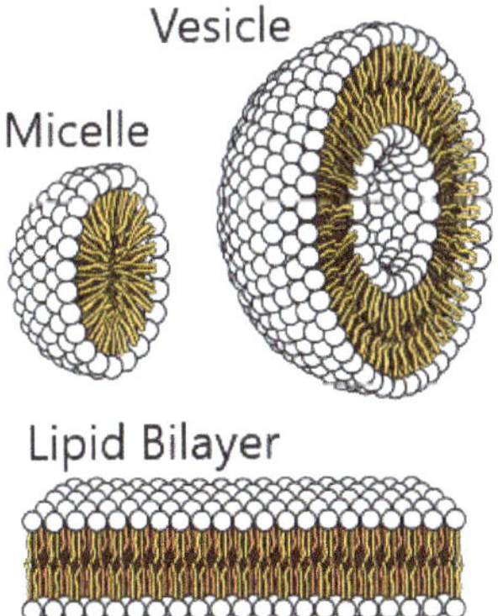

Figure 1. Different aggregates formed by lipids: micelle, bilayer, and closed bilayer, forming a vesicle. The preference of the lipids to aggregate in one structure or another is determined by the shape of the lipid; phospholipids form bilayers. Credits: Mariana Ruiz Villareal available under Public Domain.

A eukaryotic cell is typically composed of 500–1000 different species of lipids; however, the major components reduce to the phospholipids, which are asymmetrically distributed in the bilayers. A discussion on the polymorphism of lipids can found in the work of Cullis

(1986) [25]. In addition to the phospholipids, animal cell membranes also contain cholesterol and glycolipids. Cholesterol is a small molecule with a polar hydroxyl group and a short hydrocarbon chain. Cholesterol occupies the space between phospholipid tails in the inner region of the bilayer, with its head oriented close to the phospholipid head. Mammalian cell membranes are rich in cholesterol, which plays an important role in the control of bilayer fluidity, and it also affects the membrane rigidity when present at abnormally high densities [26,27]. Another important constituent of the cell membrane is transmembrane proteins, responsible for the main processes that take place in the membrane, and therefore they define the membrane functionality. Depending on the membrane, proteins represent 25–75% of the total mass of the membrane. Since proteins are much larger than lipids, this concentration corresponds to a protein per ≈50–100 lipids. Transmembrane proteins are also amphiphilic and orient their polar groups to the aqueous environment (cytosol and exterior of the cell), whereas the hydrophobic groups interact with the lipid tails. The bilayers of mammalian cells are complex structures with a bewildering number of proteins working on and through them. They have a typical thickness of 4 nm, while most eukaryotic cells are ≈5 μm–8 μm in length. Thus, the membrane thickness is three orders of magnitude smaller than the overall cell length. Although the bilayer is usually fluid, this property presents a strong dependence on the temperature and lipid composition [28].

Most cells have a complex mesh formed by actin filaments that occupies most of the inner cytosolic volume and connects the different organelles and microstructures of the cell. This structural element provides mechanical strength to the cell and it often participates in determining the cell shape and cell mobility. This structure, known as a cortical cytoskeleton, is connected with the membrane in order to coordinate the response to external perturbations [29]. The cells also contains a much simpler cytoskeletal structure, the so-called membrane cytoskeleton, which lies underneath the lipid bilayer. The membrane cytoskeleton has a structural functionality, providing strength and preventing from certain shape deformations, such as vesiculation or the pinching-off of the bilayer. The membrane cytoskeleton is a two-dimensional spectrin network anchored to the inner (cytosolic) monolayer of the plasma bilayer of certain cells [30], such as human erythrocytes. The presence of ATP is crucial for maintaining the cytoskeleton properties, and when this molecule is depleted, the cell experiences drastic shape changes. Although this phenomenon is not completely understood, the fluid gel hypothesis assumes that the network is subjected to continuous remodeling, which allows the relaxation of cytoskeleton tensions [31]. Hence, when active processes cease, the cytoskeleton loses its fluidic behavior and stiffens.

Eukaryotic cells present an extensive variety of shapes, as an adaptation to their specific function and location within the different tissues. The cortical cytoskeleton and the plasma membrane are the two main elements responsible for the cell shape and mechanical response. Still, the different organelles occupy an important portion of the cell volume, and their presence implies that the cell must accommodate them. Hence, while studying the mechanical properties of the cell, it is difficult to discern between the different effects, obscuring the understanding of the specific properties of the membrane. Taking into account this problem, the RBC represents an interesting case. Mammalian RBCs lack a nucleus and any internal structure, so that their unique components are the plasma membrane with its underlying cytoskeleton [32]. Accordingly, the shape of the RBC can be directly understood as the result of its membrane properties. The RBC is therefore studied as a model system in order to understand plasma membrane properties and, indeed, many of the studies that have elucidated key insights on membrane biology focused on RBCs. Nevertheless, RBCs are interesting not only as a model system but also due to their crucial role in our lives, as they are the main component of blood and the unique carriers of oxygen.

3. Cell Membrane Modeling

To develop a physical approach to membrane modeling, the use of mesoscopic theories is beneficial. Considering the membrane as locally homogeneous and introducing a continuum description, each small part of the membrane is characterized by some certain local properties. These properties must be consistent with the local molecular structure of the membrane, so that a connection between the micro and meso scales can be derived. The molecular complexity of biological membranes only affects a few essential aspects of relevance in a physical description of membranes: length scale separation, fluidity, hydrophobicity of the lipid tails, bilayer architecture, membrane cytoskeleton and active processes [33].

In this context, the Helfrich bending energy represents the fundamental theory of membrane elasticity. Helfrich adapted the general theory of elasticity to the particular characteristics of membranes, accounting for the structural membrane properties [5]. The main assumption of this approach is that the cell membrane can be described as a two-dimensional sheet, based on its small thickness compared to the cell length. Helfrich proposed that, from the main types of deformations that a layer can undergo—shear, tilt, stretch and bending—only the latter plays a relevant role in the membrane elasticity to characterize the shape of the RBC. He generalized the bending energy to describe the elasticity of lipid membranes, proposing a free energy, which depends on a bending rigidity modulus κ. For a bilayer, the bending modulus depends on the area-compression modulus K_A, which represents the energetic cost of expand/compress the area of the a single layer. Hence, assuming a homogeneous layer and considering a pure bending deformation, the general elastic energy reduces to the bending contribution.

3.1. Cell Membrane Dynamics

In recent years, several numerical models to understand and replicate the elastic properties of cells have been developed [34]. Various numerical techniques have been reported to model a single RBC's mechanics and its elastic properties, such as the finite element method [35], boundary integral models [36,37] lattice-Boltzmann method [38–40] and dissipative particle dynamics [41,42]. Most of these methods use a multiscale approach for single-cell modeling.

The representation of the membrane as a two-dimensional layer is reasonably accurate, the simplest and most direct formulation consists of defining a mesh of points which represents the membrane neutral surface, and from there extract the local mean curvature or deformation tensor necessary to compute the elastic energy. The most important examples include the immersed boundary methods [43,44], integral boundary methods [45,46] or multiparticle collision dynamics [47,48]. Methods in this direction have been successfully applied to the study of many membrane-related topics [31,49]. All these methods require of an explicit tracking of the membrane position and the calculation of the deformation variables, i.e., the curvature.

A different approach, based on a Eulerian rather than a Lagrangian description, are the phase-field models [50]. The membrane is identified from an auxiliary scalar field defined in the entire space, and the method details the dynamics of the field, instead of specifically dealing with the evolution of the interface. This formulation also avoids the problem of defining the boundary conditions at the membrane surface. Although the application of phase-field methods to amphiphilic systems was extensively investigated in the past [51], it is only recently that these models have been used in the study of cell morphology and dynamic response [39,52–54].

Combining the Helfrich free energy model, mentioned earlier in this section, and a phase field method, the dynamics of a membrane are defined as a function of an order parameter ϕ, which varies between -1 and 1, defined as $\Phi[\phi] = -\phi + \phi^3 - \epsilon^2\nabla^2\phi$ and a mean bending modulus $\bar{\kappa} = \frac{3\sqrt{2}}{4\epsilon^3}\kappa$. Here, ϵ is the interfacial width, and the order parameter is given as $\phi(x) = tanh(x/(\sqrt{2}\epsilon))$ [55]. The dynamic of the membrane is described as

$$\frac{\partial\phi}{\partial t} = \bar{\kappa}\nabla^2\{(3\phi^2 - 1)\Phi[\phi] - \epsilon^2\nabla^2\Phi[\phi] + \epsilon^2\bar{\sigma}(x)\nabla^2\phi + \epsilon^2\nabla\bar{\sigma}(x)\cdot\nabla\phi\}, \tag{1}$$

where the term $\mu_{mem} = (3\phi^2 - 1)\Phi[\phi] - \epsilon^2\nabla^2\Phi[\phi] + \epsilon^2\bar{\sigma}(x)\nabla^2\phi$ represents the chemical potential of the membrane. The parameter σ is the mean surface tension of the membrane defined as $\bar{\sigma}(x) = \frac{\sqrt{2}}{6\epsilon^3\bar{\kappa}}\sigma(x)$.

3.2. Membrane Dynamics and Hydrodynamic Coupling

The dynamics of the membrane are dictated by Equation (1), but, in many systems, the hydrodynamic effects of the aqueous environment are also crucial in the membrane evolution. A typical example is the study of lipid vesicles in shear flow [56,57], which serves as a model system for RBCs while flowing along capillaries forced by an external flow. To model the interaction of the membrane with the surrounding fluid, the Navier–Stokes equation is frequently used to describe the dynamics of the fluid, and both equations are coupled describing the membrane–fluid interaction. The complete Navier–Stokes phase-field model (NS-PF) [58,59] is

$$\frac{\partial\phi}{\partial t} + \mathbf{v}\cdot\nabla\phi = M\nabla^2\mu_{mem}, \tag{2}$$

$$\rho\left[\frac{\partial\mathbf{v}}{\partial t} + (\mathbf{v}\cdot\nabla\mathbf{v})\right] = -\nabla P - \phi\nabla\mu_{mem} + \eta\nabla^2\mathbf{v} + \mathbf{f_{ext}}, \tag{3}$$

where ϕ is the order parameter, $\mathbf{v}$ is the velocity of the fluid, μ_{mem} is the chemical potential of the membrane, ρ is the density of the suspension, P is the pressure exerted on the fluid and f_{ext} are the external forces applied to the fluid.

From the perspective of RBC elasticity, the membrane mechanics are often characterized with the bending and shear modulii. The minimization of the Helfrich free energy for an ellipsoidic shape under the appropriate values of area and volume leads to the biconcave discocyte of the RBC as the equilibrium shape. Nevertheless, to obtain an accurate model of RBC membranes, the cytoskeleton's elastic properties must be considered. The cytoskeleton presents a low resistance to bend, with a bending modulus at least two orders of magnitude lower than that of the bilayer. It does present, however, resistance to shear and compression in the membrane layer, and it is known to play a fundamental role in inhibiting budding and vesiculation processes. Different models have been formulated to model the cytoskeleton's elasticity. A simple way is to represent it as a spring mesh, relating the spring constant with the elastic modulii. A different approach is to recover the continuum mechanics description and consider the finite strain theory [60].

The elastic properties of the RBC membrane are highly dependent on the specific bilayer lipid composition, ATP concentration, age of the cell, and temperature. They are also known to vary with the morphological state of the cell, and echinocytes or spherocytes are considerably more rigid than discocytes. The bending rigidity has been measured by different experimental techniques [61], such as, micropipette [1], AFM [62] and optical tweezers [63]. Typical values fall between 10 and 50 k_BT, with slight deviations depending on the specific technique. The shear modulus of the bilayer is negligible due to its fluidic nature in the membrane plane, given that any shear stress is instantaneously relaxed by the rapid lateral rearrangement of lipids.

With the improvement of miniaturization techniques and methods in the past 20 years, microfluidics has become a fundamental aspect for studying the elastic and mechanical properties of the RBCs from an experimental point of view. As a result, several studies have been successful in relating theoretical and numerical analyses with experiments [7,64,65].

4. Human Red Blood Cells and Blood Components

4.1. Human Red Blood Cells

Mammalian RBCs have different shapes and sizes, depending on the animal's physiological requirements (e.g., oxygen consumption in animals inhabiting high-altitude mountains). Human RBCs have a disk shape with a typical diameter of 8 µm, with a concave region in the center where the cell achieves its minimum thickness of 1 µm, and a convex outer rim where it reaches a maximum thickness of 2 μm; see Figure 2. This particular shape is usually known as the biconcave discocyte, and it corresponds to the healthy state of the cell. The typical cell area and volume of a healthy individual are 140 μm^2 and 90 μm^3 [66], respectively. Cells present specific regulatory systems to maintain their area and volume constant, thus ensuring that their resting shape is fixed.

In humans, RBCs exhibit a huge intraindividual variability, with strong correlation with sex and age. Cells of men are up to 20% larger than in women, and men also present higher hematocrit in the circulatory system. Aging affects to the RBC membrane rigidity, so that old individuals present more rigid cells. The biconcave discocyte, however, represents just one of the many morphologies exhibited by RBCs, and it responds to a very specific conditions of area-to-volume ratio, bilayer and cytoskeleton elastic properties, membrane internal asymmetry and pH of the surrounding aqueous environment, among others. Other well-known morphologies are the stomatocyte, when the cell acquires a cup-like shape, and the echynocyte, when the cell becomes spherical and it develops many spikes around its contour. The entire shape deformation comprises a sequence of different morphologies usually known as stomato-discoechynocyte [67], and it is triggered by the disruption of the membrane microstructure which changes the membrane asymmetry.

Figure 2. Photograph of RBCs. In this picture, the biconcave disc shape of RBCs is observed. The RBCs are from a healthy 18-year-old male and were imaged on a SEM microscope as quickly as possible so the blood cells did not shrink and distort. The image has been digitally modified to add the red color typical of RBCs. Credits: Annie Cavanagh available under Creative Commons by-nc-nd 4.0 from http://wellcomeimages.org/, accessed on 30 January 2020.

The origins of the peculiar discocyte shape have been subject to debate for decades. It seems reasonable that the large cell area compared to volume (compared to that of a sphere, the so-called reduced volume $v_{red} = V/(4\pi R^2/3) = 0.6$, where $R = \sqrt{A/(4\pi)}$ is the radius of a sphere with equal area to the cell; thus, for a sphere, $v_{red} = 1$), responds to the necessity of optimizing the diffusion of oxygen across the membrane. Alternatively, it has been proposed that the disk has a low inertial momentum, so that it does not rotate when flowing in the main arteries, minimizing the formation of turbulent flows [68]. Another hypothesis postulates that the discocyte is an appropriate shape to undergo strong deformations and pass through the smallest capillaries, after recovering the normal relaxed shape [69].

Three fundamental effects, derived from the characteristic geometry of RBCs, affect blood flow: (1) The geometry gives them the capacity to align with the direction of flow. (2) The cellular membrane of a healthy RBC is flexible, which means that it can change its shape and deform under different flow conditions. (3) RBCs' shape facilitates their adhesion together, forming aggregates. All these properties of the RBCs act together to give blood a viscosity that is substantially higher than blood plasma and contribute to its non-Newtonian properties.

4.2. Blood Components

Human blood is a two-phase fluid system and consists mainly of an aqueous polymeric and ionic solution of low viscosity, the plasma, in which is suspended a 0.45–0.50 concentrated cellular fraction [70]. The plasma is a liquid-phase mixture of metabolites, proteins and lipoproteins suspended in a salt solution composed mostly of water. The cellular fraction is a complex mixture of erythrocytes (RBCs), leukocytes (white blood cells), and thrombocytes (platelets). Nearly 99% of the cellular fraction in blood is represented by RBCs [71]. The complete set of blood components is usually referred to as whole blood. Given the complex constitution of blood, it is considered as a non-Newtonian fluids, which presents a shear-thinning behavior. The rheological properties of blood are primarily due to the diversity and particular features of its constituents.

Human blood plasma is known to behave as a Newtonian fluid; however, some recent studies have observed viscoelastic behavior in human blood plasma [72]. Plasma proteins play an important role in the hemorheological properties of whole blood. First, even though blood plasma is ≈92% water, due to plasma proteins, its viscosity at 37 °C is around 1.7 times the viscosity of water at the same temperature [73]. Second, plasma proteins (especially fibrinogen) cause RBCs to stick together, forming aggregates, such as piles of coins, known as rouleaux. Rouleaux formation is important because it causes the viscosity of blood to be very dependent on the shear rate to which it is exposed [74].

The erythrocytes' volume fraction in blood is commonly referred as hematocrit. The normal range of hematocrit differs between men and women, 40 to 50% and 36 to 46%, respectively. Leukocytes and thrombocytes together only comprise about 1% of the cellular fraction. This high concentration of RBCs is the main reason that they are hemorheologically important. Additionally, the physical and morphological properties of RBCs also contribute to the non-Newtonian behavior of blood.

White blood cells (WBCs) and platelets do not have a significant hemorheological role, mainly due to their low concentration in blood in comparison with RBCs. Despite WBCs being bigger in size, presenting viscoelastic properties by themselves, and playing an important role in microcirculation resistance, their volume concentration is approximately three orders of magnitude lower than RBCs. Thus, their effects are less relevant in general circulation. In the case of platelets, they are much smaller than RBCs (2–4 μm) and their volume in blood is even smaller than the leukocytes' volume. As a consequence, they neither influence whole-blood viscosity directly nor microvascular resistance. However, recent studies have considered the biomechanics of platelets to be fundamental in clinical diagnostics [75,76].

5. Hemodynamics and Hemorheology

The circulatory system is an organ system that circulates blood along all the cells and tissues, facilitating the transport of oxygen and nutrients, which allows the nourishment of the cells [77]. It also serves as a carrier of other molecules or matter, and is used in processes such as the transport of waste products towards the excretory system, or a fast transport of hormones from one part of the body to another in response to a certain environmental condition [78]. Generally, the main function of the circulatory system is to provide the molecules that the body tissues need at each moment.

The circulatory circuit is composed of a collection of blood vessels. These blood vessels decrease in size from the arteries and veins, through arterioles and venules, to capillaries where they reach the organ tissues and nutrient exchange takes place. The circulation of blood in these microvessels: arterioles, venules and capillaries is know as microcirculation [79].

Blood circulates constantly around the body, and therefore the study of blood flow and its rheological properties is crucial to understand the processes underlying microcirculation. Moreover, several studies have demonstrated that the alteration of hemodynamics and hemorheology is associated with various diseases that affect the normal circulation of blood [16,80–82]. In this section, we will review the general features of hemodynamics and hemorheology, and the contribution of recent numerical and experimental microfluidic techniques, to study the hemodynamical and hemorheological properties of blood from a single-cell effect to their collective behavior.

5.1. Hemodynamics and Hemorheology for a Single Cell

Hemodynamics is the area of biophysics and physiology that studies the fluid dynamics of blood flow inside the different structures of the circulatory system: arteries, veins, arterioles, venules and capillaries. Blood flow in the human body is affected by several factors, such as the driving pressure of the flow, the flow characteristics of blood and the geometric structure and mechanical properties of blood vessels [83]. Hemodynamics research has a long history and is an attractive topic, with several theoretical, experimental and computational studies having been developed in the past 50 years [84–88]. The field continues to expand due to recent advancements in numerical and experimental techniques at the microscale. These new techniques have enabled the prediction and observation of blood flow in vitro, emulating in vivo conditions. The combination of computational hydrodynamics and microfluidics have become key elements to approximating the blood flow in the microcirculatory system.

Blood circulates the human body pumped by the heart, which generates a pressure difference in the system. As the blood flows through the circulatory system, the pressure falls progressively by the time it reaches the termination of the venae cavae where they empty into the right atrium of the heart [78]. The heart pumping is pulsatile, and therefore the arterial pressure alternates between a systolic pressure level and a diastolic pressure level. This pressure difference allows blood to flow through the different blood vessels in our body, enabling microcirculation.

Microcirculation flow is characterized by a low Reynolds number $Re = \frac{\eta v D}{\rho} < 1$, where v is the velocity of the flow, D is the diameter of the microvessel, η is the dynamic viscosity of the fluid and ρ is the fluid density. The Reynolds number is defined as the ratio between the inertial and viscous forces, and in the microvessels it ranges between $0.001 < Re < 0.1$. Hence, the viscous effects are more significant than the inertial effects, and the flow is laminar. In microvessels over 200 μm diameter, it can be assumed that blood is a homogeneous continuous fluid. This assumption is not true for smaller microvessels and here the individual motion of RBCs becomes important. When small objects, such as droplets or cell, enter a microchannel, the hydraulic resistance along the channel is given as the addition of the resistance of the channel in the absence of the particle and the resistance developed across the length of the object. The resistance of the object will depend on the local characteristics of the flow and the viscoelastic properties of the object [89].

To measure the factors that affect hemodynamics, several numerical and experimental techniques are used, such as dielectrophoresis, magnetic interaction, optical traps and biomarkers [14]. Using these techniques, researchers have been able to study blood flow behavior from a single RBC to their collective behavior and blood as a homogeneous fluid. When studying blood flow in confined geometries for a single cells, the effect of the system walls are relevant, enabling RBCs to form a single train at the center of the microchannel. However, when studying the collective behavior of RBCs, a focusing phenomenon arise due to the presence of walls and cells to cells interactions. The effects of focusing, or RBC migration, affects the rheological properties of blood, affecting its viscosity and therefore the blood flow.

From a numerical point of view, several techniques have been reported to model RBCs in blood flow [90–94]. Most of these methods have in common a multiscale approach for single-cell modeling in confined geometries. Confinement is crucial to exploring the

RBCs' elasticity effect in microcirculation, since it replicates the in vivo conditions of blood circulation.

To model the interaction between RBC membrane elasticity and flow, a dimensionless quantity, C_κ (usually referred to as capillary number) is defined as the ratio between the elastic relaxation time, τ_m, of the membrane of cells suspended in the fluid, and a viscous time τ_η associated with the viscous forces of the fluid [58,59]

$$C_\kappa = \frac{\tau_\kappa}{\tau_\eta} = \frac{\eta_0 \bar{v}_z d^2}{\kappa}\left(\frac{d}{b}\right), \tag{4}$$

where $\bar{v}_z$ is the mean velocity in the direction of the flow, η_0 is the viscosity of the surrounding fluid, d is the diameter of the RBC $d \approx 8$ µm, b is a geometrical parameter which accounts for the confinement and κ is the bending modulus. For human RBCs, the typical values of the bending modulus are $\kappa \approx 50 k_B T = 2 \times 10^{-19}$ J. The ratio d/b is a parameter that relates the RBC size to the size of the system to establish the confinement. The dynamics of the cell membrane are controlled by the viscosity ratio between the inner and outer regions of the cell, and the capillary number which characterizes the shear rate of the force relative to the membrane rigidity. The effective (or apparent) viscosity of the whole suspension (i.e., liquid and cells) is computed from the relation of the applied force, f_0, and the outcome flow given by the mean velocity $\bar{v}_z$,

$$\eta_{eff} = \frac{f_0}{12\bar{v}_z} b^2. \tag{5}$$

Figure 3 shows three different RBC morphologies in a Poiseuille flow, modeled using the Navier–Stokes phase field models discussed in Section 3, as a function of the capillary number [58,59]. The effects of the hydrodynamics on the viscosity of the solution is shown in Figure 4.

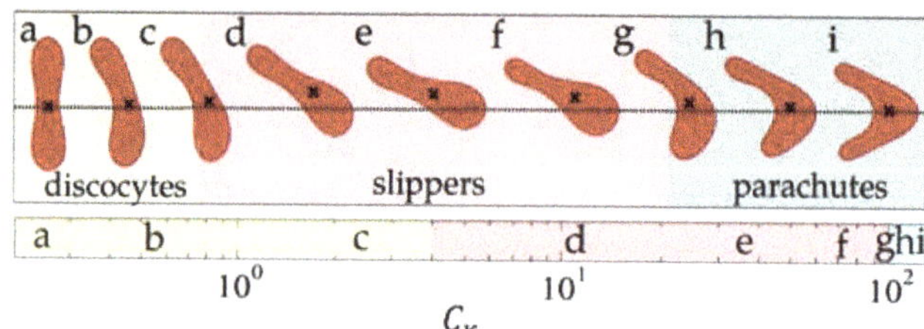

Figure 3. Red blood cell morphologies in a Poiseuille flow, modelled through Equations (4) and (5) for an increasing capillary number, C_κ. The letters associated to the red blood cells represent different stages of the cell morphology. The parameter used to model the RBC was a reduced volume $v_{red} = 0.48$ and a confinement $d/b = 0.71$, defined as the ratio between the RBC size and the system size. The colored regions represent the three main morphological regimes, namely the discocyte (yellow), the slipper (red), and the parachute (blue). The dotted line represents the channel axis, and the crosses are the center of mass of each RBC. Image reproduced from Lázaro et al. (2014) [58].

From an experimental point of view, early hemodynamical experiments only provided a qualitative understanding of blood flow. Quantitative information, such as rheological effects and blood cell deformability were difficult to obtain due to lack of time and spatial resolution. Eventually, high-speed and high-resolution cameras, with an enhanced sensitivity and mounted to an optical microscope, enabled velocity measurements of such small-scale flows. In this aspect, several techniques have been developed to measure the velocity fields of blood and RBCs at the microscale, such as μPIV (microparticles image velocimetry) or PTV (particle tracking velocimetry) and wavelet-based optical flow velocimetry (wOFV) [95–99].

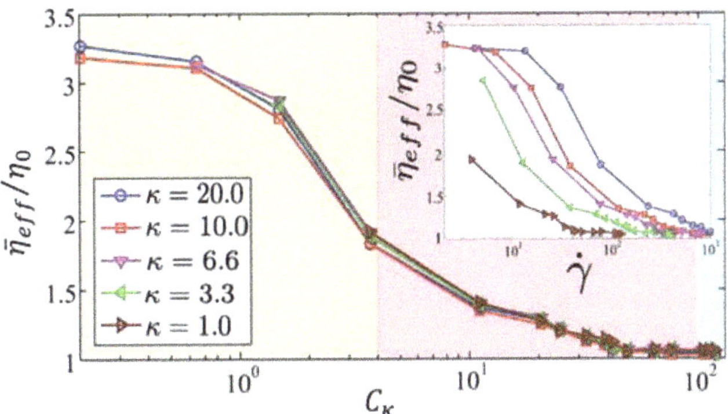

Figure 4. Effective viscosity of an RBC suspension as a function of the capillary number for different bending rigidities, κ, at confinement is $d/b \approx 0.71$. The value of η_{eff}, obtained from Equation (5), is averaged for different initial conditions of the RBC. The coloured regions correspond to the three morphological regimes shown in Figure 3. The curves for different rigidities as a function of the shear rate show the sensitivity of the viscosity and RBC morphology to the rigidity of its membrane; however, the curves collapse when the relative effect between the viscous and elastic forces is considered. Image reproduced from Lázaro et al. (2014) [58].

The rise of microfluidics in the last two decades has enabled the increase in experimental options to study RBCs' properties and their effect on blood flow [89,100–102]. The easiness of replicating small structures in microfluidics allowed the development of various designs and structures to observe and analyze the deformation of RBCs [103–105]. Typical microfluidics approaches consider a forced, or gradual, constriction of RBCs, as they circulate through very narrow slits. The viscoelastic properties of the RBC membrane are obtained, establishing a relation between the shear flow and the pressure gradient applied to them [106]. Moreover, the combination of numerical and experimental models has enabled a deeper study of the biomechanical properties of RBCs [107,108] and their sensitivity to blood flow [109]. These have been successful in capturing several changes in the morphology of the RBCs, replicating the slipper and parachute shapes of RBCs under shear flow [65,110]; see Figure 5. Other studies on the elasticity of RBCs have submitted them to extreme deformation circulating through submicrons slits [111], to simulate the filtration of RBCs in the spleen.

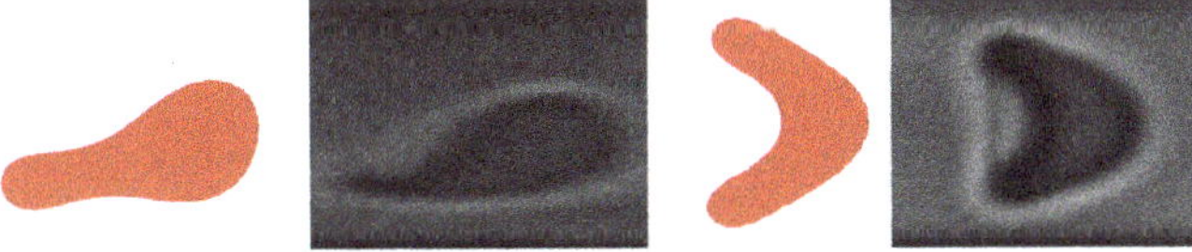

Figure 5. Comparison between slipper (**left**) and parachutes (**right**) obtained from numerical and experimental results. Numerical images reproduced from G. R. Lazaro et al. (2014) [58]. Experimental snapshots adapted by permission from RSC, G. Tomaiuolo et al. (2009) [7] under license 1181911-1.

5.2. Experimental Hemorheology: Collective Behavior of Red Blood Cells

Hemorheology is the study of the rheological properties that affect blood flow. These properties are mostly related to the non-Newtonian nature of blood, due to its composition and to the biomechanics of its erythrocytes. In vivo, blood flow is determined by multiple factors, including hematocrit levels, RBCs deformability, the elasticity of venules and arteries, and blood pressure. The rheological properties of blood have been studied for many years and it has been demonstrated that it presents a shear-thinning behavior [112–115], which means that, as the flow velocity increases, the viscosity of blood decreases. This rheological properties of blood highly depend on the properties of its RBCs, which affect the viscosity of blood, as well as its shear-thinning behavior. Typical values for the viscosity of healthy blood at a low shear rate (0.28 s^{-1}) are 39 $\pm$ 4 mPas for females and 48 $\pm$ 6 mPas for males. At high shear rates (128 s^{-1}), the viscosity values are 4.3 $\pm$ 0.2 mPas and 4.7 $\pm$ 0.2 mPas for females and males, respectively [116].

From a macrorheological point of view, the viscosity of blood is directly related to the fraction of RBCs suspended in plasma. Given the particular properties of RBCs, the increase or decrease in its concentration in plasma (hematocrit) will affect the behavior of blood, where the erythrocyte concentration is directly proportional to the viscosity [117]. Therefore, increased hematocrit levels will lead to an increased viscosity of blood and decreased hematocrit levels will lead to a decreased viscosity. Thus, the concentration of RBCs in plasma affects the whole-blood viscosity values, as well as its non-Newtonian behavior, which is lost at low hematocrit levels [118]. From a microscopical point of view, two properties of RBCs are particularly important in the effort to understand the shear-thinning behavior of blood: their deformability [119] and their tendency to form aggregates [74,120,121]. Both properties cause the highly non-Newtonian behavior observed for RBC suspensions in plasma [122]. At low shear rates, the viscosity of blood is high, whereas, at a high shear rate, red cell disaggregation and deformation reduces the viscosity of blood.

The viscosity of fluids is measured using viscometers. Complex fluids', such as blood, rheological properties are studied using a rheometer, capable of measuring their behavior under different flow conditions. Rheometers differentiate from the type of flow they induce on a material; these may be drag or pressure-induced flows. Typical drag flow rheometers are cone-plate and cylindrical Couette rheometers. On the other hand, the most typical pressure-driven flow is the capillary rheometer. The capillary rheometer was the first rheometer, and is still the most common method to measure viscosity, due its low cost and easy operation. In comparison with rotational rheometers (cone-plate and Couette), they can be closed devices, which avoid the evaporation of solvents and the expulsion of samples. Capillary viscometers and rheometers have been used since the beginning of hemorheology in the 1960s, to measure the viscosity of blood plasma and blood [123]. However, the rise of microfluidics at the end of the 1990s brought new applications and innovation in this area. In recent years, a variety of microfluidics devices and methods have been developed with the objective of measuring the viscosity of blood plasma [124–127] and blood [128], using optical detection techniques [75,129–134], pressure sensors [135,136] and electrical sensors [137–139]; see Figure 6.

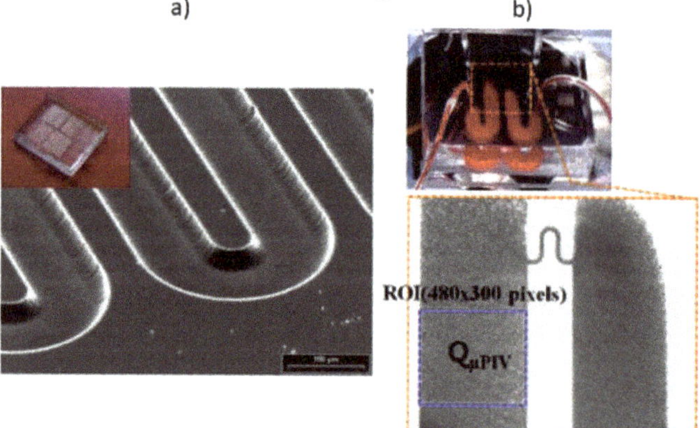

Figure 6. Images of two different microfluidics devices developed to measure blood viscosity. (**a**) Image reprinted by permission from Springer Nature, Morhell and Pastoriza (2013) [135] under license 5235930238113. (**b**) Image adapted by permission from MDPI Kang (2018) [132] under Creative Commons CC by 4.0 license.

In a capillary, the viscosity of the fluid is measured by establishing the relation between the pressure difference exerted on the fluid and the flow velocity. The pressure difference, which moves the fluid inside the microchannel, is generated through gravity, gas compression, pistons or suction. For Newtonian fluids, the viscosity is determined as the ratio between the shear stress, σ, defined as a function of the pressure exerted on the fluid, and the shear rate, $\dot{\gamma}$, defined as a function of the velocity. However, for non-Newtonian fluids, the relation between these parameter becomes non-linear and the viscosity of blood is measured through a local relation between the shear stress and the shear rate. Typical non-Newtonian viscosity models used for blood are the power law, the Carreau, the Carreau–Yasuda and the Casson models [140]. Nonetheless, the simplicity of the two parameters of the power law model makes it the most popular model used to estimate blood viscosity. This model states that the viscosity of the fluid is defined as a function of the shear rate through:

$$\eta = m\dot{\gamma}^{n-1}, \tag{6}$$

where m is a consistency factor that depends on the fluid and n is the behavior factor that defines the character of the fluid. When $n = 1$ the fluid is Newtonian, for $n < 1$ the fluid is shear thinning and for $n > 1$ the fluids is shear thickening.

A basic method to determine the viscosity of blood is the Front Microrheology method, which consists of inducing a pressure difference in the fluid through hydrostatic pressure $P_{hyd} = \rho g H$. The pressure is controlled through a fluid column inside a reservoir set at different heights H and connected to a bio-compatible tube with uniform internal cross-sections of radius r and length l_t. The tube connects the reservoir with a rectangular microchannel of width $w = 1$ mm, depth $b = 0.3$ mm and length $l_c = 4$ cm, fabricated in PDMS over a glass substrate using typical microfabrication techniques [141–143]. Figure 7a shows a schematic view of the experimental setup described. The observation of the blood–air interface (blood front) inside the microchannel is made using a microscope and a high-speed camera; see Figure 7b. The velocity of the blood front is measured tracking the mean front position as a function of time between several contiguous images. A full description of the microfluidic device and details of the experimental method are reported by Trejo-Soto et al. (2016) [127].

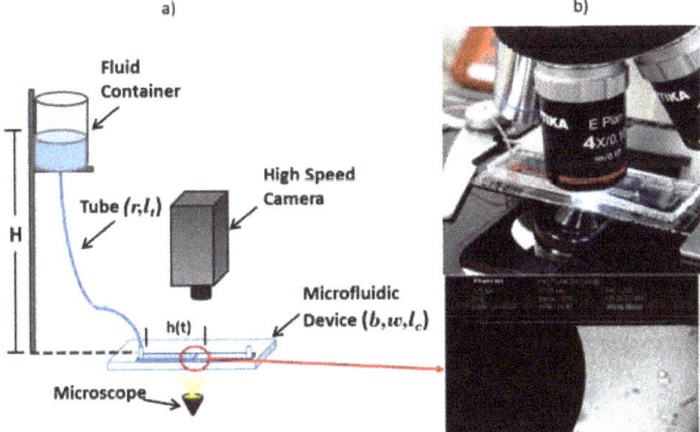

Figure 7. (**a**) Schematic representation of the experimental set up to perform blood viscosity measurements using microfluidics. The pressure difference is generated through hydrostatic pressure $P_{hyd} = \rho g H$, where H is the height from the fluid in the reservoir to a microchannel of width w, depth b and length l_c. These are connected through a tube of radius r and length l_t. (**b**) Photograph of the experimental set up. Top image: A view of the microdevice under a microscope. Bottom image: View of the blood–air interface inside the microfluidic channel, taken with an Optika XDS-3 microscope and a high-speed camera Photron Fastcam Viewer 3. Images reproduced from the work of Trejo-Soto et al. (2017) [130].

According to this experimental setup, an effective pressure $\Delta P_{eff} = \rho g H - P_L$ is defined, where P_L is the Laplace pressure due to the curvature of the fluid interface. This effective pressure is related to the stress through [130]:

$$\sigma = \frac{r}{2l_t}(\rho g H - P_L), \tag{7}$$

where r and l_t are the internal radius and the length of the tube, respectively. The shear rate of the system is defined as a function of the velocity of the interface, and the geometrical parameter of the coupled system tube-microchannel, through the following expression:

$$\dot{\gamma}_F(n) = \frac{b^2 w}{\pi r^3}\left(3 + \frac{1}{n}\right)\frac{v}{b}, \tag{8}$$

where b and w are the geometrical parameters of the experimental microchannel. The parameter n is the behavior exponent obtained using a power law model to describe the viscosity of blood. The viscosity of blood and its shear-thinning behavior have been measured and observed using several methods [144]; see Figure 8. Using the power law model, typical values of the exponent for blood are around $n = 0.80$ [129,136].

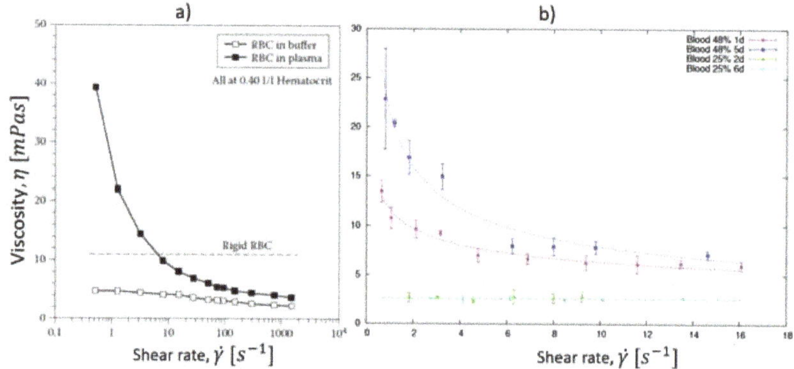

Figure 8. Viscosity of blood as function of the shear rate. In both images, the non-Newtonian nature of whole blood is shown. At a low shear rate, RBCs form aggregates which are responsible for an increase in viscosity. As the shear rate increases, cells disaggregate and move freely through blood vessels. If the shear rate keep increasing, RBCs deform, elongate and align with the direction of the flow, which happen in microcapillary vessels. The image on the **left** (**a**) shows one of the first viscosity measures of blood obtained using a typical rheometer, image reproduced from Baskurt et al. (2007) [116]. On the **right** (**b**), shows the viscosity of a fresh 48% hematocrit blood sample (magenta) and the same sample 5 days from extraction (blue), we observe how the aging of the sample affects the viscosity of the sample. Image reproduced from Trejo-Soto et al. (2017) [130].

Although this standard procedure provides important information about the bulk behavior of the fluid, it is of limited interest for understanding the flow in very confined systems, when the rheological behavior can be severely affected. For a single cell, elastic properties are more relevant, and RBCs as an ensemble are mainly affected by confinement and focusing.

5.3. Comparison with Numerical Results of the Collective Behavior of RBCs

As mentioned earlier, the flow of RBCs in tubes and channels is critically controlled by the hematocrit. The interactions between the RBCs, involving hydrodynamic interactions, purely geometrical constraints, or aggregation, play a fundamental role in the collective dynamics of the RBC suspension. At low concentrations, vesicles and hard spheres flowing in thick tubes migrate from the center line and reach a stable trajectory at ≈0.6*r* from the axis, forming an annulus of high density at this radial distance, the so-called Segre–Silberberg effect [145]. At high concentrations, however, RBCs distribute along the tube core, avoiding the region close to the wall. The transition from the single-cell to the high hematocrit behavior is still poorly understood in spite of its importance in the rheological behavior of the fluid.

When blood measures are at the microscale, other effects may be observed, namely, the Fåhraeus [146] and the Fåhraeus–Lindqvist effects [147]. The latter, characterized by a dependence of the blood viscosity with the channel thickness, are perhaps the most important example [148]. In the range between roughly 300 μm and 10 μm of the tube diameter, the effective viscosity decreases up to 4–5 times. This effect occurs as a consequence of the strong repulsion from the walls that forces the blood cells to concentrate on the central region of the channel. The formation of layers free of cells close to the walls allows a rapid flow in these regions, enhancing the overall fluidity. At high confinements, the walls' proximity enforces a more concentrated distribution of cells in the center and consequently broader layers of free flow are present [149]. In larger channels, the free layers are proportionally thinner until their effect becomes eventually negligible. Although, in the narrowest channels (<10 μm), RBCs are ordered in a single row for flow at low concentrations, and thus interactions between cells are disregardable. At an intermediate channel size (≈20 μm), RBCs present a more complex behavior and collective effects must be considered [150].

While flowing in thicker channels, where cells typically flow at higher concentrations, RBCs do interact, and collective effects substantially change the flow properties. From a theoretical point of view, the organization in trains (observed for single cells) also offers an interesting way to study the hydrodynamic interactions between neighboring cells, and how it affects the RBCs' dynamics. The membrane stiffness dictates the flow disruption induced by the RBCs. Rigid cells induce stronger perturbations of the incoming flow than softer ones. Even if deviations from the imposed flow are small when RBCs are distant, interactions strengthen for lower distances between cells, favoring RBCs' collective behavior. If RBCs are initially placed very close to each other, even at high capillary numbers, they do not migrate towards the walls but flow whilst maintaining a centered position.

RBCs are very sensitive to the hydrodynamic interactions with other cells, and the competition between these interactions and the wall effects dictates different RBC flow properties when several cells are flowing at high and moderate concentrations. For instance, in the inertial regime, the limit of single-cell behavior is characterized by the Segré–Silberberg effect, when cells migrate towards a specific lateral position, whereas at higher concentrations the collective behavior dominates and cells are located at the tube core, the Fåhraeus–Lindqvist effect. The dynamics of several RBCs at moderate concentrations have proven to differ in several aspects from the single-cell case, and this affects the rheological behavior of the suspension. Recent numerical analyses have computed the effective viscosity for three configurations (one ordered and two disordered initial conditions), at a low volume fraction and concluded that the viscosity curves show the expected shear-thinning behavior, though two main differences were found with respect to the single-cell case: the magnitude of the effective viscosity obtained and the C_κ, Equation (4), required to observe the shear-thinning decay [151]; see Figure 9.

Other important characteristics of RBCs in their collective interaction is their aggregation. RBCs have a tendency to form stacked structures (aggregation), commonly known as rouleaux. The formation of aggregates affects blood flow and its rheological properties, increasing blood viscosity, and therefore slowing down the flow. These structures have several characteristics, such as, the number of RBCs per rouleau being variable and side-to-side formations being possible, due to the particular discocyte shape of the RBCs [74]. Figure 10 shows two images of rouleaux formation of two blood samples with different hematocrit levels, where some of these characteristics are observed.

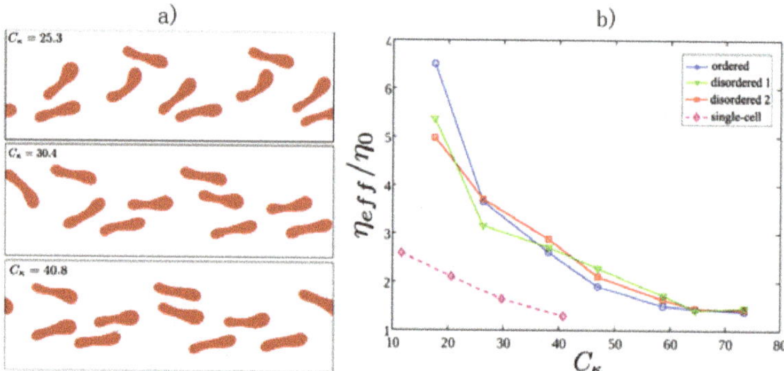

Figure 9. (**a**) Simulation of RBCs during flow in a confined channel. (**b**) Numerical results of the effective viscosity for an RBC suspension at a low concentration of RBCs and low confinement as a function of the capillary number, Equation (5). Three initial conditions, one ordered and two disordered, are calculated, obtaining similar results. The curve shows the expected shear-thinning behavior, and the differences with the single-cell case. Image reproduced from Lázaro et al. (2019) [151].

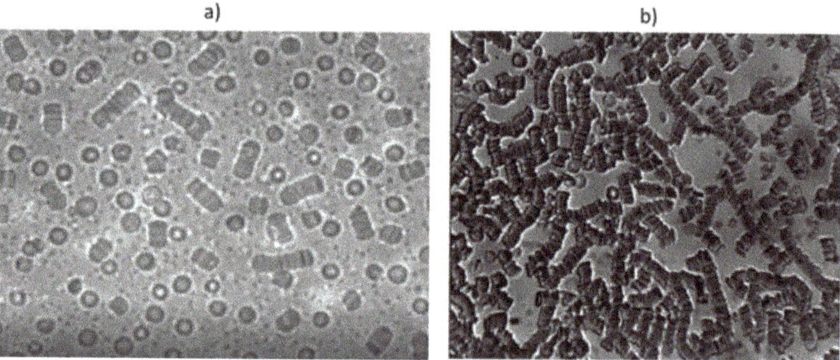

Figure 10. (**a**) Red blood cells aggregates from a blood sample at 5% hematocrit. The image was taken with an inverted Optika XDS-3 microscope using a 50× magnitude objective. (**b**) Large RBC aggregate from a blood sample at 38% hematocrit. The image was taken with an Optika B-353LDI microscope using a 40× magnitude objective.

Aggregation has rheological consequences in blood and it determines its non- Newtonian behavior at low shear rates. When RBCs are aggregated, more shear is required to move the fluid, but as shear increases, RBCs start to disaggregate, making it easier to change the state of motion of blood. If we keep increasing the shear, cells start to align with the flow, deform, and elongate. Therefore, if aggregation increases, then blood viscosity increases as well and the shear-thinning behavior of blood is altered. Numerical studies have demonstrated the effects of aggregation in blood viscosity [120,122,152–154] and experimental studies have observed that RBCs' deformability induces cell aggregation during flow in microcapillaries, allowing the formation of clusters of cells [98,155,156]. Moreover, the aging of storaged RBCs also contributes to the increase in aggregation and affects its viscosity. However, when scaling according to adhesion energies, a collapse in the viscosity curves defines a single universal behavior for blood viscosity [130]. To analyze the effects of aging, the hematocrit levels were fixed and the behavior of the blood sample as it ages was studied, showing that as the sample aged, the aggregate formation increased.

By introducing a non-linear scaling parameter, the adhesion scaling number, A, the effects of aging on RBC aggregation was quantified. This quantity is defined as

$$A = \frac{\eta_0 \dot{\gamma} d^3}{k_a E_0}, \tag{9}$$

where η_0 is the viscosity of plasma, d is the average diameter of an RBC, E_0 is the energy scale associated with the aggregation energy between RBCs [122], and k_a is a scaling factor that accounts for the relative increase in the adhesion energy when the blood ages. Then, the parameter A can be interpreted as the ratio between the characteristic viscous energy scale and the aggregation energy. When the viscosity of the fluid is normalized according to the hematocrit levels and blood plasma $\eta_{eff} = \eta/\eta_0$, the changes in the viscosity depend only on the adhesion scaling number. Figure 11a shows the effect on blood viscosity due to RBC aggregation, induced by aging.

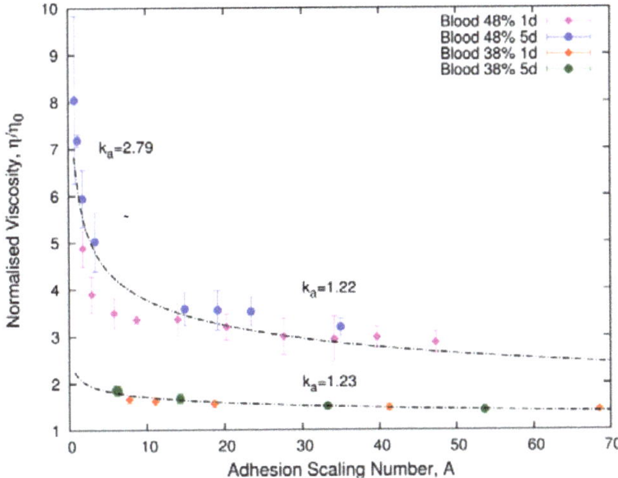

Figure 11. The plot shows how the normalized viscosity, η/η_0, follows a universal curve as the adhesion scaling number changes. Image reproduced from Trejo-Soto et al. (2017) [130].

In general, blood rheology has been proven to be experimentally difficult to measure. Thus, inaccurate interpretations have frequently been made [157]. Many microfluidics techniques have been able to reduce difficulties using capillary rheometry and pressure-driven flow. The advantages of microfluidics in this area stand out, mainly portability and the need for only a small sample. Additionally, some microfluidic techniques related experimentally blood viscosity with the properties of RBCs [158]. Still, many considerations need to be taken into account to obtain feasible results.

6. Hemorheological Pathologies and Emergent Microfluidics Diagnostics Techniques

As mentioned in previous sections, human blood is an unusual fluid. While blood plasma alone behaves as a Newtonian fluid, the complete set of blood components are non-Newtonian, meaning that, its viscosity varies according to the speed at which it circulates. This characteristic presents two important issues in clinical hemorheology. First, in vitro measurements of the viscosity of plasma alone never reflect the totality of events occurring in vivo in patient circulation. Second, the cellular elements, by acting as particles in suspension, are mainly responsible for the non-Newtonian behavior of blood. This is why, instead of considering only abnormal plasma proteins in diseases, we should also consider the rheological abnormalities of the erythrocytes. The dynamics and the elastic mechanics of RBCs in confined systems are subjects of fundamental interest due to their enormous application potential in biomedical engineering, as they affect hemorheology during blood handling and storage, or manipulate cells in pathology diagnosis. Altered blood due to abnormal RBC concentrations or stiffening of the cells can lead to a reduction in the oxygen delivered or obstruction of the blood vessels. The healthy running of RBC circulation and oxygen transport can be affected by different disorders.

Plasma proteins are responsible for the elevation of blood plasma viscosity in comparison to water; a change in their composition may as well alter the hemorheological properties of blood. Some diseases, such as Wasldenstrom's macroglobulinaemia, induce an increase in macrogobulins, which increases the viscosity of blood plasma [159]. Furthermore, in this condition, it is more likely that proteins will form rouleaux, which increases blood viscosity at low shear rates. An elevated concentration of fibrinogen in plasma generates an abnormal increase in RBC aggregation, changing the rheological behavior of blood. Aggregation and its effect in blood rheology have been related to several diseases [160–162], such as inflammation, diabetes [80,163,164], hypertension [165], obesity [166] and coronary syndromes [167–169].

Although plasma affects blood viscosity, RBCs are the most prominent hematological factor influencing hemorheology. Among circulating blood cells, erythrocytes interact most significantly with plasma, mainly as a function of the hematocrit levels. Most typical diseases related to blood viscosity are related to the percentage of RBC concentration (hematocrit), such as anemia (low % hematocrit) or polycythemia (high % hematocrit) [116]. Elevated values of blood viscosity are characteristic of hyperviscosity syndromes. Hyperviscosity may occur due to different properties of blood: an increase in the viscosity of blood plasma, a high production of fibrinogen, an increased numbers of cells (polycythemia or leukemia) or a increased resistance of cells to deformation (sicklemia or spherocytosis) [170,171]. In the case of whole blood, the most influential factor to increase its viscosity is hematocrit. If the hematocrit levels of blood exceed 65% (which is the case of polycythemia), various rheological abnormalities arise, for example, the sedimentation rates decrease significantly as a result of RBC crowding. Additionally, in severe cases, the elasticity and deformation properties of RBCs become crucial to achieving smooth driven flows; otherwise, microcirculation may be severely compromised. On the contrary, anemias present low hematocrit levels, less than 35%. Anemia may have different origins: iron deficiency, hemolysis due to particular diseases or heredity. Hematocrit levels lower than 30% tend to neglect almost every non-Newtonian characteristic and usually displays a Newtonian behavior. However, in some cases of hemolytic anemias, low hematocrit levels lead to high viscosity, due to the alterations of the RBCs' properties.

Other disorders concern inherited pathologies which affect the RBC membrane, producing abnormalities in RBC shape or deformability, which potentially reduce the healthy functioning of blood circulation. These membrane alterations provide important information about the membrane's structural balance, and their main causes and consequences are disorders such as sickle cell anemia, hemolytic anemias and thalasemic syndromes, which are directly related to the RBC elasticity, deformation or aggregation properties [16,172]. In addition, some infectious diseases, such as malaria (which does not have a genetic origin), are also known to impair the membrane microstructure, leading to cell stiffening, affecting whole-blood viscosity [173].

Sickle cell anemia (drepanocytosis) is characterized by the formation of sickle cells (ISC) that lose their capability to deform and recover the discocyte shape, altering oxygen delivery. The molecular basis for this is an abnormal phosphorilation of hemogoblin that promotes a massive aggregation of this molecule under low concentrations. The formation of these molecular aggregates affects the concentration of the protein band 3, and the cell membrane is damaged in a process similar to aging, becoming rigid [174]. Patients affected by this pathology present a reduced life expectancy, although modern medical treatments allow a normal life. Due to the presence of ISC, in oxygenated conditions, the "htc/viscosity" ratio is lower than for normal blood samples. If the natural hematocrit levels of the sample are raised to the typical levels of non-anemic blood, an increased viscosity is observed at all shear rates [82]. The plasma viscosity of subjects with sickle cell anemia is also higher than the viscosity of healthy subjects. In severe cases, ISCs obstruct microvessels, altering the normal circulation of blood. In the last decade, microfluidic devices have played the important role of determining the biophysical characteristics of sickle red cells [175], measuring the mechanical stresses on erythrocytes in sickle cell disease [176], studying vaso-oclusion [104,177,178], identifying biophysical markers [179,180], segregating sickle cells [181], and developing point-of-care diagnostic technologies for low-resource settings [182] and possible treatments [183].

Hemolytic anemias are diseases characterized by the reduction in RBC life expectancy (120 days) and an increased destruction of RBCs (trought hemolysis). Hemolytic disorders originated due to the hereditary defects of three RBC components: the membrane, enzymes and hemoglobin. Hemolysis occurs via two mechanisms, extravascular hemolysis, where RBCs are eliminated prematurely from circulation through the microcirculation fagocitic system (liver and spleen), and intravascular hemolysis, where RBC membranes rupture during blood circulation [184]. In hereditary spherocytosis, patients present a high concentration of spheroidal-shaped RBCs, as a consequence of defects in several proteins of the membrane (mainly from the bilayer–cytoskeleton links), which cause fragility of the membrane. The alteration of membrane properties allows vesiculation and a loss of the membrane surface, triggering cell-shape deformation. Spherocytes rapidly retire from circulation due to the spleenic system, leading to hemolysis. Patients must be treated with blood transfusions for critical levels of anemia [185]. Hereditary elliptocytosis is characterized by abnormalities in the spectrin dimers, causing weakness of the cytoskeleton, which impairs membrane stability. RBCs deform into ellipsoidal (or cigarshaped) cells. The RBC functionality might not be severely affected, as most patients are asymptomatic and only 10% present anemia. Thalassemia syndromes such as αthalassemia and βthalassemia are genetic hematological disorders caused by defects in the synthesis of one or more hemoglobin chains. α-thalassemia, also known as HbH disease, is caused by a reduced or absent synthesis of the α-globin chains, and an excess of β-globin chains in the cytoskeleton. It has been reported [186,187], through measurements of cellular deformability, that α-thalassemic and β-thalassemic erythrocytes exhibit increased surface areas in relation to cell volume, increased membrane rigidity and increased membrane viscosity. Although the stability of α-thalassemic erythrocytes membranes are normal, they are uniformly less dense than healthy erythrocytes. A typical diagnosis for hemolytic anemias is made using a technique known as ektacytometry [188,189]. However, in recent years, microfluidic and lab-on-a-chip devices have presented new alternatives to study RBCs deformation using diverse techniques such as deformability cytometry [103,190], magnetic measurement [191], electrical measurement [192], single-cell chamber arrays [193], combination of microfluidics with machine learning [194] and pressure-driven microrheometry [195].

Malaria is caused by the infection of a parasite of the genus Plasmodium. Infected RBCs develop advanced proteinic machinery, including the formation of organelles similar to the Golgi apparatus, which are used for nutrient transport and storage, and allow enzymatic activity. The parasite is hosted in a vacuole, and during its maturation it reaches the size of a nucleus in a typical eukaryotic cell. Apart from this new internal structure, the parasite produces changes in the membrane proteins that affect the deformability of the cell [196]. RBCs also adopt a more spherical shape, and proteins allocated in the external face of the membrane promote aggregation with other infected cells, avoiding hemolysis in the spleen. All these conformational changes strongly affect the cells' mechanical properties, modifying the rheological properties of blood [197,198]. As in the case of non-infectious diseases, in recent years, microfluidic devices have been developed to support Malaria diagnosis mostly directed to low-resource locations [199–201] or to find possible treatments [202–204].

Severe hemorheological disorders are usually related to alterations in the RBCs' mechanical properties. RBCs have physical properties of their own, and are capable of directly influencing blood flow regardless of hematocrit levels, hence the importance of taking

RBCs properties into account when studying its effects on diseases that affects whole blood viscosity. Therefore, understanding these properties from theoretical, numerical, and experimental points of views is the key to improving diagnostics techniques and to developing successful treatments. Microfluidic technologies play a remarkable role in biomedical research, and in combination with biomimetics, lab-on-a-chip and organ-on-a-chip technologies are the cornerstone of future medical diagnoses and treatments.

7. Conclusions

In this review, we have described the general features of RBC membranes and their effects on blood flow, from numerical and experimental perspectives, highlighting the achievements of microfluidics technology in performing in vitro studies assessing RBC membranes' elasticity, hemodynamics and hemorheology.

First, we described the main composition of mammalian cell membranes and refer to the RBC membranes, due to the their special characteristics, specifically their lack of a nucleus and simple structure, which makes them easier to model. We mentioned several methods to model the RBC membranes' elasticity; however, we discussed the Helfrich free energy model combined with a phase field model to study the cells and vesicles from a single RBC to their collective behavior. In this matter, we found that the dynamics of a single isolated cell immersed in a Poiseuille flow can be modeled using a phase field model coupled with the Navier–Stokes equations. Through this model, we were able to observe the deformation of the RBCs in shear flow.

We later discussed blood and its constitution to relate the elastic properties of the RBCs to the bulk behavior of blood, describing its effect in hemodynamics and hemorheology, again from single-cell to their collective behavior. We presented numerical and experimental points of view considering the latest advances in microfluidics technology to achieve new observations in vitro. From an experimental point of view, we were able to determine the viscosity of blood using microfluidic technology and to determine the effect of RBC aggregation and erythrocyte concentration on whole-blood rheological properties.

Finally, we reviewed and described hematological disorders associated with whole blood and the elastic properties of red blood cells, and how their alterations affect the hemodynamics and rheological properties of blood. We addressed these diseases taking into account the new microfluidic methods that are being developed for diagnostics and future treatment of blood pathologies and their RBC membrane abnormalities. Even though the development of these devices has increased significantly in the past decade, new applications and improvements are being created and discovered every year. Hence, microfluidics applications to diagnostics through the analysis of whole-blood properties or RBCs properties are still and will remain a hot topic in the future.

Author Contributions: Writing—original draft preparation, C.T.-S. and G.R.L.; writing—review and editing, C.T.-S., I.P. and A.H.-M. All authors have read and agreed to the published version of the manuscript.

Funding: This research was funded by VRIEA/PUCV (Grant No. DIII039.425/2021); ANID/PCI (Grant No. MEC80180021); Ministerio de Ciencia e Innovación (Grant No. PID2019-106063GB-100); MINECO (Grant No. PGC2018-098373-B-100); DURSI (Grant No. 2017 SGR 884), and SNF Project no. 200021-175719.

Institutional Review Board Statement: The experimental study mentioned in this article was conducted according to the guidelines of the Declaration of Helsinki, and approved by the University of Barcelona's Bioethics Commission (CBUB) (Project 160016 and date of approval March 1st 2016).

Informed Consent Statement: Informed consent was obtained from all subjects involved in the study.

Data Availability Statement: All the data is contained within the article.

Acknowledgments: Figure 5, adapted by permission from the Royal Society from Chemistry:Soft Matter [7] Tomaiuolo et al. (2009) under license 1181911-1; Figure 6a, reprinted by permission from Springer Nature: Microfluidics and Nanofluidics, [135] Morhell & Pastoriza (2013) under license 5235930238113. Figure 6b adapted by permission from MDPI: Micromachines [132] Kang et al. under Creative Commons CC BY 4.0 license

Conflicts of Interest: The authors declare no conflict of interest.

References

1. Evans, E.A. Bending elastic modulus of red blood cell membrane derived from buckling instability in micropipet aspiration tests. *Biophys. J.* **1983**, *43*, 27–30. [CrossRef]
2. Evans, E.A. Structure and deformation properties of red blood cells: Concepts and quantitative methods. In *Methods in Enzymology*; Elsevier: Amsterdam, The Netherlands, 1989; Volume 173, pp. 3–35.
3. Singer, S.J.; Nicolson, G.L. The fluid mosaic model of the structure of cell membranes. *Science* **1972**, *175*, 720–731. [CrossRef] [PubMed]
4. Canham, P.B. The minimum energy of bending as a possible explanation of the biconcave shape of the human red blood cell. *J. Theor. Biol.* **1970**, *26*, 61–81. [CrossRef]
5. Helfrich, W. Elastic properties of lipid bilayers: Theory and possible experiments. *Zeitschrift für Naturforschung C* **1973**, *28*, 693–703. [CrossRef]
6. Deuling, H.; Helfrich, W. Red blood cell shapes as explained on the basis of curvature elasticity. *Biophys. J.* **1976**, *16*, 861–868. [CrossRef]
7. Tomaiuolo, G.; Simeone, M.; Martinelli, V.; Rotoli, B.; Guido, S. Red blood cell deformation in microconfined flow. *Soft Matter* **2009**, *5*, 3736–3740. [CrossRef]
8. Goldsmith, H.; Marlow, J.; MacIntosh, F.C. Flow behaviour of erythrocytes-I. Rotation and deformation in dilute suspensions. *Proc. R. Soc. London. Ser. B Biol. Sci.* **1972**, *182*, 351–384.
9. Fischer, T.M.; Stohr-Lissen, M.; Schmid-Schonbein, H. The red cell as a fluid droplet: Tank tread-like motion of the human erythrocyte membrane in shear flow. *Science* **1978**, *202*, 894–896. [CrossRef]
10. Abkarian, M.; Faivre, M.; Viallat, A. Swinging of red blood cells under shear flow. *Phys. Rev. Lett.* **2007**, *98*, 188302. [CrossRef]
11. Dupire, J.; Socol, M.; Viallat, A. Full dynamics of a red blood cell in shear flow. *Proc. Natl. Acad. Sci. USA* **2012**, *109*, 20808–20813. [CrossRef]
12. Melzak, K.A.; Lázaro, G.R.; Hernández-Machado, A.; Pagonabarraga, I.; de Espada, J.M.C.D.; Toca-Herrera, J.L. AFM measurements and lipid rearrangements: Evidence from red blood cell shape changes. *Soft Matter* **2012**, *8*, 7716–7726. [CrossRef]
13. Mohandas, N.; Gallagher, P.G. Red cell membrane: Past, present, and future. *Blood* **2008**, *112*, 3939–3948. [CrossRef]
14. Toner, M.; Irimia, D. Blood-on-a-chip. *Annu. Rev. Biomed. Eng.* **2005**, *7*, 77. [CrossRef]
15. Viallat, A.; Abkarian, M. *Dynamics of Blood Cell Suspensions in Microflows*; CRC Press: Boca Raton, FL, USA, 2019.
16. Tomaiuolo, G. Biomechanical properties of red blood cells in health and disease towards microfluidics. *Biomicrofluidics* **2014**, *8*, 051501. [CrossRef]
17. Gervais, L.; De Rooij, N.; Delamarche, E. Microfluidic chips for point-of-care immunodiagnostics. *Adv. Mater.* **2011**, *23*, H151–H176. [CrossRef]
18. Sackmann, E.K.; Fulton, A.L.; Beebe, D.J. The present and future role of microfluidics in biomedical research. *Nature* **2014**, *507*, 181–189. [CrossRef]
19. Sebastian, B.; Dittrich, P.S. Microfluidics to mimic blood flow in health and disease. *Annu. Rev. Fluid Mech.* **2018**, *50*, 483–504. [CrossRef]
20. Lipowsky, R.; Sackmann, E. *Structure and Dynamics of Membranes: I. from Cells to Vesicles/II. Generic and Specific Interactions*; Elsevier: Amsterdam, The Netherlands, 1995.
21. Alberts, B. The cell as a collection of protein machines: Preparing the next generation of molecular biologists. *Cell* **1998**, *92*, 291–294. [CrossRef]
22. Alberts, B.; Bray, D.; Lewis, J.; Raff, M.; Roberts, K.; Watson, J.D. *Molecular Biology of the Cell*; Garland Science: New York, NY, USA, 1994.
23. Van Meer, G.; Voelker, D.R.; Feigenson, G.W. Membrane lipids: Where they are and how they behave. *Nat. Rev. Mol. Cell Biol.* **2008**, *9*, 112. [CrossRef]
24. Yeagle, P.L. *The Membranes of Cells*; Academic Press: Cambridge, MA, USA, 2016.
25. Cullis, P.R.; Hope, M.J.; Tilcock, C.P. Lipid polymorphism and the roles of lipids in membranes. *Chem. Phys. Lipids* **1986**, *40*, 127–144. [CrossRef]
26. Finegold, L.X. *Cholesterol in Membrane Models*; CRC Press: Boca Raton, FL, USA, 1992.
27. Maxfield, F.R.; van Meer, G. Cholesterol, the central lipid of mammalian cells. *Curr. Opin. Cell Biol.* **2010**, *22*, 422–429. [CrossRef] [PubMed]
28. Yeagle, P.L. *The Structure of Biological Membranes*; CRC Press: Boca Raton, FL, USA, 2004.
29. Boal, D.; Boal, D.H. *Mechanics of the Cell*; Cambridge University Press: Cambridge, UK, 2012.

30. Bennett, V. The spectrin-actin junction of erythrocyte membrane skeletons. *Biochim. Biophys. Acta (BBA)-Rev. Biomembr.* **1989**, *988*, 107–121. [CrossRef]
31. Li, J.; Dao, M.; Lim, C.; Suresh, S. Spectrin-level modeling of the cytoskeleton and optical tweezers stretching of the erythrocyte. *Biophys. J.* **2005**, *88*, 3707–3719. [CrossRef] [PubMed]
32. Mohandas, N.; Evans, E. Mechanical properties of the red cell membrane in relation to molecular structure and genetic defects. *Annu. Rev. Biophys. Biomol. Struct.* **1994**, *23*, 787–818. [CrossRef] [PubMed]
33. Stillwell, W. *An Introduction to Biological Membranes: From Bilayers to Rafts*; Newnes: Oxford, UK, 2013.
34. Závodszky, G.; van Rooij, B.; Azizi, V.; Hoekstra, A. Cellular level in-silico modeling of blood rheology with an improved material model for red blood cells. *Front. Physiol.* **2017**, *8*, 563. [CrossRef] [PubMed]
35. Peng, Z.; Mashayekh, A.; Zhu, Q. Erythrocyte responses in low-shear-rate flows: Effects of non-biconcave stress-free state in the cytoskeleton. *J. Fluid Mech.* **2014**, *742*, 96–118. [CrossRef]
36. Salehyar, S.; Zhu, Q. Deformation and internal stress in a red blood cell as it is driven through a slit by an incoming flow. *Soft Matter* **2016**, *12*, 3156–3164. [CrossRef] [PubMed]
37. Freund, J.B. The flow of red blood cells through a narrow spleen-like slit. *Phys. Fluids* **2013**, *25*, 110807. [CrossRef]
38. Gompper, G.; Schick, M.; Milner, S. Self-assembling amphiphilic systems. *Phys. Today* **2008**, *48*, 91–93. [CrossRef]
39. Lázaro, G.R.; Pagonabarraga, I.; Hernández-Machado, A. Phase-field theories for mathematical modeling of biological membranes. *Chem. Phys. Lipids* **2015**, *185*, 46–60. [CrossRef]
40. Pontrelli, G.; Halliday, I.; Melchionna, S.; Spencer, T.J.; Succi, S. Lattice Boltzmann method as a computational framework for multiscale haemodynamics. *Math. Comput. Model. Dyn. Syst.* **2014**, *20*, 470–490. [CrossRef]
41. Pivkin, I.V.; Karniadakis, G.E. Accurate coarse-grained modeling of red blood cells. *Phys. Rev. Lett.* **2008**, *101*, 118105. [CrossRef]
42. Fedosov, D.A.; Caswell, B.; Karniadakis, G.E. A multiscale red blood cell model with accurate mechanics, rheology, and dynamics. *Biophys. J.* **2010**, *98*, 2215–2225. [CrossRef]
43. Peskin, C.S. The immersed boundary method. *Acta Numer.* **2002**, *11*, 479–517. [CrossRef]
44. Kaoui, B.; Krüger, T.; Harting, J. How does confinement affect the dynamics of viscous vesicles and red blood cells? *Soft Matter* **2012**, *8*, 9246–9252. [CrossRef]
45. Pozrikidis, C. *Boundary Integral and Singularity Methods for Linearized Viscous Flow*; Cambridge University Press: Cambridge, UK, 1992.
46. Pozrikidis, C. Finite deformation of liquid capsules enclosed by elastic membranes in simple shear flow. *J. Fluid Mech.* **1995**, *297*, 123–152. [CrossRef]
47. Malevanets, A.; Kapral, R. Mesoscopic model for solvent dynamics. *J. Chem. Phys.* **1999**, *110*, 8605–8613. [CrossRef]
48. McWhirter, J.L.; Noguchi, H.; Gompper, G. Flow-induced clustering and alignment of vesicles and red blood cells in microcapillaries. *Proc. Natl. Acad. Sci. USA* **2009**, *106*, 6039–6043. [CrossRef]
49. Peng, Z.; Li, X.; Pivkin, I.V.; Dao, M.; Karniadakis, G.E.; Suresh, S. Lipid bilayer and cytoskeletal interactions in a red blood cell. *Proc. Natl. Acad. Sci. USA* **2013**, *110*, 13356–13361. [CrossRef]
50. Du, Q.; Liu, C.; Wang, X. A phase field approach in the numerical study of the elastic bending energy for vesicle membranes. *J. Comput. Phys.* **2004**, *198*, 450–468. [CrossRef]
51. Gompper, G.; Shick, M. *Phase Transitions and Critical Phenomena*; Elsevier: Amsterdam, The Netherlands, 1994; Volume 76, p. 16. .
52. Campelo, F.; Hernandez-Machado, A. Shape instabilities in vesicles: A phase-field model. *Eur. Phys. J. Spec. Top.* **2007**, *143*, 101–108. [CrossRef]
53. Rosolen, A.; Peco, C.; Arroyo, M. An adaptive meshfree method for phase-field models of biomembranes. Part I: Approximation with maximum-entropy basis functions. *J. Comput. Phys.* **2013**, *249*, 303–319. [CrossRef]
54. Peco, C.; Rosolen, A.; Arroyo, M. An adaptive meshfree method for phase-field models of biomembranes. Part II: A Lagrangian approach for membranes in viscous fluids. *J. Comput. Phys.* **2013**, *249*, 320–336. [CrossRef]
55. Campelo, F.; Hernandez-Machado, A. Dynamic model and stationary shapes of fluid vesicles. *Eur. Phys. J. E* **2006**, *20*, 37–45. [CrossRef] [PubMed]
56. Deschamps, J.; Kantsler, V.; Segre, E.; Steinberg, V. Dynamics of a vesicle in general flow. *Proc. Natl. Acad. Sci. USA* **2009**, *106*, 11444–11447. [CrossRef] [PubMed]
57. Deschamps, J.; Kantsler, V.; Steinberg, V. Phase diagram of single vesicle dynamical states in shear flow. *Phys. Rev. Lett.* **2009**, *102*, 118105. [CrossRef] [PubMed]
58. Lázaro, G.R.; Hernández-Machado, A.; Pagonabarraga, I. Rheology of red blood cells under flow in highly confined microchannels: I. effect of elasticity. *Soft Matter* **2014**, *10*, 7195–7206. [CrossRef] [PubMed]
59. Lázaro, G.R.; Hernández-Machado, A.; Pagonabarraga, I. Rheology of red blood cells under flow in highly confined microchannels. II. Effect of focusing and confinement. *Soft Matter* **2014**, *10*, 7207–7217. [CrossRef]
60. Mofrad, M.R. Rheology of the cytoskeleton. *Annu. Rev. Fluid Mech.* **2009**, *41*, 433–453. [CrossRef]
61. Dimova, R. Recent developments in the field of bending rigidity measurements on membranes. *Adv. Colloid Interface Sci.* **2014**, *208*, 225–234. [CrossRef]
62. Scheffer, L.; Bitler, A.; Ben-Jacob, E.; Korenstein, R. Atomic force pulling: Probing the local elasticity of the cell membrane. *Eur. Biophys. J.* **2001**, *30*, 83–90. [CrossRef]

63. Betz, T.; Lenz, M.; Joanny, J.F.; Sykes, C. ATP-dependent mechanics of red blood cells. *Proc. Natl. Acad. Sci. USA* **2009**, *106*, 15320–15325. [CrossRef]
64. Abkarian, M.; Viallat, A. Vesicles and red blood cells in shear flow. *Soft Matter* **2008**, *4*, 653–657. [CrossRef]
65. Mauer, J.; Mendez, S.; Lanotte, L.; Nicoud, F.; Abkarian, M.; Gompper, G.; Fedosov, D.A. Flow-induced transitions of red blood cell shapes under shear. *Phys. Rev. Lett.* **2018**, *121*, 118103. [CrossRef]
66. Zarda, P.; Chien, S.; Skalak, R. Elastic deformations of red blood cells. *J. Biomech.* **1977**, *10*, 211–221. [CrossRef]
67. Besis, M. Red Cell Shapes, An Illustrated Classification and Its Rationale. In *Red Cell Shape*; Besis, M., Weed, R.I., Leblond, P.F., Eds.; Springer: Berlin/Heidelberg, Germany, 1973.
68. Uzoigwe, C. The human erythrocyte has developed the biconcave disc shape to optimise the flow properties of the blood in the large vessels. *Med. Hypotheses* **2006**, *67*, 1159–1163. [CrossRef]
69. Reinhart, W.H.; Chien, S. Red cell rheology in stomatocyte-echinocyte transformation: Roles of cell geometry and cell shape. *Blood* **1986**, *67*, 1110–1118. [CrossRef]
70. Hoffman, R.; Benz, E.J., Jr.; Silberstein, L.E.; Heslop, H.; Anastasi, J.; Weitz, J. *Hematology: Basic Principles and Practice*; Elsevier Health Sciences: Amsterdam, The Netherlands, 2013.
71. Mathew, J.; Sankar, P.; Varacallo, M. *Physiology, Blood Plasma*; StatPearls Publishing: Treasure Island, FL, USA 2018.
72. Brust, M.; Schaefer, C.; Doerr, R.; Pan, L.; Garcia, M.; Arratia, P.; Wagner, C. Rheology of human blood plasma: Viscoelastic versus Newtonian behavior. *Phys. Rev. Lett.* **2013**, *110*, 078305. [CrossRef]
73. Schaller, J.; Gerber, S.; Kaempfer, U.; Lejon, S.; Trachsel, C. *Human Blood Plasma Proteins: Structure and Function*; John Wiley & Sons: Hoboken, NJ, USA, 2008.
74. Baskurt, O.; Neu, B.; Meiselman, H.J. *Red Blood Cell Aggregation*; CRC Press: Boca Raton, FL, USA, 2011.
75. Yeom, E.; Park, J.H.; Kang, Y.J.; Lee, S.J. Microfluidics for simultaneous quantification of platelet adhesion and blood viscosity. *Sci. Rep.* **2016**, *6*, 24994. [CrossRef]
76. George, M.J.; Bynum, J.; Nair, P.; Cap, A.P.; Wade, C.E.; Cox, C.S., Jr.; Gill, B.S. Platelet biomechanics, platelet bioenergetics, and applications to clinical practice and translational research. *Platelets* **2018**, *29*, 431–439. [CrossRef]
77. Fung, Y. *Biomechanics*; Springer Science+Business Media: New York, NY, USA, 1985.
78. Hall, J.E.; Hall, M.E. *Guyton and Hall Textbook of Medical Physiology E-Book*; Elsevier Health Sciences: Amsterdam, The Netherlands, 2020.
79. Secomb, T.W. Blood flow in the microcirculation. *Annu. Rev. Fluid Mech.* **2017**, *49*, 443–461. [CrossRef]
80. Cho, Y.I.; Mooney, M.P.; Cho, D.J. Hemorheological disorders in diabetes mellitus. *J. Diabetes Sci. Technol.* **2008**, *2*, 1130–1138. [CrossRef] [PubMed]
81. Cowan, A.Q.; Cho, D.J.; Rosenson, R.S. Importance of blood rheology in the pathophysiology of atherothrombosis. *Cardiovasc. Drugs Ther.* **2012**, *26*, 339–348. [CrossRef] [PubMed]
82. Connes, P.; Alexy, T.; Detterich, J.; Romana, M.; Hardy-Dessources, M.D.; Ballas, S.K. The role of blood rheology in sickle cell disease. *Blood Rev.* **2015**, *330*, 111–118 [CrossRef] [PubMed]
83. Secomb, T.W. Hemodynamics. *Compr. Physiol.* **2011**, *6*, 975–1003.
84. Goldsmith, H.; Skalak, R. Hemodynamics. *Annu. Rev. Fluid Mech.* **1975**, *7*, 213–247. [CrossRef]
85. Lipowsky, H.H. Microvascular rheology and hemodynamics. *Microcirculation* **2005**, *12*, 5–15. [CrossRef] [PubMed]
86. Popel, A.S.; Johnson, P.C. Microcirculation and hemorheology. *Annu. Rev. Fluid Mech.* **2005**, *37*, 43. [CrossRef]
87. Omori, T.; Imai, Y.; Kikuchi, K.; Ishikawa, T.; Yamaguchi, T. Hemodynamics in the Microcirculation and in Microfluidics. *Ann. Biomed. Eng.* **2015**, *43*, 238–257. [CrossRef]
88. Ju, M.; Ye, S.S.; Namgung, B.; Cho, S.; Low, H.T.; Leo, H.L.; Kim, S. A review of numerical methods for red blood cell flow simulation. *Comput. Methods Biomech. Biomed. Eng.* **2015**, *18*, 130–140. [CrossRef]
89. Abkarian, M.; Faivre, M.; Horton, R.; Smistrup, K.; Best-Popescu, C.A.; Stone, H.A. Cellular-scale hydrodynamics. *Biomed. Mater.* **2008**, *3*, 034011. [CrossRef]
90. Kaoui, B.; Biros, G.; Misbah, C. Why do red blood cells have asymmetric shapes even in a symmetric flow? *Phys. Rev. Lett.* **2009**, *103*, 188101. [CrossRef]
91. Lei, H.; Fedosov, D.A.; Caswell, B.; Karniadakis, G.E. Blood flow in small tubes: Quantifying the transition to the non-continuum regime. *J. Fluid Mech.* **2013**, *722*, 214–239. [CrossRef]
92. Fedosov, D.A.; Noguchi, H.; Gompper, G. Multiscale modeling of blood flow: From single cells to blood rheology. *Biomech. Model. Mechanobiol.* **2014**, *13*, 239–258. [CrossRef]
93. Lázaro, G.R.; Pagonabarraga, I.; Hernández-Machado, A. Elastic and dynamic properties of membrane phase-field models. *Eur. Phys. J. E* **2017**, *40*, 77. [CrossRef]
94. Arroyo, M.; DeSimone, A. Relaxation dynamics of fluid membranes. *Phys. Rev. E* **2009**, *79*, 031915. [CrossRef]
95. Sugii, Y.; Okuda, R.; Okamoto, K.; Madarame, H. Velocity measurement of both red blood cells and plasma of in vitro blood flow using high-speed micro PIV technique. *Meas. Sci. Technol.* **2005**, *16*, 1126. [CrossRef]
96. Pitts, K.L.; Mehri, R.; Mavriplis, C.; Fenech, M. Micro-particle image velocimetry measurement of blood flow: Validation and analysis of data pre-processing and processing methods. *Meas. Sci. Technol.* **2012**, *23*, 105302. [CrossRef]
97. Pitts, K.L.; Fenech, M. High speed versus pulsed images for micro-particle image velocimetry: A direct comparison of red blood cells versus fluorescing tracers as tracking particles. *Physiol. Meas.* **2013**, *34*, 1363. [CrossRef]

98. Pasias, D.; Passos, A.; Constantinides, G.; Balabani, S.; Kaliviotis, E. Surface tension driven flow of blood in a rectangular microfluidic channel: Effect of erythrocyte aggregation. *Phys. Fluids* **2020**, *32*, 071903. [CrossRef]
99. Kucukal, E.; Man, Y.; Gurkan, U.A.; Schmidt, B. Blood Flow Velocimetry in a Microchannel During Coagulation Using Particle Image Velocimetry and Wavelet-Based Optical Flow Velocimetry. *J. Biomech. Eng.* **2021**, *143*, 091004. [CrossRef] [PubMed]
100. Abkarian, M.; Faivre, M.; Stone, H.A. High-speed microfluidic differential manometer for cellular-scale hydrodynamics. *Proc. Natl. Acad. Sci. USA* **2006**, *103*, 538–542. [CrossRef] [PubMed]
101. Tomaiuolo, G.; Barra, M.; Preziosi, V.; Cassinese, A.; Rotoli, B.; Guido, S. Microfluidics analysis of red blood cell membrane viscoelasticity. *Lab A Chip* **2011**, *11*, 449–454. [CrossRef]
102. Kang, Y.J. Continuous and simultaneous measurement of the biophysical properties of blood in a microfluidic environment. *Analyst* **2016**, *141*, 6583–6597. [CrossRef] [PubMed]
103. Guruprasad, P.; Mannino, R.G.; Caruso, C.; Zhang, H.; Josephson, C.D.; Roback, J.D.; Lam, W.A. Integrated automated particle tracking microfluidic enables high-throughput cell deformability cytometry for red cell disorders. *Am. J. Hematol.* **2019**, *94*, 189–199. [CrossRef] [PubMed]
104. Man, Y.; Maji, D.; An, R.; Ahuja, S.P.; Little, J.A.; Suster, M.A.; Mohseni, P.; Gurkan, U.A. Microfluidic electrical impedance assessment of red blood cell-mediated microvascular occlusion. *Lab A Chip* **2021**, *21*, 1036–1048. [CrossRef] [PubMed]
105. Reichenwallner, A.K.; Vurmaz, E.; Battis, K.; Handl, L.; Üstün, H.; Mach, T.; Hörnig, G.; Lipfert, J.; Richter, L. Optical Investigation of Individual Red Blood Cells for Determining Cell Count and Cellular Hemoglobin Concentration in a Microfluidic Channel. *Micromachines* **2021**, *12*, 358. [CrossRef] [PubMed]
106. Guo, Q.; Duffy, S.P.; Matthews, K.; Santoso, A.T.; Scott, M.D.; Ma, H. Microfluidic analysis of red blood cell deformability. *J. Biomech.* **2014**, *47*, 1767–1776. [CrossRef]
107. Li, X.; Peng, Z.; Lei, H.; Dao, M.; Karniadakis, G.E. Probing red blood cell mechanics, rheology and dynamics with a two-component multi-scale model. *Philos. Trans. R. Soc. A Math. Phys. Eng. Sci.* **2014**, *372*, 20130389. [CrossRef]
108. Pivkin, I.V.; Peng, Z.; Karniadakis, G.E.; Buffet, P.A.; Dao, M.; Suresh, S. Biomechanics of red blood cells in human spleen and consequences for physiology and disease. *Proc. Natl. Acad. Sci. USA* **2016**, *113*, 7804–7809. [CrossRef]
109. Quinn, D.J.; Pivkin, I.; Wong, S.Y.; Chiam, K.H.; Dao, M.; Karniadakis, G.E.; Suresh, S. Combined simulation and experimental study of large deformation of red blood cells in microfluidic systems. *Ann. Biomed. Eng.* **2011**, *39*, 1041–1050. [CrossRef]
110. Guckenberger, A.; Kihm, A.; John, T.; Wagner, C.; Gekle, S. Numerical–experimental observation of shape bistability of red blood cells flowing in a microchannel. *Soft Matter* **2018**, *14*, 2032–2043. [CrossRef]
111. Lu, H.; Peng, Z. Boundary integral simulations of a red blood cell squeezing through a submicron slit under prescribed inlet and outlet pressures. *Phys. Fluids* **2019**, *31*, 031902.
112. Cokelet, G.; Merrill, E.; Gilliland, E.; Shin, H.; Britten, A.; Wells , R., Jr. The rheology of human blood—Measurement near and at zero shear rate. *Trans. Soc. Rheol.* **1963**, *7*, 303–317. [CrossRef]
113. Merrill, E.W. Rheology of blood. *Physiol. Rev* **1969**, *49*, 863–888. [CrossRef]
114. Chien, S. Shear dependence of effective cell volume as a determinant of blood viscosity. *Science* **1970**, *168*, 977–979. [CrossRef]
115. Thurston, G.B. Viscoelasticity of human blood. *Biophys. J.* **1972**, *12*, 1205. [CrossRef]
116. Baskurt, O.K. *Handbook of Hemorheology and Hemodynamics*; IOS Press: Amsterdam, The Netherlands 2007; Volume 69.
117. Eckmann, D.M.; Bowers, S.; Stecker, M.; Cheung, A.T. Hematocrit, volume expander, temperature, and shear rate effects on blood viscosity. *Anesth. Analg.* **2000**, *91*, 539–545. [CrossRef]
118. Thurston, G.B.; Henderson, N.M. Effects of flow geometry on blood viscoelasticity. *Biorheology* **2006**, *43*, 729–746.
119. Chien, S. Red cell deformability and its relevance to blood flow. *Annu. Rev. Physiol.* **1987**, *49*, 177–192. [CrossRef]
120. Liu, Y.; Liu, W.K. Rheology of red blood cell aggregation by computer simulation. *J. Comput. Phys.* **2006**, *220*, 139–154. [CrossRef]
121. McWhirter, J.L.; Noguchi, H.; Gompper, G. Deformation and clustering of red blood cells in microcapillary flows. *Soft Matter* **2011**, *7*, 10967–10977. [CrossRef]
122. Fedosov, D.A.; Pan, W.; Caswell, B.; Gompper, G.; Karniadakis, G.E. Predicting human blood viscosity in silico. *Proc. Natl. Acad. Sci. USA* **2011**, *108*, 11772–11777. [CrossRef] [PubMed]
123. Lowe, G.D.O. *Clinical Blood Rheology*; CRC Press: Boca Raton, FL, USA, 1988.
124. Srivastava, N.; Davenport, R.D.; Burns, M.A. Nanoliter viscometer for analyzing blood plasma and other liquid samples. *Anal. Chem.* **2005**, *77*, 383–392. [CrossRef] [PubMed]
125. Guillot, P.; Panizza, P.; Salmon, J.B.; Joanicot, M.; Colin, A.; Bruneau, C.H.; Colin, T. Viscosimeter on a microfluidic chip. *Langmuir* **2006**, *22*, 6438–6445. [CrossRef] [PubMed]
126. Burns, M.A.; Srivastava, N.; Davenport, R.D. Nanoliter Viscometer for Analyzing Blood Plasma and Other Liquid Samples. U.S. Patent 7,188,515, 13 March 2007.
127. Trejo-Soto, C.; Costa-Miracle, E.; Rodriguez-Villarreal, I.; Cid, J.; Alarcón, T.; Hernández-Machado, A. Capillary Filling at the Microscale: Control of Fluid Front Using Geometry. *PLoS ONE* **2016**, *11*, e0153559. [CrossRef] [PubMed]
128. Gupta, S.; Wang, W.S.; Vanapalli, S.A. Microfluidic viscometers for shear rheology of complex fluids and biofluids. *Biomicrofluidics* **2016**, *10*, 043402. [CrossRef]
129. Kang, Y.J.; Yoon, S.Y.; Lee, K.H.; Yang, S. A highly accurate and consistent microfluidic viscometer for continuous blood viscosity measurement. *Artif. Organs* **2010**, *34*, 944–949. [CrossRef]

130. Trejo-Soto, C.; Costa-Miracle, E.; Rodriguez-Villarreal, I.; Cid, J.; Castro, M.; Alarcon, T.; Hernandez-Machado, A. Front microrheology of the non-Newtonian behaviour of blood: Scaling theory of erythrocyte aggregation by aging. *Soft Matter* **2017**, *13*, 3042–3047. [CrossRef]
131. Kim, B.J.; Lee, S.Y.; Jee, S.; Atajanov, A.; Yang, S. Micro-viscometer for measuring shear-varying blood viscosity over a wide-ranging shear rate. *Sensors* **2017**, *17*, 1442. [CrossRef]
132. Kang, Y.J. Microfluidic-based technique for measuring RBC aggregation and blood viscosity in a continuous and simultaneous fashion. *Micromachines* **2018**, *9*, 467. [CrossRef]
133. Khnouf, R.; Karasneh, D.; Abdulhay, E.; Abdelhay, A.; Sheng, W.; Fan, Z.H. Microfluidics-based device for the measurement of blood viscosity and its modeling based on shear rate, temperature, and heparin concentration. *Biomed. Microdevices* **2019**, *21*, 80. [CrossRef]
134. Carvalho, V.; Gonçalves, I.M.; Souza, A.; Souza, M.S.; Bento, D.; Ribeiro, J.E.; Lima, R.; Pinho, D. Manual and Automatic Image Analysis Segmentation Methods for Blood Flow Studies in Microchannels. *Micromachines* **2021**, *12*, 317. [CrossRef]
135. Morhell, N.; Pastoriza, H. A single channel capillary microviscometer. *Microfluid. Nanofluidics* **2013**, *15*, 475–479. [CrossRef]
136. Morhell, N.; Pastoriza, H. Power law fluid viscometry through capillary filling in a closed microchannel. *Sens. Actuators B Chem.* **2016**, *227*, 24–28. [CrossRef]
137. Delamarche, E.; Temiz, Y. Continuous, Capacitance-Based Monitoring of Liquid Flows in a Microfluidic Device. U.S. Patent 10,369,567, 6 August 2019.
138. Farrarons, J.C.; Machado, A.H.; Cor, T.A.; Villarreal, A.I.R.; Catala, P.L.M. Method, Apparatus and Micro-Rheometer for Measuring Rheological Properties of Newtonian and Non-Newtonian Fluids. U.S. Patent 10,386,282, 20 August 2019.
139. Méndez-Mora, L.; Cabello-Fusarés, M.; Ferré-Torres, J.; Riera-Llobet, C.; Lopez, S.; Trejo-Soto, C.; Alarcón, T.; Hernandez-Machado, A. Microrheometer for Biofluidic Analysis: Electronic Detection of the Fluid-Front Advancement. *Micromachines* **2021**, *12*, 726. [CrossRef]
140. Larson, R.G. *The Structure and Rheology of Complex Fluids*; Oxford University Press: New York, NY, USA, 1999; Volume 150.
141. Xia, Y.; Whitesides, G.M. Soft lithography. *Annu. Rev. Mater. Sci.* **1998**, *28*, 153–184. [CrossRef]
142. Vulto, P.; Glade, N.; Altomare, L.; Bablet, J.; Del Tin, L.; Medoro, G.; Chartier, I.; Manaresi, N.; Tartagni, M.; Guerrieri, R. Microfluidic channel fabrication in dry film resist for production and prototyping of hybrid chips. *Lab A Chip* **2005**, *5*, 158–162. [CrossRef]
143. Qin, D.; Xia, Y.; Whitesides, G.M. Soft lithography for micro-and nanoscale patterning. *Nat. Protoc.* **2010**, *5*, 491–502. [CrossRef]
144. Yilmaz, F.; Gundogdu, M.Y. A critical review on blood flow in large arteries; relevance to blood rheology, viscosity models, and physiologic conditions. *Korea-Aust. Rheol. J.* **2008**, *20*, 197–211.
145. Matas, J.P.; Glezer, V.; Guazzelli, É.; Morris, J.F. Trains of particles in finite-Reynolds-number pipe flow. *Phys. Fluids* **2004**, *16*, 4192–4195. [CrossRef]
146. Fåhræus, R. VOL. IX APRIL, 1929 No. 2. *Physiol. Rev.* **1929**, *9*, 241–274 .
147. Fåhræus, R.; Lindqvist, T. The viscosity of the blood in narrow capillary tubes. *Am. J. Physiol.- Content* **1931**, *96*, 562–568. [CrossRef]
148. Stergiou, Y.G.; Keramydas, A.T.; Anastasiou, A.D.; Mouza, A.A.; Paras, S.V. Experimental and Numerical Study of Blood Flow in μ-vessels: Influence of the Fahraeus–Lindqvist Effect. *Fluids* **2019**, *4*, 143. [CrossRef]
149. Geislinger, T.M.; Franke, T. Hydrodynamic lift of vesicles and red blood cells in flow—From Fåhræus & Lindqvist to microfluidic cell sorting. *Adv. Colloid Interface Sci.* **2014**, *208*, 161–176.
150. Iss, C.; Midou, D.; Moreau, A.; Held, D.; Charrier, A.; Mendez, S.; Viallat, A.; Helfer, E. Self-organization of red blood cell suspensions under confined 2D flows. *Soft Matter* **2019**, *15*, 2971–2980. [CrossRef]
151. Lazaro, G.R.; Hernández-Machado, A.; Pagonabarraga, I. Collective behavior of red blood cells in confined channels. *Eur. Phys. J. E* **2019**, *42*, 46. [CrossRef]
152. Zhang, J.; Johnson, P.C.; Popel, A.S. Red blood cell aggregation and dissociation in shear flows simulated by lattice Boltzmann method. *J. Biomech.* **2008**, *41*, 47–55. [CrossRef]
153. Wang, T.; Pan, T.W.; Xing, Z.; Glowinski, R. Numerical simulation of rheology of red blood cell rouleaux in microchannels. *Phys. Rev. E* **2009**, *79*, 041916. [CrossRef]
154. Xu, D.; Kaliviotis, E.; Munjiza, A.; Avital, E.; Ji, C.; Williams, J. Large scale simulation of red blood cell aggregation in shear flows. *J. Biomech.* **2013**, *46*, 1810–1817. [CrossRef]
155. Tomaiuolo, G.; Lanotte, L.; D'Apolito, R.; Cassinese, A.; Guido, S. Microconfined flow behavior of red blood cells. *Med. Eng. Phys.* **2016**, *38*, 11–16. [CrossRef] [PubMed]
156. Mehri, R.; Mavriplis, C.; Fenech, M. Red blood cell aggregates and their effect on non-Newtonian blood viscosity at low hematocrit in a two-fluid low shear rate microfluidic system. *PLoS ONE* **2018**, *13*, e0199911. [CrossRef] [PubMed]
157. Thiébaud, M.; Shen, Z.; Harting, J.; Misbah, C. Prediction of anomalous blood viscosity in confined shear flow. *Phys. Rev. Lett.* **2014**, *112*, 238304. [CrossRef] [PubMed]
158. Viallat, A.; Abkarian, M. Red blood cell: From its mechanics to its motion in shear flow. *Int. J. Lab. Hematol.* **2014**, *36*, 237–243. [CrossRef] [PubMed]
159. Vijay, A.; Gertz, M.A. Waldenström macroglobulinemia. *Blood J. Am. Soc. Hematol.* **2007**, *109*, 5096–5103. [CrossRef] [PubMed]

160. Luquita, A.; Urli, L.; Svetaz, M.; Gennaro, A.M.; Volpintesta, R.; Palatnik, S.; Rasia, M. Erythrocyte aggregation in rheumatoid arthritis: Cell and plasma factor's role. *Clin. Hemorheol. Microcirc.* **2009**, *41*, 49–56. [CrossRef] [PubMed]
161. Flormann, D.; Schirra, K.; Podgorski, T.; Wagner, C. On the rheology of red blood cell suspensions with different amounts of dextran: Separating the effect of aggregation and increase in viscosity of the suspending phase. *Rheol. Acta* **2016**, *55*, 477–483. [CrossRef]
162. Gyawali, P.; Ziegler, D.; Cailhier, J.F.; Denault, A.; Cloutier, G. Quantitative measurement of erythrocyte aggregation as a systemic inflammatory marker by ultrasound imaging: A systematic review. *Ultrasound Med. Biol.* **2018**, *44*, 1303–1317. [CrossRef]
163. Le Devehat, C.; Vimeux, M.; Khodabandehlou, T. Blood rheology in patients with diabetes mellitus. *Clin. Hemorheol. Microcirc.* **2004**, *30*, 297–300.
164. Li, Q.; Li, L.; Li, Y. Enhanced RBC aggregation in type 2 diabetes patients. *J. Clin. Lab. Anal.* **2015**, *29*, 387–389. [CrossRef]
165. Presti, R.L.; Hopps, E.; Caimi, G. Hemorheological abnormalities in human arterial hypertension. *Korea-Aust. Rheol. J.* **2014**, *26*, 199–204. [CrossRef]
166. Wiewiora, M.; Piecuch, J.; Glűck, M.; Slowinska-Lozynska, L.; Sosada, K. The effects of weight loss surgery on blood rheology in severely obese patients. *Surg. Obes. Relat. Dis.* **2015**, *11*, 1307–1314. [CrossRef]
167. Lowe, G.; Lee, A.; Rumley, A.; Price, J.; Fowkes, F. Blood viscosity and risk of cardiovascular events: The Edinburgh Artery Study. *Br. J. Haematol.* **1997**, *96*, 168–173. [CrossRef]
168. Bilgi, M.; Güllü, H.; Kozanoğlu, İ.; Özdoğu, H.; Sezgin, N.; Sezgin, A.T.; Altay, H.; Erol, T.; Müderrisoğlu, H. Evaluation of blood rheology in patients with coronary slow flow or non-obstructive coronary artery disease. *Clin. Hemorheol. Microcirc.* **2012**, *53*, 317–326. [CrossRef]
169. Sloop, G.; Holsworth Jr, R.E.; Weidman, J.J.; St Cyr, J.A. The role of chronic hyperviscosity in vascular disease. *Ther. Adv. Cardiovasc. Dis.* **2015**, *9*, 19–25. [CrossRef]
170. Kwaan, H.C.; Bongu, A. The hyperviscosity syndromes. *Seminars in Thrombosis and Hemostasis*; Thieme Medical Publishers, Inc.: Stuttgart, Germany, 1999; Volume 25, pp. 199–208.
171. Gertz, M.A. Acute hyperviscosity: Syndromes and management. *Blood* **2018**, *132*, 1379–1385. [CrossRef]
172. Gallagher, P.G. Red cell membrane disorders. *ASH Educ. Program Book* **2005**, *2005*, 13–18. [CrossRef]
173. Sloop, G.D.; De Mast, Q.; Pop, G.; Weidman, J.J.; Cyr, J.A.S. The role of blood viscosity in infectious diseases. *Cureus* **2020**, *12*, e7090. [CrossRef]
174. Chien, S.; Usami, S.; Bertles, J.F. Abnormal rheology of oxygenated blood in sickle cell anemia. *J. Clin. Investig.* **1970**, *49*, 623. [CrossRef] [PubMed]
175. Aich, A.; Lamarre, Y.; Sacomani, D.P.; Kashima, S.; Covas, D.T.; De la Torre, L.G. Microfluidics in Sickle Cell Disease Research: State of the Art and a Perspective Beyond the Flow Problem. *Front. Mol. Biosci.* **2021**, *7*, 252. [CrossRef]
176. Iragorri, M.A.L.; El Hoss, S.; Brousse, V.; Lefevre, S.D.; Dussiot, M.; Xu, T.; Ferreira, A.R.; Lamarre, Y.; Pinto, A.C.S.; Kashima, S.; et al. A microfluidic approach to study the effect of mechanical stress on erythrocytes in sickle cell disease. *Lab A Chip* **2018**, *18*, 2975–2984. [CrossRef] [PubMed]
177. Higgins, J.; Eddington, D.; Bhatia, S.; Mahadevan, L. Sickle cell vasoocclusion and rescue in a microfluidic device. *Proc. Natl. Acad. Sci. USA* **2007**, *104*, 20496–20500. [CrossRef] [PubMed]
178. Man, Y.; Kucukal, E.; An, R.; Watson, Q.D.; Bosch, J.; Zimmerman, P.A.; Little, J.A.; Gurkan, U.A. Microfluidic assessment of red blood cell mediated microvascular occlusion. *Lab A Chip* **2020**, *20*, 2086–2099. [CrossRef] [PubMed]
179. Wood, D.K.; Soriano, A.; Mahadevan, L.; Higgins, J.M.; Bhatia, S.N. A biophysical indicator of vaso-occlusive risk in sickle cell disease. *Sci. Transl. Med.* **2012**, *4*, 123ra26. [CrossRef]
180. Kucukal, E.; Man, Y.; Hill, A.; Liu, S.; Bode, A.; An, R.; Kadambi, J.; Little, J.A.; Gurkan, U.A. Whole blood viscosity and red blood cell adhesion: Potential biomarkers for targeted and curative therapies in sickle cell disease. *Am. J. Hematol.* **2020**, *95*, 1246–1256. [CrossRef]
181. Clavería, V.; Connes, P.; Lanotte, L.; Renoux, C.; Joly, P.; Fort, R.; Gauthier, A.; Wagner, C.; Abkarian, M. In vitro red blood cell segregation in sickle cell anemia. *Front. Phys.* **2021**, *9*, 712. [CrossRef]
182. Ilyas, S.; Simonson, A.E.; Asghar, W. Emerging point-of-care technologies for sickle cell disease diagnostics. *Clin. Chim. Acta* **2020**, *501*, 85–91. [CrossRef]
183. Lu, X.; Chaudhury, A.; Higgins, J.M.; Wood, D.K. Oxygen-dependent flow of sickle trait blood as an in vitro therapeutic benchmark for sickle cell disease treatments. *Am. J. Hematol.* **2018**, *93*, 1227–1235. [CrossRef]
184. Osorio, G. Libro Hematología. Diagnóstico y Terapéutica. Adultos y niños. *Rev. Chil. Pediatr.* **2019**, *90*, 458–459. [CrossRef]
185. Xu, T.; Lizarralde-Iragorri, M.A.; Roman, J.; Ghasemi, R.; Lefevre, J.P.; Martincic, E.; Brousse, V.; Français, O.; El Nemer, W.; Le Pioufle, B. Characterization of red blood cell microcirculatory parameters using a bioimpedance microfluidic device. *Sci. Rep.* **2020**, *10*, 9869. [CrossRef]
186. Advani, R.; Sorenson, S.; Shinar, E.; Lande, W.; Rachmilewitz, E.; Schrier, S.L. Characterization and comparison of the red blood cell membrane damage in severe human alpha-and beta-thalassemia. *Blood* **1992**, *79*, 1058–1063. [CrossRef]
187. Krishnevskaya, E.; Payan-Pernia, S.; Hernández-Rodríguez, I.; Sevilla, Á.F.R.; Serra, Á.A.; Morales-Indiano, C.; Ferrer, M.S.; Vives-Corrons, J.L. Distinguishing iron deficiency from beta-thalassemia trait by new generation ektacytometry. *Int. J. Lab. Hematol.* **2021**, *43*, e58–e60. [CrossRef]

188. Bessis, M.; Mohandas, N.; Feo, C. Automated ektacytometry: A new method of measuring red cell deformability and red cell indices. In *Automation in Hematology*; Springer: Berlin/Heidelberg, Germany, 1981; pp. 153–165.
189. Vives-Corrons, J.L.; Krishnevskaya, E.; Rodriguez, I.H.; Ancochea, A. Characterization of hereditary red blood cell membranopathies using combined targeted next-generation sequencing and osmotic gradient ektacytometry. *Int. J. Hematol.* **2021**, *113*, 163–174. [CrossRef]
190. Gossett, D.R.; Henry, T.; Lee, S.A.; Ying, Y.; Lindgren, A.G.; Yang, O.O.; Rao, J.; Clark, A.T.; Di Carlo, D. Hydrodynamic stretching of single cells for large population mechanical phenotyping. *Proc. Natl. Acad. Sci. USA* **2012**, *109*, 7630–7635. [CrossRef]
191. Reisbeck, M.; Helou, M.J.; Richter, L.; Kappes, B.; Friedrich, O.; Hayden, O. Magnetic fingerprints of rolling cells for quantitative flow cytometry in whole blood. *Sci. Rep.* **2016**, *6*, 32838. [CrossRef]
192. Zheng, Y.; Nguyen, J.; Wang, C.; Sun, Y. Electrical measurement of red blood cell deformability on a microfluidic device. *Lab A Chip* **2013**, *13*, 3275–3283. [CrossRef]
193. Lee, D.W.; Doh, I.; Kuypers, F.A.; Cho, Y.H. Sub-population analysis of deformability distribution in heterogeneous red blood cell population. *Biomed. Microdevices* **2015**, *17*, 102. [CrossRef]
194. Rizzuto, V.; Mencattini, A.; Álvarez-González, B.; Di Giuseppe, D.; Martinelli, E.; Beneitez-Pastor, D.; del Mar Mañú-Pereira, M.; Lopez-Martinez, M.J.; Samitier, J. Combining microfluidics with machine learning algorithms for RBC classification in rare hereditary hemolytic anemia. *Sci. Rep.* **2021**, *11*, 13553. [CrossRef]
195. Méndez-Mora, L.; Cabello-Fusarés, M.; Ferré-Torres, J.; Riera-Llobet, C.; Krishnevskaya, E.; Trejo-Soto, C.; Payán-Pernía, S.; Hernández-Rodríguez, I.; Morales-Indiano, C.; Alarcón, T.; et al. Blood Rheological Characterization of β-Thalassemia Trait and Iron Deficiency Anemia Using Front Microrheometry. *Front. Physiol.* **2021**, *12, 761411* . [CrossRef]
196. Picot, J.; Ndour, P.A.; Lefevre, S.D.; El Nemer, W.; Tawfik, H.; Galimand, J.; Da Costa, L.; Ribeil, J.A.; de Montalembert, M.; Brousse, V.; et al. A biomimetic microfluidic chip to study the circulation and mechanical retention of red blood cells in the spleen. *Am. J. Hematol.* **2015**, *90*, 339–345. [CrossRef]
197. Dondorp, A.M.; Kager, P.A.; Vreeken, J.; White, N.J. Abnormal blood flow and red blood cell deformability in severe malaria. *Parasitol. Today* **2000**, *16*, 228–232. [CrossRef]
198. Hosseini, S.M.; Feng, J.J. How malaria parasites reduce the deformability of infected red blood cells. *Biophys. J.* **2012**, *103*, 1–10. [CrossRef]
199. Warkiani, M.E.; Tay, A.K.P.; Khoo, B.L.; Xiaofeng, X.; Han, J.; Lim, C.T. Malaria detection using inertial microfluidics. *Lab A Chip* **2015**, *15*, 1101–1109. [CrossRef]
200. Tay, A.; Pavesi, A.; Yazdi, S.R.; Lim, C.T.; Warkiani, M.E. Advances in microfluidics in combating infectious diseases. *Biotechnol. Adv.* **2016**, *34*, 404–421. [CrossRef]
201. Reboud, J.; Xu, G.; Garrett, A.; Adriko, M.; Yang, Z.; Tukahebwa, E.M.; Rowell, C.; Cooper, J.M. based microfluidics for DNA diagnostics of malaria in low resource underserved rural communities. *Proc. Natl. Acad. Sci. USA* **2019**, *116*, 4834–4842. [CrossRef]
202. Hou, H.W.; Bhagat, A.A.S.; Chong, A.G.L.; Mao, P.; Tan, K.S.W.; Han, J.; Lim, C.T. Deformability based cell margination—A simple microfluidic design for malaria-infected erythrocyte separation. *Lab A Chip* **2010**, *10*, 2605–2613. [CrossRef]
203. Wu, W.T.; Martin, A.B.; Gandini, A.; Aubry, N.; Massoudi, M.; Antaki, J.F. Design of microfluidic channels for magnetic separation of malaria-infected red blood cells. *Microfluid. Nanofluidics* **2016**, *20*, 41. [CrossRef]
204. Elizalde-Torrent, A.; Trejo-Soto, C.; Méndez-Mora, L.; Nicolau, M.; Ezama, O.; Gualdrón-López, M.; Fernández-Becerra, C.; Alarcón, T.; Hernández-Machado, A.; Del Portillo, H.A. Pitting of malaria parasites in microfluidic devices mimicking spleen interendothelial slits. *Sci. Rep.* **2021**, *11*, 22099. [CrossRef]

 membranes

MDPI

Article

In Silico Drug Design of Benzothiadiazine Derivatives Interacting with Phospholipid Cell Membranes

Zheyao Hu † **and Jordi Marti** *,†

Department of Physics, Polytechnic University of Catalonia-Barcelona Tech, B4-B5 Northern Campus UPC, 08034 Barcelona, Spain; zheyao.hu@upc.edu

* Correspondence: jordi.marti@upc.edu

† These authors contributed equally to this work.

Abstract: The use of drugs derived from benzothiadiazine, a bicyclic heterocyclic benzene derivative, has become a widespread treatment for diseases such as hypertension, low blood sugar or the human immunodeficiency virus, among others. In this work we have investigated the interactions of benzothiadiazine and four of its derivatives designed in silico with model zwitterionic cell membranes formed by dioleoylphosphatidylcholine, 1,2-dioleoyl-*sn*-glycero-3-phosphoserine and cholesterol at the liquid–crystal phase inside aqueous potassium chloride solution. We have elucidated the local structure of benzothiadiazine by means of microsecond molecular dynamics simulations of systems including a benzothiadiazine molecule or one of its derivatives. Such derivatives were obtained by the substitution of a single hydrogen site of benzothiadiazine by two different classes of chemical groups, one of them electron-donating groups (methyl and ethyl) and another one by electron-accepting groups (fluorine and trifluoromethyl). Our data have revealed that benzothiadiazine derivatives have a strong affinity to stay at the cell membrane interface although their solvation characteristics can vary significantly—they can be fully solvated by water in short periods of time or continuously attached to specific lipid sites during intervals of 10–70 ns. Furthermore, benzothiadiazines are able to bind lipids and cholesterol chains by means of single and double hydrogen-bonds of characteristic lengths between 1.6 and 2.1 Å.

Keywords: benzothiadiazine derivatives; drug design; molecular dynamics; phospholipid membrane

Citation: Hu, Z.; Marti, J. In Silico Drug Design of Benzothiadiazine Derivatives Interacting with Phospholipid Cell Membranes. *Membranes* **2022**, *12*, 331. https://doi.org/10.3390/membranes12030331

Academic Editor: Slawomir Sek

Received: 25 February 2022
Accepted: 15 March 2022
Published: 17 March 2022

Publisher's Note: MDPI stays neutral with regard to jurisdictional claims in published maps and institutional affiliations.

1. Introduction

Plasma membranes are fundamental in the behavior of human cells, being not only responsible for the interactions between the cell and its environment but also for processes such as cellular signaling [1], enzyme catalysis [2], endocytosis [3] and transport, among others. The main structure of the cell membrane is composed of bilayer phospholipids including sterols, proteins, glycolipids and a wide variety of other biological molecules. High compositional complexity and versatility of membranes are closely related to the environment and the physiological state of cells [4,5] so that many diseases such as cancer, cardiopathies, diabetes, atherosclerosis, infectious diseases or neurodegenerative pathologies are accompanied by changes in the composition of cell membranes [6–9]. For such a reason, the knowledge of the behavior of drugs interacting with different membrane components and their distribution in damaged tissues maybe key to improving drug efficiency and the therapy of the diseases and it has become a topic of greatest scientific interest.

It is well known that the composition of cell membranes in different tissues and organs of the human body exhibits large variations. In the treatment of diseases, an efficient drug design could enhance the interaction of active pharmaceutical ingredients with membrane components in specific tissues helping to reach the target site successfully. Thus, there is a great demand for a full understanding of the rules of drug–membrane interactions, which may help us predict the distribution and curative effect of drugs in the body when it comes to the designing and testing of new drug molecules. Generally, medicinal chemists tended to overcome the difficulty of drugs in entering cells or crossing biological barriers, such as the blood–brain barrier [10–12] by modifying their structures to enhance the lipophilicity of drugs. However, little research has been performed on the influence of drug structure on the rule of drug–membrane interaction, notably the direct information on atomic interactions of drug-membrane systems at the all-atom level. In this work we establish a procedure for the in silico design of derivatives of the well-known family of benzothiadiazines.

Heterocyclic compounds are ubiquitous in the structure of drug molecules [13,14] playing an important role in human life [15,16]. Such compounds are common parts of commercial drugs having multiple applications based on the control of lipophilicity, polarity and molecular hydrogen bonding capacity. Among them, benzothiadiazine and its derivatives have wide pharmacological applications, such as diuretic [17], anti-viral [18], anti-inflammatory [19], regulating the central nervous system [20] and, more recently, as anti-cancer agents [21–23]. In addition to the above-mentioned biopharmacological activities, benzothiadiazine derivatives also have bio-activity such as Factor Xa inhibition [24], anti-Mycobacterium [25,26] and anti-benign prostatic hyperplasia [27]. 3,4-dihydro-1,2,4-benzothiadiazine-1,1-dioxide (DBD) being the main common structure of the benzothiadiazine family was investigated in a previous work [28] to elucidate the mechanisms responsible for the interactions of DBD with the basic components of cell membranes in all-atom level for the first time. In the present work, our aim was to design in silico DBD derivatives that may be employed with the purpose of inhibiting a limited variety of tumors produced by the oncogenic protein KRas-4B (such as pancreatic, lung or colorectal [29,30]), work currently in progress in our lab. For such a purpose, it has been found convenient to model the substrate cell membrane with DOPC/DOPS lipids, since these particular components are most relevant for the absorption of the oncogene at the cell membrane's interface (see for instance [31,32]). Further, in an effort to produce a more realistic setup, we decided to include cholesterol in the membrane model. Cholesterol constitutes about 33.3% of the outer leaflet in healthy colorectal cells [33], which is in good agreement with the 30% of cholesterol adopted in this work. We have already observed [28] that DBD has a strong affinity to the DOPC species of lipids and that it is also able to bind other membrane components by single and double hydrogen bonds. In this paper, we modified DBD and evaluated the effect of different substitutes on the affinity of the DBD to cell membrane components.

2. Methods

Five models of lipid bilayer membranes in aqueous solution have been constructed using the CHARMM-GUI web-based tool [34,35]. The membrane components and the amount of particles of each class are as follows: all systems include one single DBD derivative, 112 neutral DOPC lipids, 28 DOPS associated with K^+ (DOPS-K) lipids, 60 cholesterol molecules, 49 potassium ions, 21 chlorine ions and 10,000 water molecules. The lipids have been distributed in two symmetric leaflets embedded inside an electrolyte potassium–chloride solution at 0.15 M concentration. We have considered five different setups, where only the benzothiadiazine derivative is different in each case. We considered a previously investigated [28] standard DBD species as the reference (DBD1) and four more DBD derivatives (DBD2, DBD3, DBD4 and DBD5), designed by ourselves using in silico techniques. The way we designed the new DBD species followed the fact that medicinal chemists modify the chemical structure of the drug for the purpose of improving its therapeutic effect, reducing toxicity and side effects. The modification method depends on the structure of the drug. Generally, when performing the structure modification, the basic structure of the drug will remain unchanged and only some functional group will change. When the drug acts, the binding methods of drug and receptor form a reversible complex and are generally by ionic bond, hydrogen bond or covalent bond. In a previous work [28] we observed that DBD can form hydrogen bonds (HB) and become absorbed by the cell membrane with DBD having strong affinity for DOPC. 'H2' and 'H4' sites of DBD are important for the formation of such HB with membrane components. The 'R' site (shown in Figure 1) is very close to the 'H2' and 'H4' sites so that the size, electronegativity and other properties of the R substituent will affect the ability of 'H2'/'H4' to form hydrogen bonds with cell membrane components. So, with the tool of CHARMM-GUI platform "Ligand Reader & Modeller", we introduced methyl, ethyl, fluorine and trifluoromethyl into this site in order to assess the effect of new drug structures on the behavior of DBD in cell membranes.

Sketches of all species are reported in Figure 1. Each DBD species and each phospholipid was described with atomic resolution (DBD1 and DBD4 have 20 sites, DBD2 and DBD5 have 23 sites, DBD3 has 26 sites, DOPC has 138 sites, DOPS has 131 sites and cholesterol has 74 sites). In all simulations water has been represented by rigid 3-site TIP3P [36] molecules, included in the CHARMM36 force field [37,38], that was adopted for lipid—lipid and lipid—protein interactions. In particular, we selected the version CHARMM36m [39], which is able to reproduce the area per lipid for the most relevant phospholipid membranes in excellent agreement with experimental data. The parameterization of the DBD species was performed by means of the "Ligand Reader & Modeller" tool in CHARMM-GUI platform (https://charmm-gui.org/?doc=input/ligandrm, accessed on 31 December 2021). All bonds involving hydrogen atoms were set to fixed lengths, allowing fluctuations of bond distances and angles for the remaining atoms. Van der Waals interactions were cut off at 12 Å with a smooth switching function starting at 10 Å. Finally, long-ranged electrostatic forces were computed using the particle mesh Ewald method [40], with a grid space of 1 Å, updating electrostatic interactions every time step of the simulation runs.

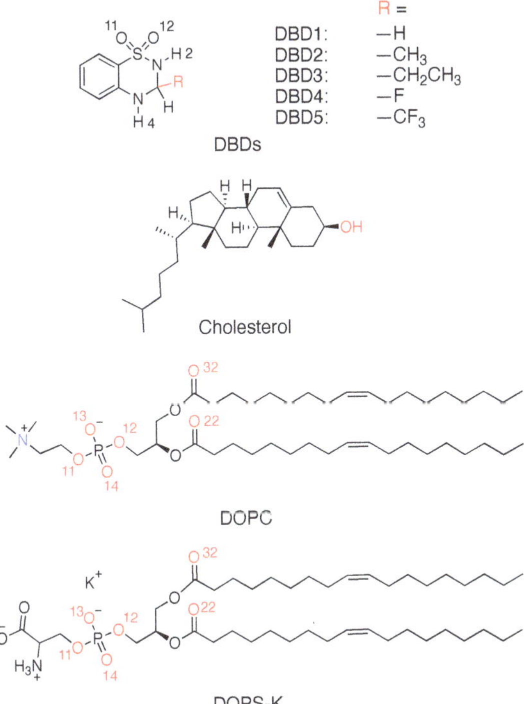

Figure 1. Chemical structures of benzothiadiazine derivatives, phospholipids and cholesterol. Site 'R' stands for the five DBD derivatives considered in the present work.

Molecular dynamics (MD) simulations have been revealed to be a very reliable tool for the simulation of the microscopic structure and dynamics of all sorts of condensed systems, such as aqueous solutions in bulk or under confinement [41–47] towards model cell membranes in electrolyte solution [48–50] and, more recently, small-molecule and protein systems attached to phospholipid membranes [51–53]. Five sets of MD runs were performed by means of the GROMACS2021 simulation package [54–58]. We run all the simulations at the fixed pressure of 1 atm and at the temperature of 310.15 K, typical of the human body and also well above the crossover temperatures for pure DOPC and DOPS needed to be at the liquid crystal phase (253 and 262 K, respectively) [59]. In all cases, the temperature was controlled by a Nose–Hoover thermostat [60] with a damping coefficient of 1 ps^{-1}, whereas the pressure was controlled by a Parrinello–Rahman barostat [61] with a damping time of 5 ps. In the isobaric—isothermal ensemble, i.e., under the condition of a constant number of particles, pressure and temperature, equilibration periods for all simulations were around 200 ns. In all cases, we recorded statistically meaningful trajectories of 600 ns. The simulation boxes had the same size in all cases, i.e., $78.1 \times 78.1 \times 95.7$ Å^3. We have considered periodic boundary conditions in the three directions of space. The simulation time step was fixed to 2 fs in all cases.

3. Results and Discussion

3.1. Characteristics of the Bilayer Systems

The phospholipid bilayer considered in this work was previously simulated and its main characteristics were reported [28,30]. We found reliable values of the area per lipid A and the thickness Δz of the membranes to be in qualitative agreement with available experimental data. In order to corroborate these results in the present work where the system contains DBD derivatives, we computed A and Δz as usual, considering the total surface along the XY plane (plane along the bilayer surface) divided by the number of lipids and cholesterol in one single leaflet [62] and the difference between the z-coordinates of the phosphorus atoms of the two leaflets, respectively. The results of the averaged values obtained from the 600 ns production runs are reported in Table 1, whereas the time evolution of both properties is displayed in Figure 2.

Table 1. Area/lipid A and thickness Δz of the systems simulated in this work, given in Å^2 and Å units, respectively. Estimated errors based on standard deviations correspond to the last significant figures, i.e., ± 0.01 in each case.

DBD Derivative	A	Δz
DBD1	52.18	43.04
DBD2	52.20	43.02
DBD3	52.23	42.98
DBD4	52.19	43.02
DBD5	52.23	42.97

The results shown in Figure 2 indicate that the simulated trajectories were well equilibrated in all cases. The comparison with previous results indicates that the effect of DBD derivatives on the area per lipid and thickness of the membrane is totally marginal. Firstly, the averaged result of $A = 52.2$ Å^2 in all cases matches perfectly the previous reported value of 52.0 Å^2 [30] (where a large protein was embedded in the system) and also the experimental value of 54.4 Å^2 reported by Nagle et al. [63]. Area/lipid shows fluctuations around 5% of the averaged values. Secondly, thickness of the membranes are also in good qualitative agreement with previous works: from Figure 2 we observe fluctuations less than 5% of the averaged values, of around 43.0 Å, as expected. Such value is in qualitative agreement with the experimental measurement of $\Delta z = 40$ Å for the DOPC-cholesterol (30%) bilayer, as reported by Nagle et al. [63] and it matches the previously found $\Delta z = 43.0$ Å obtained in previous simulations for the DOPC/DOPS/cholesterol membrane [30].

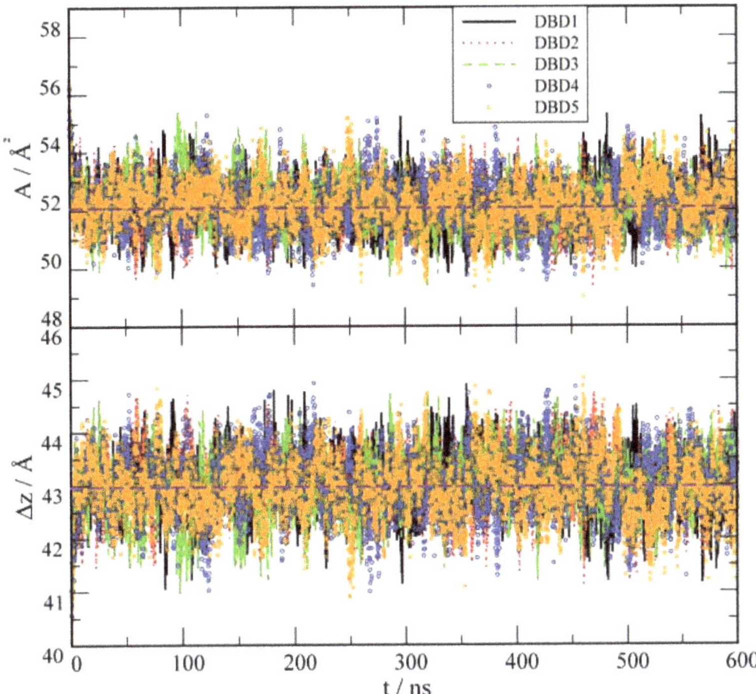

Figure 2. Area per lipid A and thickness Δz of the membrane systems including DBD derivatives as a function of simulation time t. DBD1 (continuous line); DBD2 (dotted line); DBD3 (dashed line); DBD4 (circles); DBD5 (squares). Long-dashed (purple) lines indicate the average values reported in Table 1.

3.2. *Local Structure of Benzothiadiazine Derivatives*

3.2.1. Radial Distribution Functions

We considered the so-called atomic pair radial distribution functions (RDF) $g_{AB}(r)$, defined, in a multicomponent system, for a species B close to a tagged species A as:

$$g_{AB}(r) = \frac{V \langle n_B(r) \rangle}{4\, N_B\, \pi r^2\, \Delta r}, \tag{1}$$

where $n_B(r)$ is the number of atoms of species B surrounding a given atom of species A inside a spherical shell of width Δr. V is the total volume of the system and N_B is the total number of particles of species B. The physical meaning of the RDF stands for the probability of finding a particle B at a given distance r of a particle A. Our RDF are normalized so that tend to 1 at long distances, i.e., when the local density equals the averaged one.

We have evaluated the local structure of the DBD derivatives when solvated by lipids, cholesterol and water according to Equation (1). Only a few of all possible RDF are reported, since we have selected the most relevant ones for the purpose of highlighting the main interactions between the tagged particles. The results are presented in Figures 3–5, where we have selected the hydrogen sites 'H2', 'H4', 'O11' and 'O12' of DBD derivatives, since these are the most active sites, able to form hydrogen bonds with the surrounding partners (lipid, cholesterol species and eventually water). In all cases we can observe a clear first coordination shell associated to the binding of DBD derivatives to the membranes, with corresponding maxima indicating the typical HB distances, together with much lower second shells centered around 4–5 Å. As a general fact, the HB detected cover a noticeably wide range of distances, between 1.6 and 2.1 Å.

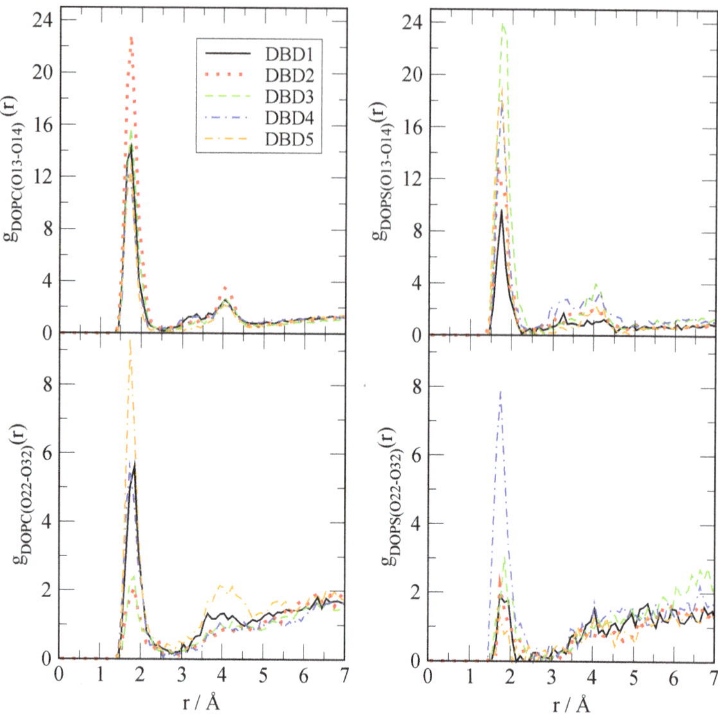

Figure 3. Radial distribution functions between site 'H2' of DBD derivatives and selected oxygen sites in DOPC and DOPS phospholipids. Sites 'O13-14' stand for head groups of the cell membrane phospholipids and sites 'O22-32' stand for tail groups located deeper in the membrane interface.

The structure of DBD derivatives described by their 'H2' site (see Figure 1) indicates the existence of HB formed by 'H2' and several sorts of lipid oxygen sites and it is represented in Figure 3. We can notice that the typical HB length is of 1.7 Å in all cases, both for the binding with oxygen atoms of the phosphoryl group 'O13-14' (located at the head groups of DOPC and DOPS, with both oxygen sites sharing a negative charge) and for the binding with sites 'O22-32' (located in the tail groups of the lipids) as well. This is the typical distance of the binding of small-molecules to cell membranes, such as tryptophan to dipalmytoilphosphatidylcholine (see for instance the review [64]). It should be pointed out that using fluorescence spectroscopy, Liu et al. [65] obtained values for the HB lengths of tryptophan-water between 1.6 and 2.1 Å, i.e., of the same range as those reported here.

This indicates that: (1) all sorts of DBD derivatives can bind the membranes at both head and tail groups and (2) depending on the oxygen sites, some derivatives are able to create HB stronger than others; however, the strength of the HB binding is not uniform and it clearly depends of the class of derivative and lipid chain involved. Despite we will qualitatively analyze the strength of the HB in Section 3.2.2, we can give some general clues here. For instance, DBD2 is able to bind DOPC more strongly than DBD1 (species that we will consider as the reference), with the remaining derivatives making bonds of similar strength. Nevertheless, when DOPS is concerned, all derivatives form stronger HB than DBD1, with DBD3 the strongest. Similar trends are observed when the internal tail group sites 'O22-32' are analyzed: DBD5 makes the strongest HB with DOPC and DBD4 makes the strongest bond with DOPS. In this latter case, the enhancement of the HB is milder than it occurred in the former case (head group bindings). Overall, we can observe that the one-site modifications proposed with the design of the new DBD-derivatives reported in this work has produced significant changes and enhancement of the HB connections to the model cell membrane. Generally speaking, electron-donating groups (DBD2-DBD3) produce similar qualitative effects on the DBD–membrane hydrogen bond connections, whereas electron-accepting types (DBD4-DBD5) tend to produce opposite effects. This can be valuable information to assess the affinity of new designed drugs to target specific oncogenes such as KRas-4B, work that it is been currently developed in our group.

Concerning hydrogens 'H4' of DBD derivatives and their binding characteristics when associated to DOPC and DOPS (Figure 4), we observed that they can be also connected either to 'O11' or 'O12' of the phospholipids (head groups), either to 'O22' or 'O32' (tail groups). Interestingly, in the case of DBD's 'H4', HB lengths are within the range of 2–2.1 Å, significantly longer than those formed by H2 (range around 1.6 to 1.8 Å). This was already observed for the reference DBD1 in a previous work [28]. In this case, the strongest HB is observed when 'H4' of DBD3 is connected to DOPS's 'O11-12' oxygens. In this particular case, the new DBD derivatives have shown to be able to bind the internal regions of the membrane, whereas the original benzothiadiazine species (DBD1) had a very low probability to penetrate these regions. Again for the 'H4' binding site, we have found a general enhancement of the binding of DBD derivatives with the main phospholipids forming our cell membrane system.

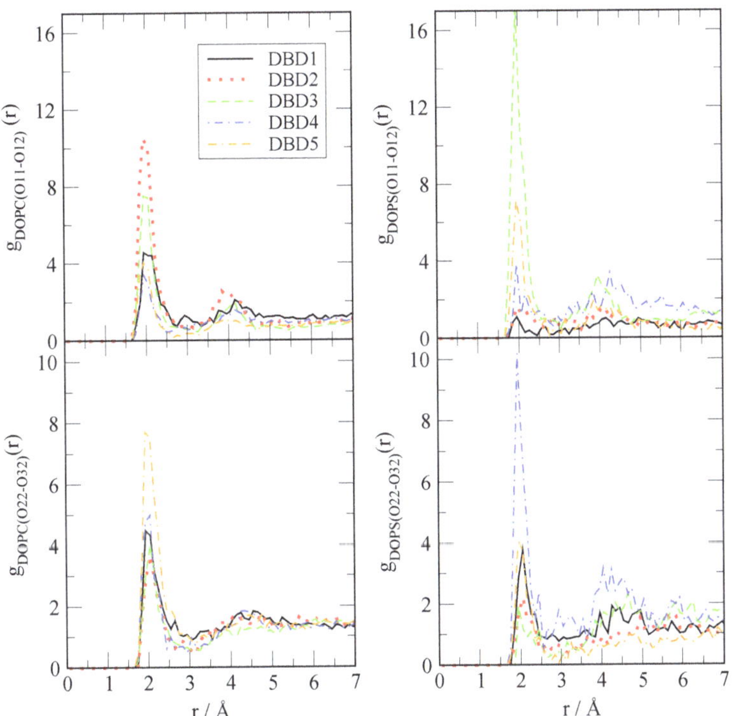

Figure 4. Radial distribution functions between site 'H4' of DBD derivatives and selected oxygen sites in DOPC and DOPS phospholipids.

In the third RDF set (Figure 5) we report interactions between sites 'H2', 'H4' and 'O11-12' of DBD with water (plots at the left column) and cholesterol (plots at the right column). In the case of water, HB can be established between 'H2' and the oxygen site of water (top) or, alternatively, between 'H4' and the oxygen of water (bottom). In both cases, the strength of the interaction is low, which suggests that DBD derivatives are strongly bound to the cell membrane and can be solvated by a few water molecules located at the interface. We have not observed long term episodes of DBD derivatives fully solvated by water.

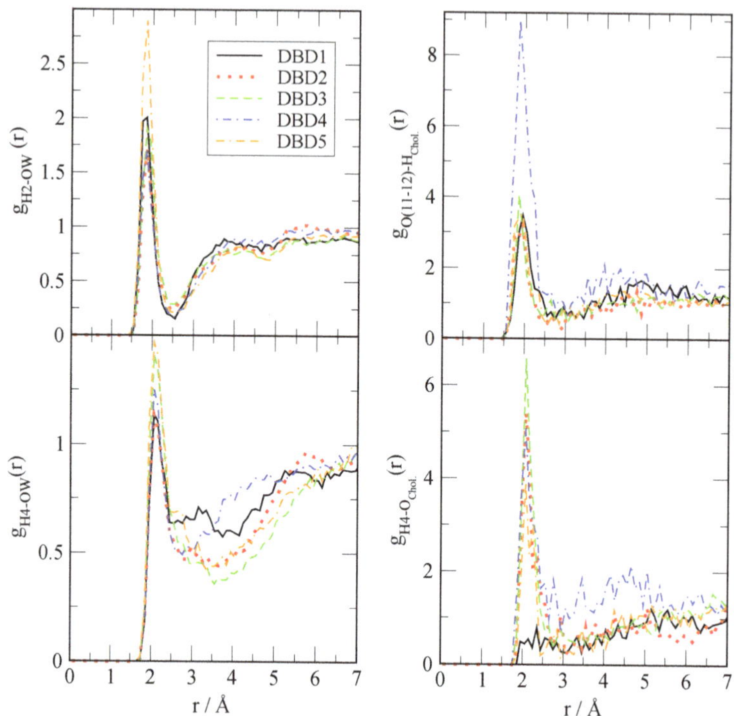

Figure 5. Radial distribution functions between sites 'H2' and 'H4' of DBD derivatives and selected sites of water (left column) and cholesterol (right column).

We have located some extent of hydrogen-bonding between DBD and cholesterol. However, no significant binding of 'H2' with cholesterol has been observed, whereas interactions of both 'H4' and 'O11-12' sites of DBD derivatives have been detected. In particular, the strongest contributions are seen for with hydroxyl's oxygens of cholesterol with 'H4' of the DBD species, which were undetected for the reference original DBD1 as well as for oxygens of the DBD derivatives with hydroxyl's hydrogen of cholesterol. In the latter case, we found a particularly strong contribution of DBD4, i.e., the derivative containing a fluoride residue instead the original hydrogen atom. The HB lengths are in the range of 2.1 Å in all cases.

3.2.2. Potentials of Mean Force between Benzothiadiazine Derivatives and Lipids

Among the wide variety of one-dimensional free-energy methods proposed to compute the potential of mean force (PMF) between two tagged particles [66] a simple but meaningful choice is to consider the radial distance r as an order parameter, able to play the role of the reaction coordinate of the process, within the framework of unbiased simulations as those reported in the present work and to proceed with a direct estimation of the reversible work as described below. This has become one of standard choices to compute free-energy barriers in MD simulations, together with constrained MD simulations [67] or the popular *umbrella sampling* procedure [68]. In case that more accurate values of the free-energy barriers are needed, the optimal choices are: (1) to use constraint-bias simulation combined with force averaging for Cartesian or internal degrees of freedom [66]; (2) the use of multi-dimensional reaction coordinates [69] such as transition path sampling [70–72] or

(3) considering collective variables, such as metadynamics [30,73] although such methods require a huge amount of computational time. Since the determination of reaction coordinates for the binding of DBD at zwitterionic membranes is out of the scope of this work, we limit ourselves to use radial distances between two species as our order parameters to perform reversible work calculations.

In this framework, a good approximation of the PMF can be obtained by means of the reversible work $W_{AB}(r)$ required to move two tagged particles (A, B) from infinite separation to a relative separation r (see for instance Ref. [74], chapter 7):

$$W_{AB}(r) = -\frac{1}{\beta} \ln g_{AB}(r), \tag{2}$$

where $\beta = 1/(k_B T)$ is the Boltzmann factor, k_B the Boltzmann constant and T the temperature. In the calculations reported here, the radial distance r is the distance used in the corresponding RDF (Section 3.2) i.e., it is not related to the atom position relative to the center of the membrane. All free-energy barriers are simply defined (in $k_B T$ units) by a neat first minimum and a first maximum of each $W(r)$, with barrier size ΔW obtained as the difference between the former. As a sort of example, we present the free-energy barriers with largest values for each DBD species in Figure 6.

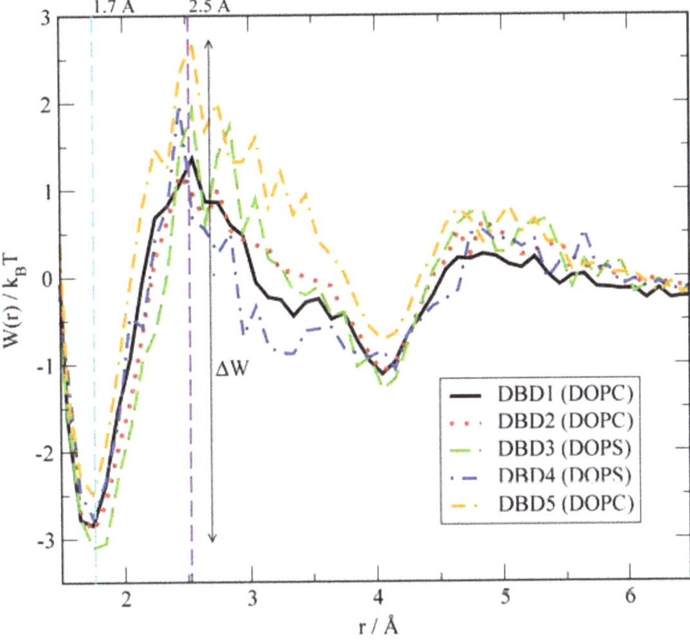

Figure 6. Potentials of mean force for the binding of 'H2' sites of DBD derivatives to the sites 'O13-14' of DOPC and DOPS.

The full set of free-energy barriers for a wide selection of bound pairs has been reported in Table 2. There we can observe overall barriers between 1.2 and 5.2 $k_B T$, which correspond to 0.7–3.1 kcal/mol, for the simulated temperature of 310.15 K. We observe stable binding distances (given by the position of the first minima of the PMF) matching the typical hydrogen-bond distances, as expected. As a reference, it is known that the typical energy of water–water hydrogen-bonds estimated from ab initio calculations is of 4.9 kcal/mol for a water dimer in vacuum [75], whereas in our model system (including TIP3P water) the barrier associated to the HB signature, given by the first maximum of water's oxygen–hydrogen RDF, is of 1.1 kcal/mol. This low value can be directly associated with two facts: (1) first, we have estimated this energy in the bulk, condensed phase of the aqueous ionic solvent, whereas the reference value of Feyereisen et al. [75] corresponds to an isolated water dimer, i.e., can be related to gas phase; (2) secondly, the TIP3P water model included in the CHARMM36 force field is well known to have significant drawbacks to describe liquid water [76].

Table 2. Free-energy barriers ΔW (in $k_B T$) from reversible work calculations for the binding of DBD to cholesterol, lipids and water. In order to quantify the height of all barriers, 1 $k_B T = 0.596$ kcal/mol. Labels as indicated in Figure 1. Estimated errors of ± 0.1 in all cases.

DBD Site	Lipid Site	ΔW				
		DBD1	**DBD2**	**DBD3**	**DBD4**	**DBD5**
H2	O13-14 DOPC	4.2	4.0	4.0	4.2	5.2
H2	O22-32 DOPC	3.4	3.2	2.5	4.4	3.7
H2	O13-14 DOPS	4.2	3.8	5.0	4.7	4.7
H2	O22-32 DOPS	2.8	3.5	3.3	3.0	2.2
H4	O11 DOPC	2.0	2.5	2.4	1.9	3.1
H4	O22-32 DOPC	1.6	2.2	1.9	2.3	2.3
H4	O11 DOPS	2.3	1.2	3.5	1.8	3.8
H4	O22-32 DOPS	1.8	1.6	2.0	4.0	3.9
H4	O Cholesterol	1.2	3.2	2.9	2.4	3.4
O11-12	H Cholesterol	1.8	2.4	2.2	2.6	2.3

In an earlier work [28] we reported by the first time DBD–membrane related free-energy barriers. For the sake of comparison with other similar systems, we can remark that the PMF of tryptophan in a di-oleoyl-phosphatidyl-choline bilayer membrane shows a barrier of the order of 4 kcal/mol [77], whereas the barrier for the movement of tryptophan attached to a poly-leucine α-helix inside a DPPC membrane was reported to be of 3 kcal/mol [78]. Finally, neurotransmitters such as glycine, acetylcholine or glutamate were reported to show small barriers of about 0.5–1.2 kcal/mol when located close to the lipid glycerol backbone [79]. These values could further indicate that our estimations match well the order of magnitude of the free-energy barriers for other small-molecules of similar size.

We designed two sets of DBD derivatives according to their characteristics: in DBD2 and DBD3 we replaced a hydrogen by electron-donating groups (methyl and ethyl, respectively) whereas in DBD4 and DBD5 we replaced a hydrogen by electron-accepting groups (fluorine and trifluoromethyl, respectively). Regardless of the type of replacement considered, our general result is that most of the barriers are in the range of 1–5 kcal/mol, regardless of the specific derivative considered. As more specific features, we can observe that the barriers corresponding to the HB formed by the residue 'H2' of the DBD derivatives are overall larger than those related to the hydrogen-bonds formed by 'H4', which suggests that 'H2' is the most stable binding site between DBD species and the model cell membranes considered in this work. Among the five DBD species analyzed we can observe that, regarding the 'H2' site of DBD, interactions of its derivatives with DOPC are about 10% stronger that those with DOPS but when 'H4' is concerned, the strength of its HB with DOPC is weaker than those with DOPS only when the tail groups 'O22-32' are considered. Nevertheless, the barriers of 'H4' to head groups are of similar size for both DOPC and DOPS. Further, we should remark a gross feature based on the class of substitution: derivatives DBD2 and DBD3 (where the -H group of the original DBD1 was replaced by electron-donating groups) show similar free-energy barriers and close to the values obtained for DBD1, whereas derivatives DBD4 and DBD5 (where the -H group of the original DBD1 was replaced by electron-accepting groups) also show similar free-energy barriers but less similar to the values obtained for DBD1. Finally, the binding of DBD with cholesterol is revealed to be sensibly weaker than that to DOPC and DOPS.

With the aim of a better understanding of the geometrical shape of the HB established between DBD and lipid species and as a sort of example, we report in Figure 7 a series of three snapshots describing the simultaneous binding of DBD4 with a few counterparts: so, we can observe that DBD4's 'H2' and 'H4' are able to bridge oxygens 'O13' and 'O14' of DOPC and 'O22-32' of DOPS (A), also 'O22-32' of DOPC and 'O' of the hydroxyl group of cholesterol and finally 'O13' and 'O14' of DOPS and 'O' of the hydroxyl group of cholesterol. This remarkable bridging properties of DBD4 are qualitatively similar to those of DBD1. Both species, and to some extent all of DBD derivatives, can also form closed-ring structures (see Ref. [28], Figure 6). The bridging bonds highlighted here are quite similar to the HB structures observed in tryptophan [80] and melatonin absorbed at cell membrane surfaces [53].

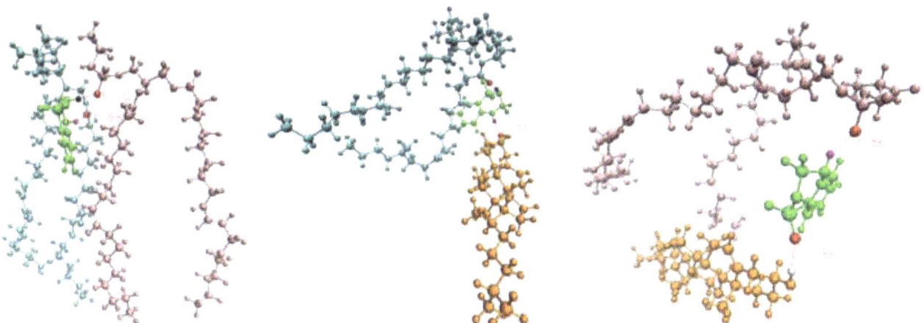

Figure 7. Snapshots of relevant configurations between benzothiadiazine derivative DBD4 (green) and their partner hydrogen-bonding sites. Lipid molecules: DOPS (pink), DOPC (cyan), cholesterol (orange). Specific sites: DBD4-H2 (black), DBD4-H4 (magenta), oxygen atoms in DBD4, DOPS, DOPC and cholesterol are depicted in red whereas hydrogen atom in cholesterol is depicted in white. Typical hydrogen-bond distances are indicated in red. This figure has been created by means of the "Visual Molecular Dynamics" package [81].

3.2.3. Time-Dependent Atomic Site–Site Distances

Once the local structures of the DBD derivatives have been fully evaluated, we make an estimation of the HB dynamics by computing the average lifetime of some of the HB reported by RDF. Other typical MD properties involving time-correlation functions such as power spectra [82,83], relaxation times or self-diffusion coefficients [84,85] that were considered in previous studies, are out of the scope of this paper and have not been considered here. We display the time evolution of selected atom–atom distances $d(t)$ in Figure 8 only for the pairings of 'H2' of DBD3 and sites 'O13' and 'O14' of DOPC and DOPS (top panel) and for 'H4' of DBD3 and sites 'O11' and 'O12' of DOPC and DOPS (bottom panel), as a sort of example. The full set of averaged values are reported in Table 3. We have selected in Figure 8 representative intervals (of more than 100 ns) from the full MD trajectory of 600 ns where the pattern of formation and breaking of HB is clearly seen, including a large extent of fluctuations. This means that such patterns have been systematically observed throughout the whole trajectory.

We can observe that typical HB distances of 1.7 and 2.05 Å are reached. Sites 'O13' and 'O14' (and 'O11' and 'O12') of DOPC and DOPS have been averaged given their equivalence. Typical HB lifetimes can vary enormously, between short lived HB of less than 1 ns (DBD1 with cholesterol) up to long-life HB of more than 70 ns (DBD2 with the head-group of DOPC, i.e., sites 'O13-14'). As general trends, we can highlight that (1) sites 'H2' of the benzothiadiazine derivatives are able to form much longer lived HB than sites 'H4', especially for the DOPS species and (2) DBD–cholesterol hydrogen bonds have rather short lifetimes in the range of 1–10 ns. A closer look indicates that the longest living HB established between DBD and cholesterol are those composed by cholesterol's hydrogen as donor and oxygens of DBD as acceptors, about twice longer that HB formed by hydroxyl's oxygen of cholesterol and hydrogen 'H4' of DBD derivatives. For the sake of comparison, we should remark that the typical lifetime of hydrogen-bonds in pure water has been estimated to be of the order of 1 ps [86]. Finally, we should indicate that the shorter lifetimes reported in a previous work where DBD1 was studied [28] must be attributed to the shorter trajectories considered there and, especially, to the fact that some lifetimes were estimated without taking into account short-lived breaking and reformation of HB, as we did in the present work.

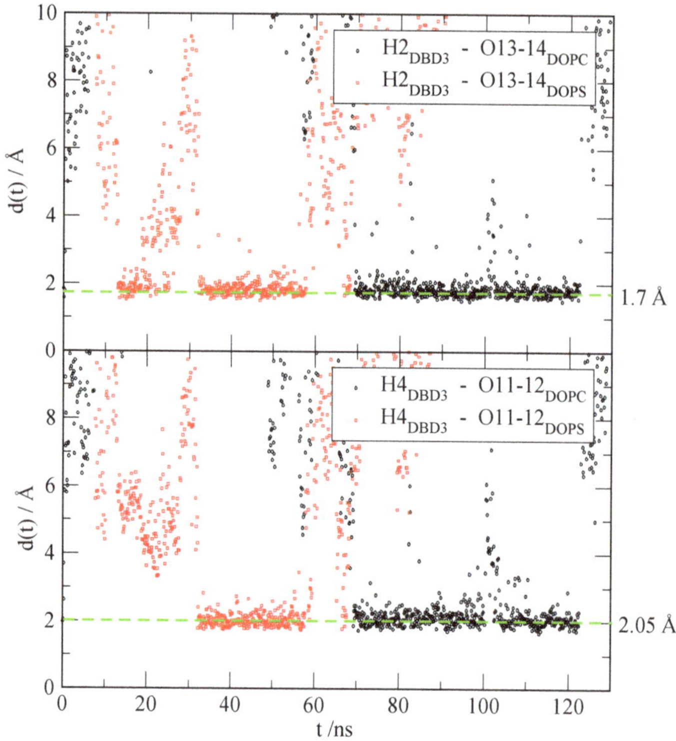

Figure 8. Time evolution of distances between selected sites of DBD3 ('H2', 'H4') and their partner oxygen sites of DOPC and DOPS.

Table 3. Averaged distances (in Å) between selected sites of DBD and the membrane. Continuous time intervals (τ, in ns) have been obtained from averaged computations along the 600 ns trajectory. Labels as indicated in Figure 1. Estimated errors of ±0.1 in all cases.

DBD Site	Lipid Site	Distance	τ_{DBD1}	τ_{DBD2}	τ_{DBD3}	τ_{DBD4}	τ_{DBD5}
H2	O13-14 DOPC	1.7	67.4	73.7	63.2	28.4	38.5
H2	O22-32 DOPC	1.8	20.4	11.8	12.3	19.4	24.6
H2	O13-14 DOPS	1.7	6.9	8.8	27.0	9.6	9.7
H2	O22-32 DOPS	1.8	1.4	1.9	2.3	3.1	0.9
H4	O11-12 DOPC	1.9	42.5	64.4	61.5	15.6	35.9
H4	O22-32 DOPC	2.0	16.5	14.0	13.0	16.6	28.4
H4	O11-12 DOPS	1.9	1.3	1.9	24.8	1.9	6.4
H4	O22-32 DOPS	2.0	1.8	1.4	0.9	2.6	1.9
H4	O Cholesterol	2.0	0.5	3.4	4.4	3.7	3.6
O11-12	H Cholesterol	1.9	4.6	4.5	6.6	10.3	6.0

4. Conclusions

We report results from molecular dynamics simulations of benzothiadiazine derivatives embedded in a phospholipid bilayer membrane formed by 200 lipid molecules with concentrations of 56% of DOPC, 14% of DOPS and 30% of cholesterol in aqueous potassium chloride solution using the CHARMM36m force field. Starting from a standard 3,4-dihydro-1,2,4-benzothiadiazine-1,1-dioxide molecule we have designed in silico four derivatives based on the replacement of a single hydrogen atom by two different classes of chemical groups, one of them electron-donating groups (methyl and ethyl) and another one by electron-accepting groups (fluorine and trifluoromethyl). The electronegativity of these two groups is very different: whereas the electronegativity of electron-donating groups is smaller than that of hydrogen atoms, the electronegativity of electron-accepting groups is larger. In this paper, the electronegativity of methyl and ethyl groups, being smaller than that of hydrogen has the inductive effect of electron donation, which will increase the electron density of the DBD molecule to a certain extent. On the contrary, fluorine and trifluoromethyl will reduce the electron density of the molecule. When the hydrogen of the C-H bond in the DBD molecule is replaced by a substituent, the electron density distribution of the molecule changes, which has significant impact on the formation of hydrogen bonds between the drug and cell membrane components, as it has been reported in the present work.

As a gross feature, the same class of chemical groups produce similar effects on the HB between DBD and cell membranes, whereas different types tend to produce overall opposite effects. With this kind of study our aim is to elucidate the effects of different chemical groups on DBD–cell interactions. Our analysis is based on the computation of the local structures of the DBD derivatives when associated to lipids, water and cholesterol molecules. After the systematic analysis of meaningful data, we have found that the location of DBD at the interface of the membrane is permanent. We have computed RDF defined for the most reactive particles, especially hydrogens 'H2' and 'H4' and oxygens 'O11-12' of DBD (see Figure 1) correlated with sites of lipids and cholesterol able to form HB with DBD. All RDF have shown a strong first coordination shell and a weak second coordination shell for all DBD–lipid structures. The first shell is the signature of HB of lengths between 1.7 and 2.1 Å, in overall good agreement with experimental measurements [65] for comparable small-molecules at interfacial membranes.

The analysis of PMF of DBD–lipid interactions has revealed free-energy barriers of the order of 1–3 kcal/mol (Table 2), with the largest barriers corresponding to hydrogen bonds between DBD's 'H2' site and oxygens sites of DOPC and DOPS; however, it has been observed that DBD derivatives are able to bind to cholesterol as well as the two classes of phospholipids, providing bridging connections that are able to locally estabilize and compactify the cell membrane, although the area per lipid and thickness of the whole membrane are not affected by the presence of the DBD species in any case. The influence of cholesterol has been especially noted in the weakening of DBD–lipid HB connections, which should be taken in consideration for the interaction of drugs with cell membranes from a pharmaceutical point of view. After a thorough analysis monitoring relative distances between tagged sites of DBD and lipids we have estimated the lifetime of HB by averaging data from the 600 ns MD trajectories to range in between 1 and 70 ns.

Author Contributions: Conceptualization, Z.H. and J.M.; methodology, Z.H. and J.M.; validation, Z.H. and J.M.; formal analysis, Z.H. and J.M.; investigation, Z.H. and J.M.; resources, Z.H. and J.M.; data curation, Z.H.; writing—original draft preparation, Z.H. and J.M.; writing—review and editing, Z.H. and J.M.; visualization, Z.H. and J.M.; supervision, J.M.; project administration, J.M.; funding acquisition, J.M. All authors have read and agreed to the published version of the manuscript.

Funding: Z.H. is the recipient of a grant from the China Scholarship Council (number 202006230070). J.M. gratefully acknowledges financial support from by the Spanish Ministry of Economy and Knowledge (grant PGC2018-099277-B-C21, funds MCIU/AEI/FEDER, UE).

Institutional Review Board Statement: Not applicable.

Informed Consent Statement: Not applicable.

Data Availability Statement: Not applicable.

Acknowledgments: We thank Huixia Lu for fruitful discussions and technical support.

Conflicts of Interest: The authors declare no conflict of interest.

References

1. Escribá, P.V.; Sastre, M.; García-Sevilla, J.A. Disruption of cellular signalling pathways by daunomycin through destabilization of nonlamellar membrane structures. *Proc. Natl. Acad. Sci. USA* **1995**, *92*, 7595. [CrossRef]
2. Tong, S.; Lin, Y.; Lu, S.; Wang, M.; Bogdanov, M.; Zheng, L. Structural Insight into Substrate Selection and Catalysis of Lipid Phosphate Phosphatase PgpB in the Cell Membrane. *J. Biol Chem.* **2016**, *291*, 18342. [CrossRef]
3. Doherty, G.J.; McMahon, H.T. Mechanisms of endocytosis. *Annu. Rev. Biochem.* **2009**, *78*, 857. [CrossRef] [PubMed]
4. Escribá, P.V.; González-Ros, J.M.; Goñi, F.M.; Kinnunen, P.K.; Vigh, L.; Sánchez-Magraner, L.; Fernández, A.M.; Busquets, X.; Horvath, I.; Barceló-Coblijn, G. Membranes: A meeting point for lipids, proteins and therapies. *J. Cell Mol. Med.* **2008**, *12*, 829. [CrossRef]
5. Noutsi, P.; Gratton, E.; Chaieb, S. Assessment of Membrane Fluidity Fluctuations during Cellular Development Reveals Time and Cell Type Specificity. *PLoS ONE.* **2016**, *11*, e0158313. [CrossRef]
6. Hakomori, S. Aberrant glycosylation in cancer cell membranes as focused on glycolipids: Overview and perspectives. *Cancer Res.* **1985**, *45*, 2405.
7. Simons, K.; Ehehalt, R. Cholesterol, lipid rafts, and disease. *J. Clin. Investig.* **2002**, *110*, 597. [CrossRef] [PubMed]
8. Vigh, L.; Escribá, P.V.; Sonnleitner, A.; Sonnleitner, M.; Piotto, S.; Maresca, B.; Horvath, I.; Harwood, J.L. The significance of lipid composition for membrane activity: New concepts and ways of assessing function. *Prog. Lipid Res.* **2005**, *44*, 303. [CrossRef]
9. Kim, Y.; Shanta, S.R.; Zhou, L.H.; Kim, K.P. Mass spectrometry based cellular phosphoinositides profiling and phospholipid analysis: A brief review. *Exp. Mol. Med.* **2010**, *42*, 1. [CrossRef] [PubMed]
10. Jolliet-Riant, P.; Tillement, J.P. Drug transfer across the blood-brain barrier and improvement of brain delivery. *Fundam. Clin. Pharmacol.* **1999**, *13*, 16. [CrossRef]
11. Bodor, N.; Buchwald, P. Barriers to remember: Brain-targeting chemical delivery systems and Alzheimer's disease. *Drug Discov. Today* **2002**, *7*, 766. [CrossRef]
12. Waterhouse, R.N. Determination of lipophilicity and its use as a predictor of blood-brain barrier penetration of molecular imaging agents. *Mol. Imaging Biol.* **2003**, *5*, 376. [CrossRef] [PubMed]
13. Rees, C.W. Polysulfur-Nitrogen Heterocyclic Chemistry. *J. Heterocycl. Chem.* **1992**, *29*, 639. [CrossRef]
14. Vitaku, E.; Smith, D.T.; Njardarson, J.T. Analysis of the structural diversity, substitution patterns, and frequency of nitrogen heterocycles among U.S. FDA approved pharmaceuticals. *J. Med. Chem.* **2014**, *57*, 10257. [CrossRef] [PubMed]
15. Horton, D.A.; Bourne, G.T.; Smythe, M.L. The combinatorial synthesis of bicyclic privileged structures or privileged substructures. *Chem. Rev.* **2003**, *103*, 893. [CrossRef] [PubMed]
16. Sharma, V.; Kamal, R.; Kumar, V. Heterocyclic Analogues as Kinase Inhibitors: A Focus Review. *Curr. Top. Med. Chem.* **2017**, *17*, 2482. [CrossRef] [PubMed]
17. Platts, M.M. Hydrochlorothiazide, a new oral diuretic. *Br. Med. J.* **1959**, *1*, 1565. [CrossRef]
18. Martínez, A.; Esteban, A.I.; Castro, A.; Gil, C.; Conde, S.; Andrei, G.; Snoeck, R.; Balzarini, J.; De Clercq, E. Novel potential agents for human cytomegalovirus infection: Synthesis and antiviral activity evaluation of benzothiadiazine dioxide acyclonucleosides. *J. Med. Chem.* **1999**, *42*, 1145. [CrossRef]
19. Tait, A.; Luppi, A.; Hatzelmann, A.; Fossa, P.; Mosti, L. Synthesis, biological evaluation and molecular modelling studies on benzothiadiazine derivatives as PDE4 selective inhibitors. *Bioorg. Med. Chem.* **2005**, *13*, 1393. [CrossRef]
20. Larsen, A.P.; Francotte, P.; Frydenvang, K.; Tapken, D.; Goffin, E.; Fraikin, P.; Caignard, D.H.; Lestage, P.; Danober, L.; Pirotte, B.; et al. Synthesis and Pharmacology of Mono-, Di-, and Trialkyl-Substituted 7-Chloro-3,4-dihydro-2H-1,2,4-benzothiadiazine 1,1-Dioxides Combined with X-ray Structure Analysis to Understand the Unexpected Structure-Activity Relationship at AMPA Receptors. *ACS Chem. Neurosci.* **2016**, *7*, 378. [CrossRef]
21. Ma, X.; Wei, J.; Wang, C.; Gu, D.; Hu, Y.; Sheng R. Design, synthesis and biological evaluation of novel benzothiadiazine derivatives as potent PI3Kδ-selective inhibitors for treating B-cell-mediated malignancies. *Eur. J. Med. Chem.* **2019**, *170*, 112. [CrossRef] [PubMed]
22. Shaik, T.B.; Malik, M.S.; Routhu, S.R.; Seddigi, Z.S.; Althagafi, I.I.; Kamal, A. Evaluation of anticancer and anti-mitotic properties of quinazoline and quinazolino-benzothiadiazine derivatives. *Anti-Cancer Agents Med. Chem.* **2020**, *20*, 599–611. [CrossRef] [PubMed]
23. Huwaimel, B.I.; Bhakta, M.; Kulkarni, C.A.; Milliken, A.S.; Wang, F.; Peng, A.; Brookes, P.S.; Trippier, P.C. Discovery of Halogenated Benzothiadiazine Derivatives with Anticancer Activity. *Chem. Med. Chem.* **2021**, *16*, 1143–1162. [CrossRef] [PubMed]
24. Hirayama, F.; Koshio, H.; Katayama, N.; Ishihara, T.; Kaizawa, H.; Taniuchi, Y.; Sato, K.; Sakai-Moritani, Y.; Kaku, S.; Kurihara, H.; et al. Design, synthesis and biological activity of YM-60828 derivatives. Part 2: Potent and orally-bioavailable factor Xa inhibitors based on benzothiadiazine-4-one template. *Bioorg. Med. Chem.* **2003**, *11*, 367–381. [CrossRef]

25. Kamal, A.; Reddy, K.-S.; Ahmed, S.-K.; Khan, M.-N.; Sinha, R.-K.; Yadav, J.-S.; Arora, S.-K. Anti-tubercular agents. Part 3. Benzothiadiazine as a novel scaffold for anti-Mycobacterium activity. *Bioorg. Med. Chem.* **2006**, *14*, 650–658. [CrossRef]
26. Kamal, A.; Shetti, R.-V.; Azeeza, S.; Ahmed, S.-K.; Swapna, P.; Reddy, A.-M.; Khan, I.-A.; Sharma, S.; Abdullah, S.-T. Anti-tubercular agents. Part 5: Synthesis and biological evaluation of benzothiadiazine 1,1-dioxide based congeners. *Eur. J. Med. Chem.* **2010**, *45*, 4545–4553. [CrossRef]
27. Tait, A.; Luppi, A.; Franchini, S.; Preziosi, E.; Parenti, C.; Buccioni, M.; Marucci, G.; Leonardi, A.; Poggesi, E.; Brasili, L. 1,2,4-Benzothiadiazine derivatives as alpha1 and 5-HT1A receptor ligands. *Bioorg. Med. Chem. Lett.* **2005**, *15*, 1185–1188. [CrossRef]
28. Hu, Z.; Martí, J.; Lu, H. Structure of benzothiadiazine at zwitterionic phospholipid cell membranes. *J. Chem. Phys.* **2021**, *155*, 154303. [CrossRef]
29. Prior, I.A.; Lewis, P.D.; Mattos, C.A comprehensive survey of Ras mutations in cancer. *Cancer Res.* **2012**, *72*, 2457–2467. [CrossRef]
30. Lu, H.; Martí, J. Long-lasting Salt Bridges Provide the Anchoring Mechanism of Oncogenic Kirsten Rat Sarcoma Proteins at Cell Membranes. *J. Phys. Chem. Lett.* **2020**, *11*, 9938–9945. [CrossRef]
31. Jang, H.; Abraham, S.J.; Chavan, T.S.; Hitchinson, B.; Khavrutskii, L.; Tarasova, N.I.; Nussinov, R.; Gaponenko, V. Mechanisms of membrane binding of small GTPase K-Ras4B farnesylated hypervariable region. *J. Biol. Chem.* **2015**, *290*, 9465–9477. [CrossRef]
32. Lu, S.; Jang, H.; Muratcioglu, S.; Gursoy, A.; Keskin, O.; Nussinov, R.; Zhang, J. Ras conformational ensembles, allostery, and signaling. *Chem. Rev.* **2016**, *116*, 6607–6665. [CrossRef] [PubMed]
33. Liu, S.-L.; Sheng, R.; Jung, J.H.; Wang, L.; Stec, E.; O'Connor, M.J.; Song, S.; Bikkavilli, R.K.; Winn, R.A.; Lee, D.; et al. Orthogonal lipid sensors identify transbilayer asymmetry of plasma membrane cholesterol. *Nat. Chem. Biol.* **2017**, *13*, 268–274. [CrossRef] [PubMed]
34. Jo, S.; Kim, T.; Iyer, V.G.; Im, W. CHARMM-GUI: A web-based graphical user interface for CHARMM. *J. Comput. Chem.* **2008**, *29*, 1859. [CrossRef]
35. Jo, S.; Lim, J.B.; Klauda, J.B.; Im, W. CHARMM-GUI Membrane Builder for mixed bilayers and its application to yeast membranes. *Biophys. J.* **2009**, *97*, 50. [CrossRef]
36. Jorgensen, W.L.; Chandrasekhar, J.; Madura, J.D.; Impey, R.W.; Klein, M.L. Comparison of simple potential functions for simulating liquid water. *J. Chem. Phys.* **1983**, *79*, 926. [CrossRef]
37. Klauda, J.B.; Venable, R.M.; Freites, J.A.; O'Connor, J.W.; Tobias, D.J.; Mondragon-Ramirez, C.; Vorobyov, I.; MacKerell, A.D., Jr.; Pastor, R.W. Update of the CHARMM all-atom additive force field for lipids: Validation on six lipid types. *J Phys. Chem. B* **2010**, *114*, 7830. [CrossRef] [PubMed]
38. Lim, J.B.; Rogaski, B.; Klauda, J.B. Update of the cholesterol force field parameters in CHARMM. *J. Phys. Chem. B* **2012**, *116*, 203. [CrossRef]
39. Huang, J.; MacKerell, A.D., Jr. CHARMM36 all-atom additive protein force field: Validation based on comparison to NMR data. *J. Comput. Chem.* **2013**, *34*, 2135. [CrossRef] [PubMed]
40. Linse, B.; Linse, P. Tuning the smooth particle mesh Ewald sum: Application on ionic solutions and dipolar fluids. *J. Chem. Phys.* **2014**, *141*, 184114. [CrossRef]
41. Heinzinger, K. *Computer Modelling of Fluids Polymers and Solids*; Springer: Berlin/Heidelberg, Germany, 1990; pp. 357–394.
42. Brodholt, J.P. Molecular dynamics simulations of aqueous NaCl solutions at high pressures and temperatures. *Chem. Geol.* **1998**, *151*, 11. [CrossRef]
43. Chowdhuri, S.; Chandra, A. Molecular dynamics simulations of aqueous NaCl and KCl solutions: Effects of ion concentration on the single-particle, pair, and collective dynamical properties of ions and water molecules. *J. Chem. Phys.* **2001**, *115*, 3732. [CrossRef]
44. Nagy, G.; Gordillo, M.C.; Guàrdia, E.; Martí, J. Liquid water confined in carbon nanochannels at high temperatures. *J. Phys. Chem. B* **2007**, *111*, 12524. [CrossRef] [PubMed]
45. Videla, P.; Sala, J.; Martí, J.; Guàrdia, E.; Laria, D. Aqueous electrolytes confined within functionalized silica nanopores. *J. Chem. Phys.* **2011**, *135*, 104503. [CrossRef]
46. Sala, J.; Guàrdia, E.; Martí, J. Specific ion effects in aqueous eletrolyte solutions confined within graphene sheets at the nanometric scale. *Phys. Chem. Chem. Phys.* **2012**, *14*, 10799. [CrossRef]
47. Calero, C.; Martí, J.; Guàrdia, E. 1H nuclear spin relaxation of liquid water from molecular dynamics simulations. *J. Phys. Chem. B* **2015**, *119*, 1966–1973. [CrossRef] [PubMed]
48. Joseph, S.; Mashl, R.J.; Jakobsson, E.; Aluru, N.R. Electrolytic transport in modified carbon nanotubes. *Nano Lett.* **2003**, *3*, 1399. [CrossRef]
49. Allen, T.W.; Andersen, O.S.; Roux, B. Molecular dynamics—Potential of mean force calculations as a tool for understanding ion permeation and selectivity in narrow channels. *Biophys. Chem.* **2006**, *124*, 251. [CrossRef]
50. Yang, J.; Calero, C.; Martí, J. Diffusion and spectroscopy of water and lipids in fully hydrated dimyristoylphosphatidylcholine bilayer membranes. *J. Chem. Phys.* **2014**, *140*, 104901. [CrossRef] [PubMed]
51. Lu, H.; Martí, J. Binding and dynamics of melatonin at the interface of phosphatidylcholine-cholesterol membranes. *PLoS ONE* **2019**, *14*, e0224624. [CrossRef]

52. Drew Bennett, W.F.; He, S.; Bilodeau, C.L.; Jones, D.; Sun, D.; Kim, H.; Allen, J.E.; Lightstone, F.C.; Ingólfsson, H.I. Predicting small molecule transfer free energies by combining molecular dynamics simulations and deep learning. *J. Chem. Inf. Model.* **2020**, *60*, 5375. [CrossRef] [PubMed]
53. Lu, H.; Martí, J. Cellular absorption of small molecules: Free energy landscapes of melatonin binding at phospholipid membranes. *Sci. Rep.* **2020**, *10*, 9235. [CrossRef] [PubMed]
54. Abraham, M.J.; Murtola, T.; Schulz, R.; Páll, S.; Smith, J.C.; Hess, B.; Lindahl, E. GROMACS: High performance molecular simulations through multi-level parallelism from laptops to supercomputers. *SoftwareX* **2015**, *1*, 19. [CrossRef]
55. Pronk, S.; Páll, S.; Schulz, R.; Larsson, P.; Bjelkmar, P.; Apostolov, R.; Shirts, M.R. Smith, J.C.; Kasson, P.M. van der Spoel, D.; et al. GROMACS 4.5: A high-throughput and highly parallel open source molecular simulation toolkit. *Bioinformatics* **2013**, *29*, 845. [CrossRef] [PubMed]
56. Van Der Spoel, D.; Lindahl, E.; Hess, B.; Groenhof, G.; Mark, A.E.; Berendsen, H.J.C. GROMACS: Fast, flexible, and free. *J. Comput. Chem.* **2005**, *26*, 1701. [CrossRef] [PubMed]
57. Lindahl, E.; Hess, B.; Van Der Spoel, D. GROMACS 3.0: A package for molecular simulation and trajectory analysis. *Mol. Model. Annu.* **2001**, *7*, 306. [CrossRef]
58. Berendsen, H.J.C.; Van der Spoel, D.; Van Drunen, R. GROMACS: A message-passing parallel molecular dynamics implementation. *Comput. Phys. Commun.* **1995**, *91*, 43. [CrossRef]
59. Chen, W.; Dusa, F.; Witos, J.; Ruokonen, S.-K.; Wiedmer, S.K. Determination of the main phase transition temperature of phospholipids by nanoplasmonic sensing. *Sci. Rep.* **2018**, *8*, 14815. [CrossRef] [PubMed]
60. Evans, D.J.; Holian, B.L. The Nose–Hoover thermostat. *J. Chem. Phys.* **1985**, *83*, 4069–4074. [CrossRef]
61. Parrinello, M.; Rahman, A. Crystal structure and pair potentials: A molecular-dynamics study. *Phys. Rev. Lett.* **1980**, *45*, 1196. [CrossRef]
62. Pandey, P.R.; Roy, S. Headgroup mediated water insertion into the DPPC bilayer: A molecular dynamics study. *J. Phys. Chem. B* **2011**, *115*, 3155–3163. [CrossRef] [PubMed]
63. Pan, J.; Tristram-Nagle, S.; Nagle, J.F. Effect of cholesterol on structural and mechanical properties of membranes depends on lipid chain saturation. *Phys. Rev. E* **2009**, *80*, 021931. [CrossRef] [PubMed]
64. Martí, J.; Lu, H. Microscopic interactions of melatonin, serotonin and tryptophan with zwitterionic phospholipid membranes. *Int. J. Mol. Sci.* **2021**, *22*, 2842. [CrossRef] [PubMed]
65. Liu, H.; Zhang, H.; Jin, B. Fluorescence of tryptophan in aqueous solution. *Spectrochim. Acta Part A Mol. Biomol. Spectrosc.* **2013**, *106*, 54–59. [CrossRef] [PubMed]
66. Trzesniak, D.; Kunz, A.-P.E.; van Gunsteren, W.F. A comparison of methods to compute the potential of mean force. *ChemPhysChem* **2007**, *8*, 162–169. [CrossRef] [PubMed]
67. Guàrdia, E.; Rey, R.; Padró, J.A. Statistical errors in constrained molecular dynamics calculations of the mean force potential. *Mol. Sim.* **1992**, *9*, 201–211. [CrossRef]
68. K'astner, J. Umbrella sampling. *Wiley Interdiscip. Rev. Comp. Mol. Sci.* **2011**, *1*, 932–942. [CrossRef]
69. Geissler, P.L.; Dellago, C.; Chandler, D.; Hutter, J.; Parrinello, M. Autoionization in liquid water. *Science* **2001**, *291*, 2121–2124. [CrossRef]
70. Martí, J.; Csajka, F.S.; Chandler, D. Stochastic transition pathways in the aqueous sodium chloride dissociation process. *Chem. Phys. Lett.* **2000**, *328*, 169–176. [CrossRef]
71. Bolhuis, P.G.; Chandler, D.; Dellago, C.; Geissler, P.L. Transition path sampling: Throwing ropes over rough mountain passes, in the dark. *Annu. Rev. Phys. Chem.* **2002**, *53*, 291–318. [CrossRef]
72. Martí, J.; Csajka, F.S. Transition path sampling study of flip-flop transitions in model lipid bilayer membranes. *Phys. Rev. E* **2004**, *69*, 061918. [CrossRef] [PubMed]
73. Barducci, A.; Bussi, G.; Parrinello, M. Well-tempered metadynamics: A smoothly converging and tunable free-energy method. *Phys. Rev. Lett.* **2008**, *100*, 020603. [CrossRef] [PubMed]
74. Chandler, D. *Introduction to Modern Statistical Mechanics;* Oxford University Press: Oxford, UK, 1987
75. Feyereisen, M.W.; Feller, D.; Dixon, D.A. Hydrogen bond energy of the water dimer. *J. Phys. Chem.* **1996**, *100*, 2993–2997. [CrossRef]
76. Vega, C.; Abascal, J.L.F.; Conde, M.M.; Aragones, J.L. What ice can teach us about water interactions: A critical comparison of the performance of different water models. *Faraday Discuss.* **2009**, *141*, 251–276. [CrossRef]
77. MacCallum, J.L.; Bennett, W.F.D.; Tieleman, D.P. Distribution of amino acids in a lipid bilayer from computer simulations. *Biophys. J.* **2008**, *94*, 3393–3404. [CrossRef]
78. De Jesus, A.J.; Allen, T.W. The role of tryptophan side chains in membrane protein anchoring and hydrophobic mismatch. *BBA Biomembr.* **2013**, *1828*, 864–876. [CrossRef]
79. Peters, G.H.; Werge, M.; Elf-Lind, M.N.; Madsen, J.J.; Velardez, G.F.; Westh, P. Interaction of neurotransmitters with a phospholipid bilayer: A molecular dynamics study. *Chem. Phys. Lipids* **2014**, *184*, 7–17. [CrossRef] [PubMed]
80. Lu, H.; Martí, J. Effects of cholesterol on the binding of the precursor neurotransmitter tryptophan to zwitterionic membranes. *J. Chem. Phys.* **2018**, *149*, 164906. [CrossRef] [PubMed]
81. Humphrey, W.; Dalke, A.; Schulten, K. VMD: Visual molecular dynamics. *J. Mol. Graph.* **1996**, *14*, 33–38. [CrossRef]

82. Martí, J.; Padró, J.A.; Guàrdia, E. Computer simulation of molecular motions in liquids: Infrared spectra of water and heavy water. *Mol. Sim.* **1993**, *11*, 321–336. [CrossRef]
83. Padró, J.A.; Martí, J. Response to "Comment on 'An interpretation of the low-frequency spectrum of liquid water'" [J. Chem. Phys. 118, 452 (2003)]. *J. Chem. Phys.* **2004**, *120*, 1659–1660. [CrossRef]
84. Padró, J.A.; Martí, Guàrdia, E. Molecular dynamics simulation of liquid water at 523 K. *J. Phys. Condens. Matter* **1994**, *6*, 2283. [CrossRef]
85. Martí, J.; Gordillo, M.C. Microscopic dynamics of confined supercritical water. *Chem. Phys. Lett.* **2002**, *354*, 227–232. [CrossRef]
86. Martí, J. Dynamic properties of hydrogen-bonded networks in supercritical water. *Phys. Rev. E* **2000**, *61*, 449. [CrossRef] [PubMed]

MDPI

Article

Investigating Structural Dynamics of KCNE3 in Different Membrane Environments Using Molecular Dynamics Simulations

Isaac K. Asare [1], Alberto Perez Galende [1], Andres Bastidas Garcia [1], Mateo Fernandez Cruz [1], Anna Clara Miranda Moura [1], Conner C. Campbell [1], Matthew Scheyer [1], John Paul Alao [2], Steve Alston [1], Andrea N. Kravats [2], Charles R. Sanders [3], Gary A. Lorigan [2,*] and Indra D. Sahu [1,2,*]

[1] Natural Science Division, Campbellsville University, Campbellsville, KY 42718, USA; ikasar294@students.campbellsville.edu (I.K.A.); apere15@students.campbellsville.edu (A.P.G.); andresbastidas4556@students.campbellsville.edu (A.B.G.); mfern512@students.campbellsville.edu (M.F.C.); ammour424@students.campbellsville.edu (A.C.M.M.); cccamp92@students.campbellsville.edu (C.C.C.); mwsche97@students.campbellsville.edu (M.S.); salston@campbellsville.edu (S.A.)

[2] Department of Chemistry and Biochemistry, Miami University, Oxford, OH 45056, USA; alaooj@miamioh.edu (J.P.A.); kravatan@miamioh.edu (A.N.K.)

[3] Center for Structural Biology, Department of Biochemistry, Vanderbilt University, Nashville, TN 37232, USA; chuck.sanders@vanderbilt.edu

* Correspondence: gary.lorigan@miamioh.edu (G.A.L.); idsahu@campbellsville.edu (I.D.S.); Tel.: +1-(513)-529-2813 (G.A.L.); +1-(270)-789-5597 (I.D.S.)

Abstract: KCNE3 is a potassium channel accessory transmembrane protein that regulates the function of various voltage-gated potassium channels such as KCNQ1. KCNE3 plays an important role in the recycling of potassium ion by binding with KCNQ1. KCNE3 can be found in the small intestine, colon, and in the human heart. Despite its biological significance, there is little information on the structural dynamics of KCNE3 in native-like membrane environments. Molecular dynamics (MD) simulations are a widely used as a tool to study the conformational dynamics and interactions of proteins with lipid membranes. In this study, we have utilized all-atom molecular dynamics simulations to characterize the molecular motions and the interactions of KCNE3 in a bilayer composed of: a mixture of POPC and POPG lipids (3:1), POPC alone, and DMPC alone. Our MD simulation results suggested that the transmembrane domain (TMD) of KCNE3 is less flexible and more stable when compared to the N- and C-termini of KCNE3 in all three membrane environments. The conformational flexibility of N- and C-termini varies across these three lipid environments. The MD simulation results further suggested that the TMD of KCNE3 spans the membrane width, having residue A69 close to the center of the lipid bilayers and residues S57 and S82 close to the lipid bilayer membrane surfaces. These results are consistent with previous biophysical studies of KCNE3. The outcomes of these MD simulations will help design biophysical experiments and complement the experimental data obtained on KCNE3 to obtain a more detailed understanding of its structural dynamics in the native membrane environment.

Keywords: KCNE3; structural dynamics; lipid bilayers; molecular dynamics simulation; membrane mimetic

Citation: Asare, I.K.; Galende, A.P.; Garcia, A.B.; Cruz, M.F.; Moura, A.C.M.; Campbell, C.C.; Scheyer, M.; Alao, J.P.; Alston, S.; Kravats, A.N.; et al. Investigating Structural Dynamics of KCNE3 in Different Membrane Environments Using Molecular Dynamics Simulations. *Membranes* **2022**, *12*, 469. https://doi.org/10.3390/membranes12050469

Academic Editors: Jordi Marti and Carles Calero

Received: 11 February 2022
Accepted: 21 April 2022
Published: 27 April 2022

1. Introduction

KCNE3 is a potassium channel accessory transmembrane protein belonging to the KCNE family that regulates the function of various voltage-gated potassium channels such as KCNQ1 and KCNQ4 [1–4]. KCNE3 has been expressed in the small intestine, colon, and human heart [5–7]. Previous studies have shown that in the presence of KCNE3, KCNQ1's voltage sensitivity shows a linear current-voltage (I-V) relationship that gives rise to a potassium ion conductivity in non-excitable cells as polarized epithelial cells of the colon, small intestine, and airways [3,8,9]. Its malfunction has been proven to contribute to health disorders such as cardiac arrhythmia, long QT syndrome, tinnitus, cystic fibrosis, and Menière's disease [3,5,10–15]. For such a biologically significant membrane protein, little information is known about the structural and dynamic properties of KCNE3 in native like membrane environment, where interactions between lipids and proteins help stabilize the structure of the protein and influence protein function within the membrane. Previous NMR studies of KCNE3 in detergent micelles and isotropic bicelles by the Sanders lab have shown KCNE3's structure consists of an extracellular N-terminus surface associated amphipathic helix connected by a loop to an alpha helical transmembrane domain [16]. A disordered C-terminus is connected to the transmembrane domain by a short juxta membrane helix [16]. Recent studies by Sun et al. using cryo-electron microscopy (Cryo-EM) showed that KCNE3 tucks its single membrane spanning helix against KCNQ1 at a point that appears to keep the voltage sensor in its depolarized confirmation [8]. However, it is not fully understood how these various sections behave structurally and dynamically in various membrane bilayer environments.

Molecular dynamics (MD) simulations serve as a structure biology tool to complement experimental studies in order to study the stability and structural dynamic properties of membrane proteins at an atomic level [17–21]. Here, we use all-atom MD simulations in the course of 105 ns to study stability and structural dynamic properties of KCNE3 in bilayers composed of POPC (1-palmitoyl-2-oleoyl-*sn*-glycero-3-phosphocholine)/POPG (1-palmitoyl-2-oleoyl-*sn*-glycero-3-phospho-(1′-*rac*-glycerol) (sodium salt)) (3:1), POPC alone, and DMPC (1,2-dimyristoyl-sn-glycero-3-phosphocholine) only. The POPC/POPG mixtures, POPC alone and DMPC are widely used lipid systems to mimic biological membrane bilayers for biophysical studies [16,17]. Previous MD simulation studies on similar membrane proteins and other protein systems have suggested that the simulation times of 10–100 ns can provide reliable analysis of protein–detergent and protein–lipid interactions [17,22–24]. We have analyzed MD simulation trajectory data to obtain several structural dynamics related parameters such as backbone root mean square deviation (RMSD), root mean square fluctuation (RMSF), lipid bilayer membrane width, Z-distances, total protein–lipid interaction energy, TMD helical tilt angle, and a heat map of the correlation between parameters, results that yield insight into the stability, molecular motion and interaction of KCNE3 in different phospholipid bilayer membranes.

2. Methods

2.1. Molecular Dynamics Modeling of Wild-Type KCNE3 in Lipid Bilayers

Nanoscale molecular dynamics (NAMD) [25] version 2.14 with the CHARMM36 force field was employed to perform molecular dynamics simulations on a full length KCNE3 (PDB ID: 2M9Z, the original pdb file is available in the Supporting Information of the ref. [16] in lipid bilayers composed of POPC/POPG (3:1), POPC alone, and DMPC alone [26–28]). The simulation set up and input files were generated by using CHARMM-GUI [29]. The visual molecular dynamics software (VMD) Xplor version 1.13 [30,31] was used for MD trajectory data analysis. The bilayer, composed of a pre-equilibrated lipid molecules with a ~12,010.5 Å^2 surface, was built using membrane builder protocol under CHARMM-GUI [29,32]. The total charge of KCNE3 was 2.0 in the simulation. The positively charged amino acid residues were protonated, and negatively charged amino acids were deprotonated. The histidine (HIS) residues were protonated to the neutral form (HSD). The protein was inserted into the membrane and the system was solvated into a TIP3 water box and ionized to add bulk water above and below the membrane and to neutralize the system with KCl using the membrane builder protocol [29,32]. The final assembled system comprised waters, phospholipids, ions and the protein (a total of ~174,071 atoms). Six equilibration steps of each protein-lipid system were performed for 50–200 ps with 2 fs timesteps using the NAMD program with input files generated by CHARMM-GUI [29,32]. The minimization equilibration utilized collective variable restraints to slowly release system motion and facilitate simulation stability. Starting from this equilibrated system,

NAMD simulations were carried out for ~105 ns using Langevin dynamics for the three membrane environments [18]. Electrostatic interactions were computed using the Particle-Mesh Ewald algorithm with a 12 Å cutoff distance [33] and Van der Waals interactions were computed with a 12 Å cutoff distance and a switching function to reduce the potential energy function smoothly to zero between 10–12 Å. Periodic-boundary conditions were used, and constant temperature (303 K) and pressure (1 atm) were maintained. Equations of motion were integrated with a timestep of 2 fs and trajectory data was recorded in 20 ps increments [18].

2.2. Analysis of the MD Simulation Data

The structures in the MD trajectory data were aligned with respect to the starting structure in the production run for each membrane bilayer environment before further analysis. The stability and structural dynamic behavior of KCNE3 was obtained from the aligned trajectory data by calculating root mean square deviation (RMSD) of all atoms of the backbone, root mean square fluctuation (RMSF), lipid bilayer membrane width, Z-distances, total protein–lipid interaction energy, and TMD helical tilt angle using the scripts available in the VMD software package [30]. The heatmaps for the correlation between different simulation parameters were graphed using Matlab (https://www.mathworks.com accessed on 10 February 2022). The images were prepared using the Igor Pro graphics program (https://www.wavemetrics.com accessed on 10 February 2022). All molecular dynamics simulations were run on the Miami Redhawk cluster computing facility at Miami University.

3. Results and Discussions

The stability and structural dynamic properties of KCNE3 in different phospholipid bilayer environments were investigated using NAMD molecular dynamics simulation trajectory data. A wild-type KCNE3 protein was incorporated into three different lipid bilayer environments including POPC/POPG (3:1), POPC alone, and DMPC alone to study how structural and dynamic properties of KCNE3 behave in different lipid bilayer environments; POPC and POPG are monounsaturated lipids and DMPC is saturated lipid. These lipids are widely used in studying membrane protein/peptides. POPG lipids contain a negative charge and hence the mixture of POPC and POPG at the molar ratio of 3 to 1 may provide more favorable condition to stabilize the TMD of KCNE3 buried into lipid bilayers while spanning the width of the bilayer membrane [34,35]. Figure 1A shows the chemical structure of the lipids used in this study. The protein lipid systems were set up beginning with the NMR structure of KCNE3. KCNE3 was inserted into the respective lipid bilayer with the transmembrane helix spanning the membrane with the C- and N-terminal helices oriented on either side of the bilayer and interacting with the solvent. Figure 1B shows the setup for the KCNE3-POPC/POPG system; the amino acid sequence of the wild-type KCNE3 is indicated in Figure 1C with the red box indicating the transmembrane helix, the blue boxes indicating the N- and C-terminal helices and the distribution of charged amino acids by color codes [16].

Molecular Motion of KCNE3 in Different Phospholipid Bilayer Environments

An all-atom molecular dynamics simulation on wild-type KCNE3 in three different lipid bilayer environments was carried out over the course of 105 ns. Figure 2 shows the snapshots of the representative MD simulation output trajectory data of KCNE3 incorporated into all three lipid bilayer systems (POPC/POPG, POPC alone, and DMPC alone) for 16 ns, 40 ns, 80 ns and 105 ns. The interaction of C- and N-termini of KCNE3 with the lipid bilayer surface is flexible and dynamic for all three lipid compositions. Interestingly, the initial few amino acid sites of N-terminal of KCNE3 showed formation of a short beta strand structure in the DMPC lipid system.

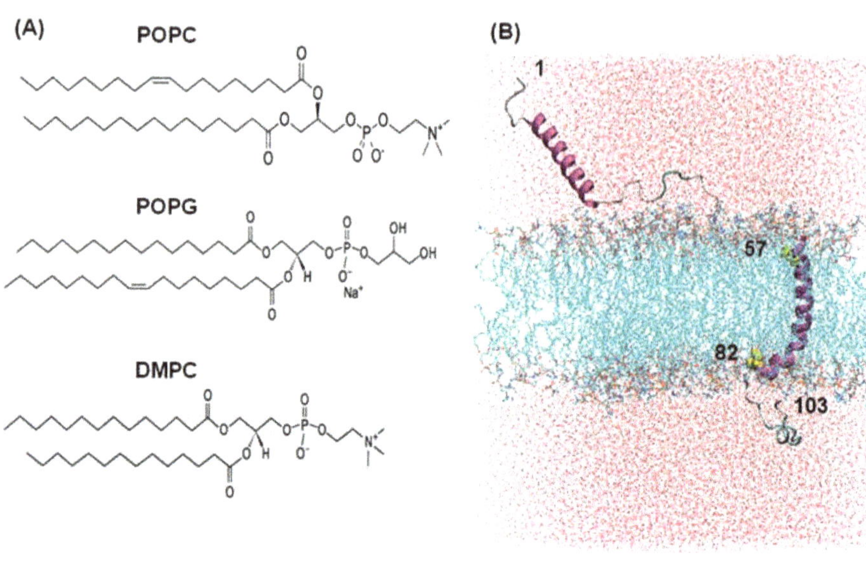

(C)

Wild-type KCNE3 amino acid sequence:

1 METTNGTET 10 WYESLHAVLKALNATLHSNLL 31 CRPGPGLGPDNQ

43 TEERRASLPGRDDN 57 SYMYI LFVMFLFAVTVGSLILGYTRS 83 RKV

DKRS 90 DPYHVY 97 KNRVSMI

Figure 1. (**A**) Chemical structure of phospholipids used in the NAMD molecular dynamics simulations. (**B**) An illustrative example of the cartoon representation of the NMR structure of KCNE3 (PDB ID: 2M9Z) incorporated into POPC/POPG lipid bilayers and solvated into water box [16]. Amino acid sites 1–56 represent N-terminus, amino acid sites 57–82 represent TMD and sites 83–103 represent C-terminus. The amino acid sites 57 and 82 are colored yellow. (**C**) Amino acid sequence of the wild-type KCNE3 with distribution of charges. Positive charges (Red), negative charges (Blue), and Histidine (Green) are color coded. The highlighted red box represents the transmembrane domain and blue boxes represent N- and C-terminal helices.

In order to analyze the conformational stability and molecular motion of the wild-type KCNE3 in membrane environments, a backbone root mean square deviation (RMSD) was calculated from the trajectory data and plotted as a function of simulation time for different segments of the protein including transmembrane domain (TMD), C-terminus, N-terminus, C-terminal helix, and N-terminal helix for POPC/POPG (3:1), POPC alone, and DMPC lipid bilayers as shown in Figure 3. We omitted analysis of the first 15 ns of each trajectory of the production run to avoid the equilibration time of the system. The RMSD measures the mean position of the amino acid residues in the structure of the subsequent simulation frames and compares them to the initial structure [22]. The RMSD is important in identifying regions of the proteins that have higher flexibility as well as regions that are stabilized. The initial trajectories for all simulations in the POPC/POPG and POPC alone systems are similar. The RMSD profile pattern for POPC/POPG (Figure 3A) shows that the RMSD values for the TMD of KCNE3 are lower than that of N-terminus and N-terminal helix until 49 ns and then increases to have similar values by 105 ns. The RMSD values for C-terminal and C-terminal helix are lower than the that of the TMD, N-terminus and N-terminal helix and vary throughout the simulation. Similarly, the RMSD profile pattern for POPC (Figure 3B) shows that the RMSD values for the TMD of KCNE3 are very close to that of N-terminus and N-terminal helix during the simulation. The RMSD values for the C-terminal helix are relatively lower than the TMD, N-terminus and N-terminal helix and C-terminus with fluctuating values. The RMSD values for the C-terminus are also close to these values, but fluctuate throughout the simulation. The RMSD profile pattern for DMPC (Figure 3C) shows that the RMSD values for the TMD, N-terminus, N-terminal helix, C-terminal helix, and C-terminus behave similarly with similar RMSD values. KCNE3 appears to be more stable in DMPC than in POPC/POPG or POPC alone, as the RMSD profiles for each segment are suppressed by comparison. In the POPC/POPG and POPC alone systems, the N-terminus and N-terminal helix have the highest RMSDs of all the segments of KCNE3. These data suggest that these regions of the protein have conformationally higher backbone fluctuations in the KCNE3 structure. This is expected, as the N-terminus contains a larger number of amino acid residues compared to the C-terminus and the TMD [16]. In the POPC/POPG and POPC alone systems, the RMSD values of the TMD begins at higher values than that of the C-terminus and C-terminal helix. However, the C-terminus and C-terminal helix have larger fluctuations as compared to the TMD, suggesting that the C-terminus is more mobile and unstable than the TMD. The overall fluctuations of the C-terminus are, however, lower than that of the N-terminus. The relatively smaller fluctuations observed for the TMD throughout the simulation suggests that it is the most stable segment of the protein and has the greatest stability of all segments studied. In the DMPC membrane mimetic system, the TMD is observed to have similar backbone fluctuations as in the POPC/POPG and POPC alone systems. However, the C-terminus segment starts out with a higher RMSD than that of the N-terminus, in contrast to the other two POPC/POPG and POPC alone systems. Similarly, higher backbone fluctuations for N- and C-termini reveal a similar level of conformational instability in the DMPC bilayer system. The average RMSD values for different segments of the KCNE3 are also calculated for all three lipid systems from the data in Figure 3 and shown in Table 1. The average RMSD values vary between 10.4–23.5 Å for POPC/POPG, 10.3–17.4 Å for POPC alone, and 9.5–15 Å for DMPC. The average RMSD value for the TMD in DMPC is the least value for TMD of all three lipid systems studied. The C-terminal helix has the lowest average RMSD value when compared to different segments of the protein in all three corresponding lipid systems. The standard deviation calculated of the average RMSD data show higher values for the outside regions of the protein compared to the TMD in all three corresponding lipid systems. The RMSD data for different regions of KCNE3 in different lipid bilayer environments suggest that the backbone flexibility for different segments of KCNE3 is different in POPC/POPG, POPC alone, and DMPC bilayer membranes. Our overall RMSD data suggest that the regions of the KCNE3 that is outside the membrane or interact with the surface are more flexible while DMPC lipid system plays a more stabilizing role.

Table 1. Average RMSD calculated from the RMSDs shown in Figure 3. The error represents standard deviation.

	Average RMSD (Å)		
	POPC/POPG	POPC	DMPC
TMD	14.7 ± 3.5	14.2 ± 3.1	9.7 ± 1.6
N-terminal helix	18.6 ± 3.5	15.9 ± 3.3	11.3 ± 4.4
N-terminus	23.5 ± 3.8	17.4 ± 3.1	13.3 ± 4.0
C-terminal helix	10.4 ± 4.0	10.3 ± 2.5	9.5 ± 1.6
C-terminus	11.3 ± 2.6	14.0 ± 3.9	15 ± 2.8

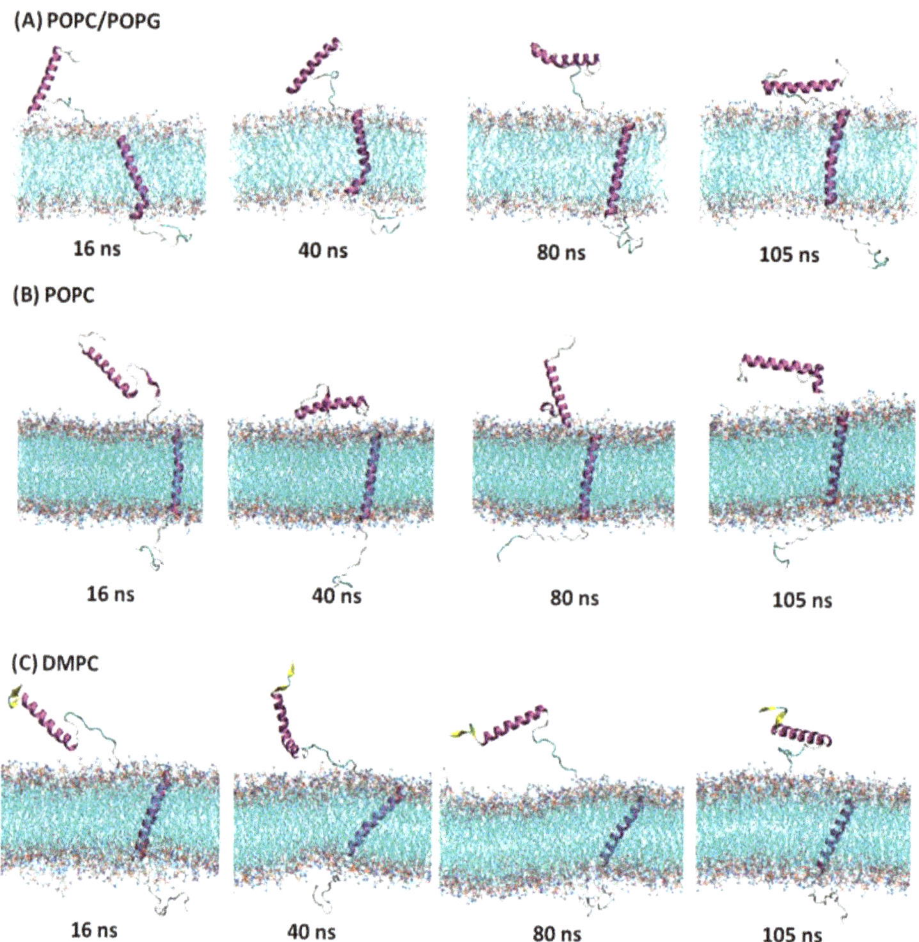

Figure 2. Snapshots of the representative MD simulation trajectory data of KCNE3 at 16 ns, 40 ns, 80 ns, and 105 ns for POPC/POPG (**A**), POPC alone (**B**), and DMPC alone (**C**) lipid bilayers. The hydrogen atom and water are omitted to make visualization simple and clear.

The RMSD data only provide the average behavior of the motion of the different segments of the protein while interacting with lipid bilayer membrane. We also wanted to understand how the flexibility of the particular regions assessed above contributed to the overall fluctuations that disturb the KCNE3's stability. The residue fluctuation profile of bilayer-integrated KCNE3 were quantitatively determined by the root mean square fluctuation (RMSF) throughout the simulation as shown in Figure 4. While the RMSD indicates positional differences of entire structures over the course of the simulation, the RMSF calculates how much a residue fluctuates during the simulation [22]. Consequently, it helps determine the flexibility of individual residues. Figure 4 shows the RMSF for KCNE3 residues in the three bilayer conditions. The profile for KCNE3 is similar for all three bilayer compositions. Overall, residues 1–9 (unstructured region) and ~25–35 (around the terminal of N-terminal helix) of the N-terminus and residues ~96–103 (unstructured region) of the C-terminus have the largest RMSF, suggesting they are the most flexible.

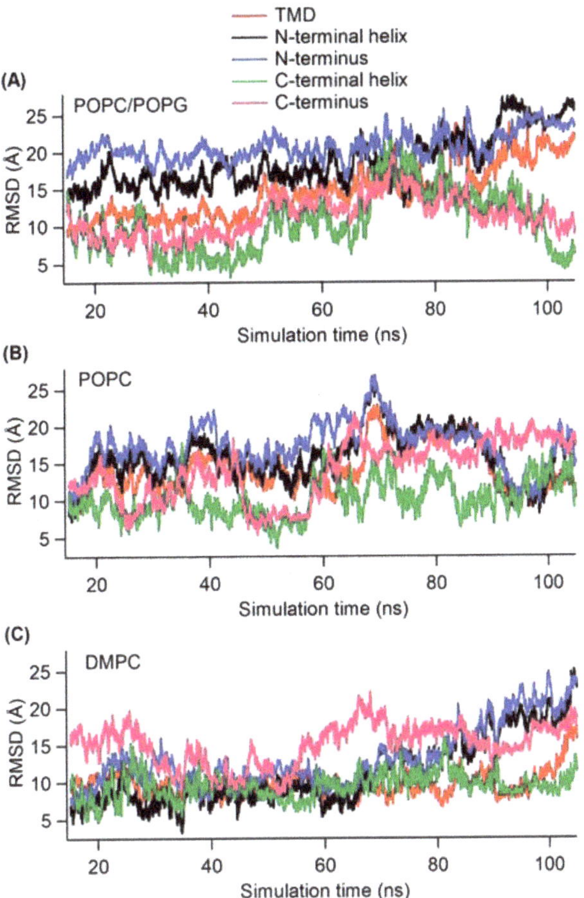

Figure 3. Root mean square deviation (RMSD) as a function of simulation time for different segments of KCNE3 in POPC/POPG (**A**), POPC alone (**B**), and DMPC (**C**).

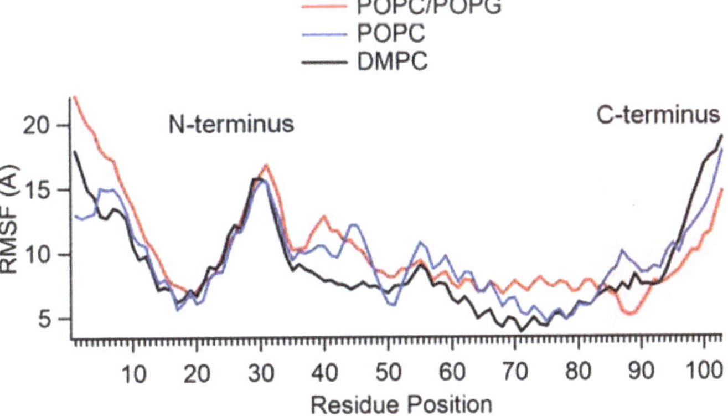

Figure 4. Plot of the root mean square fluctuation (RMSF) of KCNE3 as a function of simulation time for three different lipid compositions: POPC/POPG (Red), POPC (Blue), and DMPC (Black).

These results agree with the RMSD calculations that highlighted the highest fluctuations in the C- and N-termini. The RMSF of N-terminal residues 11–24 (helical region) and the TMD section from residue 57–82 are lower and indicate stability. The smallest RMSF fluctuations of the TMD region occur in DMPC, which is in agreement with our observations regarding the RMSD of this region. The previous NMR data-restrained molecular dynamics simulation on KCNE3 in DMPC lipid bilayers suggested the dynamic interaction of N- and C-termini helices with membrane surface [16]. These helices contain amphipathic amino acid sequences that do not deeply bury into the lipid bilayers, and hence these helices can dynamically interact with bilayer surfaces. The fluctuation of different segments of KCNE3 as suggested by the RMSF plot is consistent with the RMSD data and earlier NMR studies [16]. Our RMSF data suggest the N- and C-termini are more flexible with higher RMSD values in all three lipid compositions.

We wanted to better understand the formation of the lipid bilayer in the presence of reconstituted KCNE3, since we observed a suppressed RMSD for the KCNE3 TMD region in DMPC, in comparison to POPC/POPG and POPC alone. Both tails of the DMPC lipid only have 14 carbons, while POPC and POPG have 16 and 18 (Figure 1A). We measured the width of the membrane bilayer as a function of the simulation time for all three membrane mimic environments (POPC/POPG, POPC alone, and DMPC) to determine whether DMPC was forming compacted bilayers that stabilized the KCNE3 TMD. The membrane width was calculated by measuring the distance between the center of mass of the phosphorus of the upper and lower lipid head groups. The membrane width is shown as a time series in Figure 5A, while the probability distribution of the timeseries data is represented in Figure 5B. The membrane width of DMPC is the lowest of all three systems, as expected based on the length of hydrocarbon chains. The membrane width probability distribution plot (Figure 5B) shows the membrane width peak is centered around 37 Å for POPC/POPG, 35 Å for POPC and 31 Å for DMPC. The membrane width for POPC/POPG lipid bilayers is thicker than that of POPC lipid bilayers, despite having the same number of carbon atoms in the acyl chains.

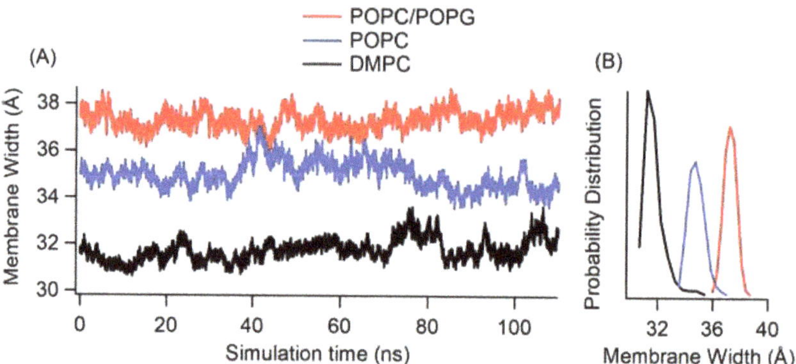

Figure 5. Membrane bilayer width incorporating KCNE3 protein as a function of simulation time (**A**) and membrane width probability distribution (**B**) for POPC/POPG (Red), POPC (Blue), and DMPC (Black) bilayers.

Next, we wanted to understand the protein topology with respect to the lipid bilayer membrane, since we observed that each bilayer had a different membrane width. The membrane thickness is oriented about the *z*-axis with the center of mass of the membrane bilayer located at Z = 0. We calculated the distance from the *z*-axis (Z-distance) of different segments of KCNE3 from the center of the mass of the lipid bilayers in all three different lipid membrane environments (POPC/POPG, POPC alone and DMPC). Previous NMR studies in micelles and isotropic bicelles suggested that amino acid residue sites 57 to 82 belong to the TMD of the KCNE3 that spans the membrane bilayer width [16]. The

Z-distances of the center of mass of the N-terminal helix, residues S57, A69 and S82, and the C-terminal helix from the center of mass of the lipid bilayers were calculated from the MD trajectory data. These Z-distance data can provide us with the information on how much various residues and different segments in the protein structure moved away from the center of the lipid bilayers when incorporated into different membrane environments. Figure 6 shows the plot of Z-distance as a function of simulation time for center of mass of different segments (N- and C-termini helices), and sites S57, A69, and S82 of TMD of KCNE3 in three different lipid bilayer environments (POPC/POPG, POPC alone, and DMPC). Figure 6A indicates that the TMD termini sites S57 and S82 are close to the surface of the lipid bilayer and span the width of the membrane for POPC/POPG lipid bilayers. The amino acid residue A69 lies close to the center of the lipid bilayers for POPC/POPG as indicated by the Z-distance around zero. The Z-distance for N- and C-termini helices varies outside the membrane width range. A similar trend of Z-distance profiles were observed for POPC alone and DMPC lipid bilayer environments. However, the Z-distance ranges for the TMD termini residues S57 and S82 for DMPC is lower than that for POPC/POPG and POPC alone. This is expected, as the DMPC bilayer width is lower than that of the POPC/POPG and POPC alone (Figure 5). The behavior of Z-distance pattern profile for these lipid bilayer environments is consistent with the membrane width profile shown in Figure 5.

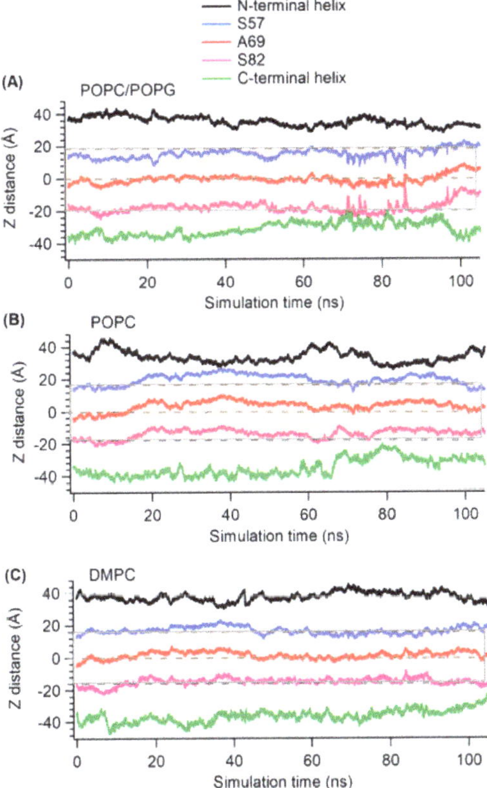

Figure 6. The plot of z-axis distance (Z-distance) as a function of simulation time for KCNE3 incorporated into POPC/POPG (**A**), POPC (**B**), and DMPC (**C**) lipid bilayers. Shaded regions represent the average width of the corresponding lipid bilayers calculated from Figure 5.

To understand the stability of the interaction of the KCNE3 reconstituted into lipid bilayer membrane environments, we calculated the internal energy of KCNE3 and plotted this energy as a function of simulation time for all three membrane bilayer environments. The corresponding histograms for total internal energy, electrostatic energy contribution and van der Waals energy contribution are shown in Figure 7. Figure 7A shows similar total energy profiles for all three systems. When the data is represented as a probability distribution (right panel), the total internal energy of the KCNE3 is the lowest, with more favorable values in the POPC/POPG lipid bilayers. The total internal energy of KCNE3 increases for POPC bilayers and is the least favorable in DMPC. Figure 7B shows the similar internal energy trends and histogram profiles for electrostatic energy contribution when compared to the total energy profile for all three systems. Figure 7C shows a lower van der Waals contribution to the total internal energy when compared to the electrostatic energy contribution. The probability distribution (Figure 7C, right panel) shows the van der Waals energy of KCNE3 has a slightly lower value in POPC/POPG lipid bilayers when compared to the POPC alone and DMPC alone systems both having similar van der Waals energy contributions. The electrostatic interactions are the dominant contribution to the total energy for all three lipid environments. The trend of the total internal energy in all three lipid environments suggests that the overall protein structure is more stable in POPC/POPG bilayer membrane compared to the cases of POPC and DMPC. Our hypothesis is that when the KCNE3 is unable to interact with the lipids, it relies on internal interactions to stabilize the structure.

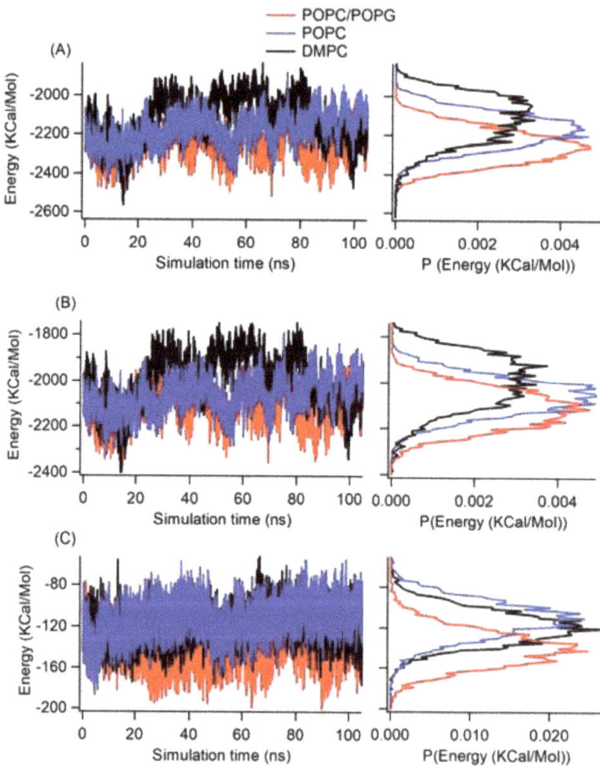

Figure 7. Internal energy of KCNE3 in lipid bilayer membranes as a function of simulation time (left panel) and corresponding histogram (right panel) for total internal energy (**A**), electrostatic energy (**B**) and van der Waals energy (**C**). The x-axis of the histogram plot represents the probability distribution.

To test this hypothesis, we computed the interaction energy of different segments of the KCNE3 with the lipid bilayer membrane as shown in Figure 8. In all three lipid environments, the interaction energy of the TMD section with the lipid is lower than that of C-terminus, C-terminal helix, and N-terminal helix with the respective lipid. The interaction energy of the N-terminus is close to interaction energy of the TMD but fluctuates throughout the simulation. While the interactions of the N- and C- termini appear to be strong and exhibit large fluctuations, the N- and C-termini helices weakly interact with the lipid. Inspection of the trajectory data suggests these helices are closely interacting with the membrane surface throughout the simulation, where the interaction energy attains the lowest values. The average interaction energies for each section of KCNE3 during the simulation were calculated for all three lipid systems from the interaction energy data (Figure 8) as shown in Table 2. The average interaction energy for the TMD of KCNE3 is not significantly different for all three lipid systems. Similarly, other segments of the protein have similar average interaction energy (within the error) in all three corresponding lipid systems. However, the standard deviation values are larger for the N- and C-termini in all three lipid systems. These data suggest that the interactions of the N- and C-termini of KCNE3 with the membrane surface are dynamic.

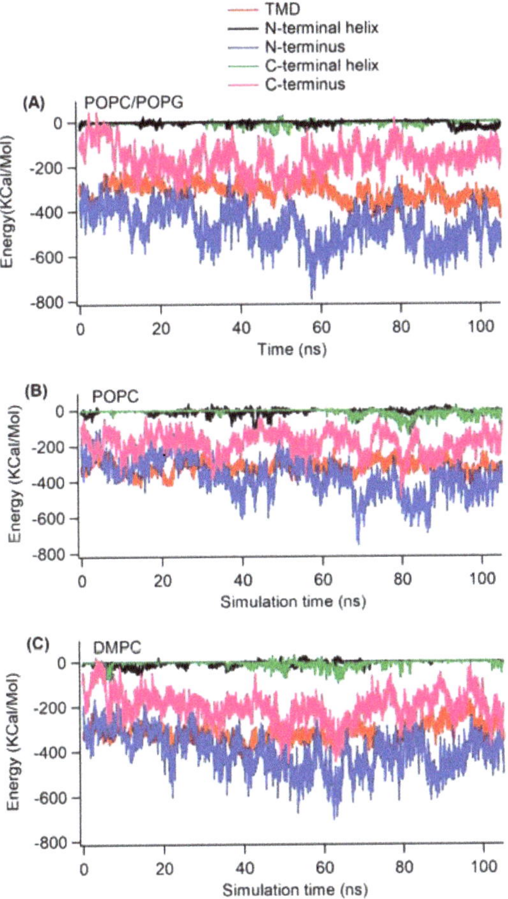

Figure 8. Interaction energy of different segments of KCNE3 with lipid bilayer membranes as a function of simulation time for POPC/POPG (**A**), POPC (**B**), and DMPC (**C**) lipid bilayers.

Table 2. Average interaction energy calculated from the interaction energy shown in Figure 8. The error represents standard deviation.

	Average Interaction Energy (KCal/Mol)		
	POPC/POPG	POPC	DMPC
TMD	−313.7 ± 42.3	−316.6 ± 37.5	−317.5 ± 42.2
N-terminal helix	−5.6 ± 10.9	−8.0 ± 14.8	−5.4 ± 9.4
N-terminus	−465.3 ± 85.3	−371.8 ± 105.3	−401.3 ± 86.8
C-terminal helix	−4.9 ± 8.8	−11.6 ± 20.2	−7.5 ± 13.7
C-terminus	−162.5 ± 64.5	−184.7 ± 63.7	−200.6 ± 68.5

These interaction energy data suggest that the N-terminus interacts most strongly with the membrane surface but the interaction is dynamic and unstable, as the standard deviation in the average energy calculation is very high for this segment of the protein. While the interaction energy of TMD is higher than that of the N-terminus, the stability of the TMD structure may be attributed to the internal energy of the protein its transient interactions with water molecules [16].

To better understand the conformational stability and the interaction of the transmembrane domain (TMD) of KCNE3 with membrane bilayers, we calculated the helical tilt of the TMD within the bilayer. In our previous results, we saw that the membrane thickness was dependent upon the lipid environment, though there were minute differences in the Z-distances of the terminal residues of the TMD helix. We wanted to establish whether the deformation of the helix occurred to accommodate the structure within the bilayer. We have plotted the probability density of the transmembrane (TM) helical tilt of KCNE3 with the membrane normal and the Z-distance of the TMD of KCNE3 from the center of the mass of lipid bilayers for all three different membrane environments (POPC/POPG, POPC alone, and DMPC) as shown in Figure 9. The initial Z-distance fluctuates around 0 Å. When KCNE3 is embedded in the POPC/POPG lipid bilayer, two conformations of the TMD helix are observed; the dominant population is centered around a Z-distance of −3 Å and a helical tilt of 45°, while the second less populated conformation is at a Z-distance of 2 Å and a helical tilt of 75°. In the case of POPC alone, similar conformations are observed as in the POPC/POPG system, though the populations are more diffuse with sampling of many intermediate states between the two dominant conformations. By comparison, only one dominant conformation exists in the DMPC lipid, which is unique to DMPC and not observed in the other two lipid environments. The highest probability is centered around a Z-distance of 3.5 Å, while the helical tilt fluctuates between 45° and 70°. These data suggest that there is a strong correlation between TM helical tilt angle and Z-distance for POPC/POPG bilayers and DMPC bilayers, and a weak correlation for POPC bilayers. Interestingly, the dominant probability density for the DMPC membrane appears for the Z-distance of the TMD of around 4 Å from the center of the mass of the bilayers and TM helical angle of around 45–70°. This suggests that the TMD helix tilts to remain embedded within the lipid bilayers. However, in DMPC the TMD helix is more stationary at the Z-distance and samples less tilt angles, suggesting that it is more stable in this lipid system. In contrast, when the TMD helix is in POPC/POPG and POPC alone, the tilting behavior of the TMD helix results in changes in Z-position, suggesting the TMD helix is more mobile within these lipids. These probability distribution patterns are also consistent with the membrane width data shown in Figure 5.

We wanted to further understand the conformational stability and interaction of different segments of KCNE3 with membrane bilayers. We plotted the correlation between the total interaction energy of different segments of KCNE3 (N-terminus, N-terminal helix, TMD, C-terminal helix and C-terminus) and the corresponding Z-distances from the center of the mass of the membrane bilayers in Figure 10. Figure 10A shows the probability density for the N-terminus. Similar trends are observed for POPC/POPG and DMPC, with one dominant population that varies between Z-distances of 35–50 Å and interacts strongly with the lipids with energies ranging from −550 to −750 kcal/mol. Interestingly, the probability density for POPC alone is more dispersed and involves much lower interaction energies. Visualization of the trajectory data suggests that the interaction of the N-terminus with the POPC membrane surface is dynamic and very unstable, with a wide spreading of its portion above the surface with occasionally anchoring to it. For the N-terminal helix of KCNE3 (Figure 10B), the probability densities are similar for all three lipid environments. However, an additional disperse density with higher interactions and Z-distances closer to the lipid head groups is observed for POPC/POPG. Visualization of the trajectory data suggests that N-terminal helix also interacts dynamically with the POPC/POPG membrane surface and develops a bending in the helix during the interaction during certain periods of simulation times. A similar trend of a dominant population for the TMD helix is observed for all three systems (Figure 10C). However, an additional dispersed density with higher interactions with same Z-distance has been observed in POPC. The probability density for the TMD helix also shows higher interactions in DMPC. For the C-terminal helix (Figure 10D) and the C-terminus (Figure 10E), similar populations are observed in all three lipid systems with a slightly weaker interactions observed in the POPC/POPG system. Together, these data suggest that the TMD of KCNE3 stably interacts with all three lipid systems, with DMPC conferring the greatest stability. While the N-terminus of KCNE3 is interacting strongly with POPC/POPG, it is more dynamic and less stable. The interaction trend of C-terminus is similar in POPC alone and DMPC. The weak interaction of the C-terminus in POPC/POPG suggests that the unanchored regions of either termini are stabilized by the interactions with water. The probability density pattern for different KCNE3 segments observed in these three environments are consistent with our RMSD, RMSF, membrane width, Z-distances, and total interaction energy data (Figures 3–7).

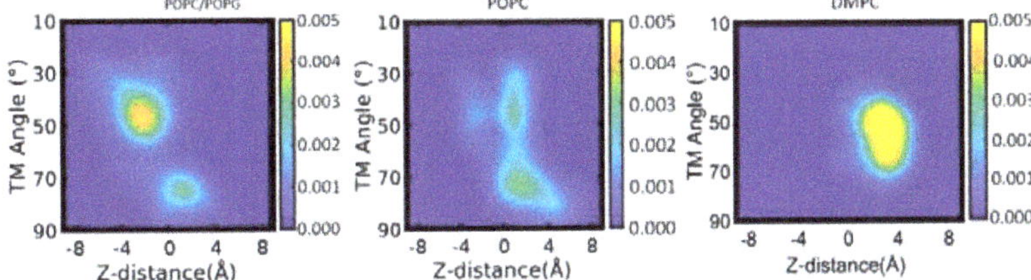

Figure 9. Probability density plot of transmembrane (TM) helical tilt angle against the Z-distance of TMD of KCNE3 in different lipid bilayer membranes. The yellow color indicates the highest probability, and the blue color represents the lowest probability.

NMR studies by the Sanders lab on KCNE3 in LMPC (lyso-myristoylphosphatidyl choline) micelles and DMPG (dimyristoylphosphatidylglycerol)/ DHPC (dihexanoylphosphatidylcholine) isotropic bicelles have suggested that KCNE3 adopts a single α-helical transmembrane domain (57–81). This is connected to a flexible loop with N-terminal surface associated amphipathic helix (10–30) and a short juxtamembrane helix (90 to 95) and a disordered C-terminus (96 to 103) [16]. The previous double electron electron resonance (DEER) electron paramagnetic resonance (EPR) data on KCNE3 in POPC/POPG bilayered vesicles suggested that the TMD helix of KCNE3 adopts a moderate curvature with residues T71, S74, and G78 facing the concave side of the curvature [16]. The TMD of KCNE3 is crucial to its function, and the curvature is important for binding to the activated-state channel [16]. A recent cryo-electron microscopic (Cryo-EM) spectroscopic study on the KCNE3-KCNQ1 complex in detergent micelles suggested that there is a deviation on the structure of KCNE3 interacting with KCNQ1 relative to the KCNE3 NMR structure model in isotropic bicelles with a root mean square deviation (RMSD) of 7.6 Å between the two structures [8]. Our all-atom molecular dynamics simulation data for 105 ns obtained on KCNE3 in POPC/POPG, POPC alone and DMPC bilayers reported in this study suggested that the center of mass of the KCNE3 TMD is slightly increased and more stable in DMPC when compared to POPC/POPG and POPC alone. In contrast, N- and C-termini were more conformationally flexible and interacted differently in all three environments. The

N- and C-termini helices dynamically interacted with the solvent or partially interacted with the membrane surface. The MD simulation results further suggested that the TMD of KCNE3 spans the membrane bilayer width with the amino acid residue A69 situated close to the center of lipid bilayers and the residues S52 and S82 are close to the surface of the membrane bilayer. The total internal energy of KCNE3 suggested that the POPC/POPG lipid bilayer membrane provides more stability in protein–membrane interactions. Our molecular dynamics simulation data are consistent with earlier experimental biophysical studies on KCNE3 [8,16,36]. Extending the MD simulation time longer than 105 ns may provide additional insight on the structural dynamic properties of KCNE3 while interacting with different mimetic systems.

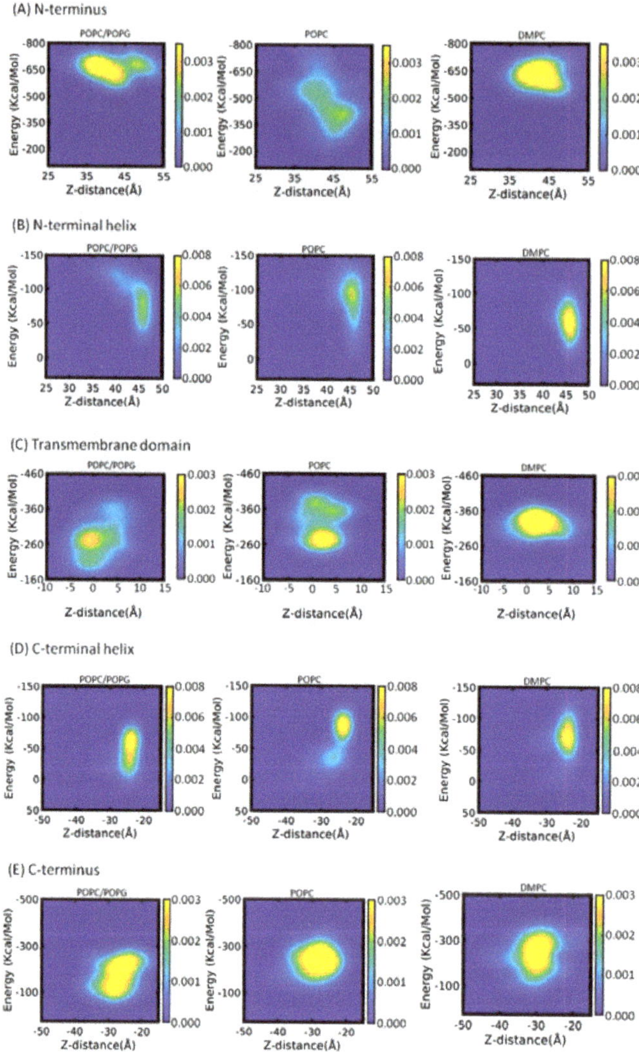

Figure 10. Probability density plot of total interaction energy of N-terminus (**A**), N-terminal helix (**B**), Transmembrane domain (**C**), C-terminal helix (**D**) and C-terminus (**E**) of KCNE3 with lipid bilayers against corresponding Z-distances from the center of mass of the lipid bilayers for different lipid bilayer membranes. The yellow color indicates the highest probability and blue color represents the lowest probability.

4. Conclusions

All atom molecular dynamics simulations for 105 ns were performed on KCNE3 incorporated into POPC/POPG, POPC alone and DMPC alone membrane bilayer environments to study the structural dynamic properties of KCNE3. The MD simulation results suggested that the TMD of the KCNE3 is less conformationally flexible and more stable when compared to the N- and C-termini in all three membrane environments. The N- and C-termini of KCNE3 are conformationally more flexible and dynamic in all these three lipid environments. The MD simulation results further suggested that the TMD of KCNE3 spans the membrane width, with residue A69 located near the center of the lipid bilayers and residues S57 and S82 located at the opposing lipid bilayer membrane surfaces. These MD simulation results complement the experimental biophysical studies of KCNE3 in lipid bilayer membranes to illuminate its structural dynamic properties in more detail.

Author Contributions: Conceptualization, I.D.S., G.A.L. and C.R.S.; methodology, I.K.A. and I.D.S.; formal analysis, I.K.A., I.D.S., A.N.K. and J.P.A.; investigation, I.K.A., A.P.G., A.B.G., M.F.C., A.C.M.M., C.C.C. and M.S.; resources, I.D.S. and G.A.L.; writing—original draft preparation, I.K.A. and I.D.S.; writing—review and editing, G.A.L., S.A., A.N.K., C.R.S. and I.D.S., supervision, I.D.S.; funding acquisition, I.D.S. All authors have read and agreed to the published version of the manuscript.

Funding: This work is generously supported by National Science Foundation NSF MCB-2040917 award. Gary A. Lorigan would like to acknowledge support from NSF (MRI-1725502) grant, NIGMS/NIH Maximizing Investigator's Research Award (MIRA) R35 GM126935 award, the Ohio Board of Regents, and Miami University. Gary A. Lorigan would also like to acknowledge support from the John W. Steube Professorship. Charles R. Sanders would like to acknowledge support from US NIH grant R01 HL122010.

Institutional Review Board Statement: Not applicable.

Informed Consent Statement: Not applicable.

Data Availability Statement: Not applicable.

Acknowledgments: We would like to appreciate Jens Mueller, the Facility Manager Redhawk Cluster Miami University, for assistance with the computational work.

Conflicts of Interest: The authors declare no conflict of interest.

Abbreviations

MD	molecular dynamics;
RMSD	root mean square deviation;
RMSF	root mean square fluctuation;
LMPC	lyso-myristoylphosphatidyl choline;
DMPC	1,2-dimyristoyl-sn-glycero-3-phosphocholine;
DHPC	dihexanoylphosphatidylcholine;
DMPG	dimyristoylphosphatidylglycerol;
POPC	1-palmitoyl-2-oleoyl-*sn*-glycero-3-phosphocholine;
POPG	1-palmitoyl-2-oleoyl-*sn*-glycero-3-phospho-(1′-*rac*-glycerol) (sodium salt);
TMD	Transmembrane Domain;
EPR	electron paramagnetic resonance;
DEER	double electron-electron resonance.

References

1. Abbott, G.W. KCNE1 and KCNE3: The yin and yang of voltage-gated K^+ channel regulation. *Gene* **2016**, *576*, 1–13. [CrossRef] [PubMed]
2. Lewis, A.; McCrossan, Z.A.; Abbott, G.W. MinK, MiRP1, and MiRP2 diversify Kv3.1 and Kv3.2 potassium channel gating. *J. Biol. Chem.* **2004**, *279*, 7884–7892. [CrossRef]
3. Schroeder, B.C.; Waldegger, S.; Fehr, S.; Bleich, M.; Warth, R.; Greger, R.; Jentsch, T.J. A constitutively open potassium channel formed by KCNQ1 and KCNE3. *Nature* **2000**, *403*, 196–199. [CrossRef] [PubMed]

4. Barro-Soria, R.; Ramentol, R.; Liin, S.I.; Perez, M.E.; Kass, R.S.; Larsson, H.P. KCNE1 and KCNE3 modulate KCNQ1 channels by affecting different gating transitions. *Proc. Natl. Acad. Sci. USA* **2017**, *114*, E7367–E7376. [CrossRef] [PubMed]
5. Ohno, S.; Toyoda, F.; Zankov, D.P.; Yoshida, H.; Makiyama, T.; Tsuji, K.; Honda, T.; Obayashi, K.; Ueyama, H.; Shimizu, W.; et al. Novel KCNE3 Mutation Reduces Repolarizing Potassium Current and Associated with Long QT Syndrome. *Hum. Mutat.* **2009**, *30*, 557–563. [CrossRef]
6. Lundquist, A.L.; Manderfield, L.J.; Vanoye, C.G.; Rogers, C.S.; Donahue, B.S.; Chang, P.A.; Drinkwater, D.C.; Murray, K.T.; George, L.A., Jr. Expression of multiple KCNE genes in human heart may enable variable modulation of I(Ks). *J. Mol. Cell. Cardiol.* **2005**, *38*, 277–287. [CrossRef] [PubMed]
7. Grahammer, F.; Warth, R.; Barhanin, J.; Bleich, M.; Hug, M.J. The small conductance K^+ channel. **2001** KCNQ1—Expression, function, and subunit composition in murine trachea. *J. Biol. Chem.* **2001**, *276*, 42268–42275. [CrossRef]
8. Sun, J.; MacKinnon, R. Structural Basis of Human KCNQ1 Modulation and Gating. *Cell* **2020**, *180*, 340–347.e9. [CrossRef]
9. Barro-Soria, R.; Perez, M.E.; Larsson, H.P. KCNE3 acts by promoting voltage sensor activation in KCNQ1. *Proc. Natl. Acad. Sci. USA* **2015**, *112*, E7286–E7292. [CrossRef]
10. Preston, P.; Wartosch, L.; Gunzel, D.; Fromm, M.; Kongsuphol, P.; Ousingsawat, J.; Kunzelmann, K.; Barhanin, J.; Warth, R.; Jentsch, T.J. Disruption of the K^+ Channel beta-Subunit KCNE3 Reveals an Important Role in Intestinal and Tracheal Cl^- Transport. *J. Biol. Chem.* **2010**, *285*, 7165–7175. [CrossRef]
11. Boucherot, A.; Schreiber, R.; Kunzelmann, K. Regulation and properties of KCNQ1 (K(v)LQT1) and impact of the cystic fibrosis transmembrane conductance regulator. *J. Membr. Biol.* **2001**, *182*, 39–47. [CrossRef] [PubMed]
12. Lundby, A.; Ravn, L.S.; Svendsen, J.H.; Haunso, S.; Olesen, S.P.; Schmitt, N. KCNE3 mutation V17M identified in a patient with lone atrial fibrillation. *Cell. Physiol. Biochem.* **2008**, *21*, 47–54. [CrossRef] [PubMed]
13. Abbott, G.W.; Butler, M.H.; Goldstein, S.A.N. Phosphorylation and protonation of neighboring MiRP2 sites: Function and pathophysiology of MiRP2-Kv3.4 potassium channels in periodic paralysis. *FASEB J.* **2006**, *20*, 293–301. [CrossRef]
14. Delpon, E.; Cordeiro, J.M.; Nunez, L.; Thomsen, P.E.B.; Guerchicoff, A.; Pollevick, G.D.; Wu, Y.S.; Kanters, J.K.; Larsen, C.T.; Burashnikov, E.; et al. Functional Effects of KCNE3 Mutation and Its Role in the Development of Brugada Syndrome. *Circ.-Arrhythmia Electrophysiol.* **2008**, *1*, 209–218. [CrossRef] [PubMed]
15. Zhang, D.F.; Liang, B.; Lin, J.; Liu, B.; Zhou, Q.S.; Yang, Y.Q. KCNE3 R53H substitution in familial atrial fibrillation. *Chin. Med. J.* **2005**, *118*, 1735–1738. [PubMed]
16. Kroncke, B.M.; Van Horn, W.D.; Smith, J.; Kang, C.B.; Welch, R.C.; Song, Y.L.; Nannemann, D.P.; Taylor, K.C.; Sisco, N.J.; George, A.L.; et al. Structural basis for KCNE3 modulation of potassium recycling in epithelia. *Sci. Adv.* **2016**, *2*, e1501228. [CrossRef]
17. Sansom, M.S.P.; Bond, P.J.; Deol, S.S.; Grottesi, A.; Haider, S.; Sands, Z.A. Molecular simulations and lipid-protein interactions: Potassium channels and other membrane proteins. *Biochem. Soc. Trans.* **2005**, *33*, 916–920. [CrossRef]
18. Ramelot, T.A.; Yang, Y.; Sahu, I.D.; Lee, H.-W.; Xiao, R.; Lorigan, G.A.; Montelione, G.T.; Kennedy, M.A. NMR structure and MD simulations of the AAA protease intermembrane space domain indicates peripheral membrane localization within the hexaoligomer. *FEBS Lett.* **2013**, *587*, 3522–3528. [CrossRef]
19. Sahu, I.D.; Craig, A.F.; Dunagum, M.M.; McCarrick, R.M.; Lorigan, G.A. Characterization of bifunctional spin labels for investigating the structural and dynamic properties of membrane proteins using EPR spectroscopy. *J. Phys. Chem. B* **2017**, *121*, 9185–9195. [CrossRef]
20. Weng, J.; Wang, W. Molecular Dynamics Simulation of Membrane Proteins. In *Protein Conformational Dynamics*; Han, K.-L., Zhang, X., Yang, M.-J., Eds.; Springer: New York, NY, USA, 2013; pp. 305–329.
21. Muller, M.P.; Jiang, T.; Sun, C.; Lihan, M.Y.; Pant, S.; Mahinthichaichan, P.; Trifan, A.; Tajkhorshid, E. Characterization of Lipid-Protein Interactions and Lipid-Mediated Modulation of Membrane Protein Function through Molecular Simulation. *Chem. Rev.* **2019**, *119*, 6086–6161. [CrossRef]
22. Pandey, B.; Grover, A.; Sharma, P. Molecular dynamics simulations revealed structural differences among WRKY domain-DNA interaction in barley (*Hordeum vulgare*). *BMC Genom.* **2018**, *19*, 132. [CrossRef] [PubMed]
23. Li, M.H.; Luo, Q.A.; Xue, X.G.; Li, Z.S. Molecular dynamics studies of the 3D structure and planar ligand binding of a quadruplex dimer. *J. Mol. Model.* **2011**, *17*, 515–526. [CrossRef] [PubMed]
24. Sonntag, Y.; Musgaard, M.; Olesen, C.; Schiott, B.; Moller, J.V.; Nissen, P.; Thogersen, L. Mutual adaptation of a membrane protein and its lipid bilayer during conformational changes. *Nat. Commun.* **2011**, *2*, 304. [CrossRef]
25. Phillips, J.C.; Braun, R.; Wang, W.; Gumbart, J.; Tajkhorshid, E.; Villa, E.; Chipot, C.; Skeel, R.D.; Kalé, L.; Schulten, K. Scalable molecular dynamics with NAMD. *J. Comput. Chem.* **2005**, *26*, 1781–1802. [CrossRef]
26. MacKerell, A.D.; Bashford, D.; Bellott, M.; Dunbrack, R.L.; Evanseck, J.D.; Field, M.J.; Fischer, S.; Gao, J.; Guo, H.; Ha, S.; et al. All-atom empirical potential for molecular modeling and dynamics studies of proteins. *J. Phys. Chem. B* **1998**, *102*, 3586–3616. [CrossRef] [PubMed]
27. MacKerell, A.D.; Feig, M.; Brooks, C.L. Improved treatment of the protein backbone in empirical force fields. *J. Am. Chem. Soc.* **2004**, *126*, 698–699. [CrossRef] [PubMed]
28. Asare, I.K.; Perez, A.; Garcia, A.B.; Cruz, M.F.; Moura, A.C.M.; Campbell, C.C.; Scheyer, M.; Alao, J.P.; Kravats, A.N.; Lorigan, G.A.; et al. Studying structural and dynamic properties of KCNE3 in different membrane mimic systems using molecular dynamics simulations. *Biophys. J.* **2022**, *121*, 196a. [CrossRef]

29. Jo, S.; Kim, T.; Iyer, V.G.; Im, W. Software news and updates—CHARNIM-GUI: A web-based grraphical user interface for CHARMM. *J. Comput. Chem.* **2008**, *29*, 1859–1865. [CrossRef]
30. Humphrey, W.; Dalke, A.; Schulten, K. VMD-Visual Molecular Dynamics. *J. Molec.Graph.* **1996**, *14*, 33–38. [CrossRef]
31. Schwieters, C.D.; Clore, G.M. The VMD-XPLOR visualization package for NMR structure refinement. *J. Magn. Reson.* **2001**, *149*, 239–244. [CrossRef]
32. Jo, S.; Kim, T.; Im, W. Automated Builder and Database of Protein/Membrane Complexes for Molecular Dynamics Simulations. *PLoS ONE* **2007**, *2*, e880. [CrossRef] [PubMed]
33. Essmann, U.; Perera, L.; Berkowitz, M.L.; Darden, T.; Lee, H.; Pedersen, L.G. A smooth particle mesh ewald method. *J. Chem. Phys.* **1995**, *103*, 8577–8593. [CrossRef]
34. Barrett, P.J.; Song, Y.; Van Horn, W.D.; Hustedt, E.J.; Schafer, J.M.; Hadziselimovic, A.; Beel, A.J.; Sanders, C.R. The Amyloid Precursor Protein Has a Flexible Transmembrane Domain and Binds Cholesterol. *Science* **2012**, *336*, 1168–1171. [CrossRef] [PubMed]
35. Sahu, I.D.; Craig, A.F.; Dunagan, M.M.; Troxel, K.R.; Zhang, R.; Meiberg, A.G.; Harmon, C.N.; MaCarrick, R.M.; Kroncke, B.M.; Sanders, C.R.; et al. Probing Structural Dynamics and Topology of the KCNE1 Membrane Protein in Lipid Bilayers via Site-Directed Spin Labeling and Electron Paramagnetic Resonance Spectroscopy. *Biochemistry* **2015**, *54*, 6402–6412. [CrossRef]
36. Kang, C.B.; Vanoye, C.G.; Welch, R.C.; Van Horn, W.D.; Sanders, C.R. Functional Delivery of a Membrane Protein into Oocyte Membranes Using Bicelles. *Biochemistry* **2010**, *49*, 653–655. [CrossRef]

MDPI
St. Alban-Anlage 66
4052 Basel
Switzerland
Tel. +41 61 683 77 34
Fax +41 61 302 89 18
www.mdpi.com

Membranes Editorial Office
E-mail: membranes@mdpi.com
www.mdpi.com/journal/membranes

www.ingramcontent.com/pod-product-compliance
Lightning Source LLC
LaVergne TN
LVHW070900170726
843515LV00005B/686

9783036549378